AF794377

Design of Shipboard Power System Grounding/Earthing

IEEE Press
445 Hoes Lane
Piscataway, NJ 08854

IEEE Press Editorial Board
Sarah Spurgeon, *Editor-in-Chief*

Moeness Amin
Jón Atli Benediktsson
Adam Drobot
James Duncan
Joydeep Mitra
Ekram Hossain
Brian Johnson
Hai Li
James Lyke
Diomidis Spinellis
Desineni Subbaram Naidu
Tony Q. S. Quek
Behzad Razavi
Thomas Robertazzi

Design of Shipboard Power System Grounding/Earthing

Norbert Doerry

Mohammed M. Islam

John Prousalidis

Published by
Standards Information Network

IEEE*Press*

WILEY

About the Authors

Dr. Norbert Doerry has over 40 years of experience in the operation, design, construction, repair, and modernization of naval ships; he is an internationally recognized expert in shipboard power system design and naval ship design. He is a retired US naval officer having attained the rank of Captain as an engineering duty officer. Dr. Doerry is a graduate of the United States Naval Academy with an electrical engineering bachelor degree and has earned three degrees from MIT: a master of science in electrical engineering and computer science, a naval engineers degree, and a PhD in naval electrical power systems. He is currently focused on ship design activity modeling, developing ship design tools, educating the ship and ship power system design workforce, and developing IEEE standards for shipboard applications. Dr. Doerry chairs standards development working groups for IEEE Std 45.1 and 45.3. He is a fellow in the Society of Naval Architects and Marine Engineers (SNAME) and has been recognized with a number of awards including the American Society of Naval Engineers (ASNE) Gold Medal, the Legion of Merit, and the Superior Civilian Service Award. He is an active member of IEEE, SNAME and ASNE and has published over 100 papers and reports. Dr. Doerry lives in Burke, Virginia, with his wife; he has two children and two grandchildren residing nearby.

Mohammed M. Islam is a retired marine electrical power system design engineer with 50 years of experience. He has participated in shipboard electrical power system standard development for 25 years. As chair of the IEEE Std 45 working group, he initiated many IEEE standards such as IEEE Std 45.1 through IEEE Std 45.8, IEEE Std 1580, IEEE Std 1713, and IEEE Std 1662. IEEE-1713 has become International standard ISO/IEC/IEEE-80005 series defining a global standard on ship-to-shore interconnections. The 80005 series of standards are paving the path in support of HV, LV, and DC ship-to-shore interconnections in support of sustainable port, green shipping, and cold ironing. He currently chairs the working group updating IEEE Std 45.7 on shipboard power system switchboards.

Mohammed M. Islam founded the Marine System Coordinating Committee (MSCC) and served as its inaugural chairperson. Born in Pakistan (now Bangladesh), he is a marine engineering graduate of 1966 (fourth batch) from Juldia Marine Academy Pakistan. He earned a bachelor of electrical engineering degree from New York State University, Fort Schuyler Maritime College, in 1974. Mohammed M. Islam's scientific interests include design and development of ship electric propulsion, All Electric Ship, green ships, alternative fuels, cold ironing, smart and sustainable ports, and HVAC/HVDC interconnections. In addition to many technical papers, he has previously authored three books on shipboard electrical power system and coauthored one book. When not performing shipboard power system design and development consultancy work, Mohammed M. Islam enjoys a retired family life with his wife, children, and grandchildren.

Dr. John Prousalidis is a full professor and the director of the Marine Engineering Laboratory of the National Technical University of Athens (NTUA). Born in Athens, Greece, he received his MSc (1991) and PhD (1997) degrees from the School of Electrical and Computer Engineering of NTUA. In 2001, he joined the Academic Staff of the School of Naval Architecture and Marine Engineering of NTUA in the scientific field of "marine electrical and electronic engineering." He is the author/coauthor of 4 books, about 100 papers, reviewer in IEEE, IET, Elsevier and IMarEST journals. He is a member of the Technical Chamber of Greece, IEEE, IMarEST, member of the board of the Hellenic Joint Branch of RINA/IMarEST, member of the Publication Supervisory Board of IMarEST Publications. He is currently serving as the vice-chairman of the IEEE Marine Power Systems Coordinating Committee (IEEE-MSCC), is a member of the IEEE/EPPC Working Group on Energy, is member of the IEC/ISO/IEEE JWG28 dealing with 80005 series of standards on ship-to-shore interconnections, and since September 2022 is head of the Industrial Connection Activity Sustainable Maritime of IEEE(ICA-22-013). He has participated in about 45 research projects and was the coordinator of 15. His scientific interests include ship electric propulsion, All Electric Ship, green ships, alternative fuels, cold ironing, smart and sustainable ports, and HVDC/HVAC submersible interconnections.

Preface

In the realm of terrestrial and shipboard power systems, the concept of "ground" (or "earth" per IEC) is a term that is often used interchangeably, implying a common understanding. However, the reality of shipboard power system grounding is quite distinct as it pertains to the unique voltage potential associated with the vast ocean. Given that for most ships, the metallic hulls and structures are in direct contact with the ocean, it is accurate to describe these elements as being at "ground" (or "earth") potential.

Shipboard power systems can be ungrounded, solidly grounded, or resistance grounded; since shipboard power systems may have multiple separately derived systems, ships can have examples of all three grounding methods depending on specific application requirements.

Ungrounded and high-resistance grounded shipboard power systems possess a remarkable feature: an inadvertent grounding of a single conductor leads to low ground fault currents. This, in turn, facilitates uninterrupted operation in the presence of such a fault. The implementation of high-resistance grounded systems has witnessed a growing prevalence within distribution networks with nominal system voltages exceeding 1000 V. More recently, this practice has found its way into systems with nominal voltages as low as 440 V. Solidly grounded systems exist where commonality with commercial equipment is desired.

Moreover, shipboard power systems equipped with power electronic conversion devices, such as variable frequency drives (VFDs), demand a rigorous effort to ensure the proper control of harmonic distortion and common-mode (CM) currents and voltages. The effective management of CM currents and voltages necessitates additional consideration in grounding strategies.

Bonding, often referred to as protective earthing (PE), encompasses the deliberate act of connecting exposed metal components, which are not designed to carry electrical currents during standard operations, to the ship's hull through a low-impedance pathway. Bonding primarily serves as a safety measure, mitigating the risk of electrical shocks to personnel arising from capacitively coupled voltages on exposed metal parts, inductively coupled currents within those parts, or insulation failures.

One of the coauthors, Mohammed M. Islam, boasts an impressive five decades of experience in designing shipboard power systems and conducting research on these systems, including their grounding aspects. Following the completion of a full electrical propulsion ship with a 6600 V power generation system, he recognized the need to seek clarification on several shipboard power generation, distribution, and grounding issues. This led to contact with the IEEE-45 working group, which revealed that the work to address these issues was indeed substantial.

Subsequently, Moni took on a leadership role within the IEEE-45 working group. Over the course of a decade, twelve standards were introduced to support the design and development of shipboard power systems. Additionally, Moni authored three books to elucidate the design and development of shipboard power systems as a complementary resource to the IEEE standards.

During the development of these books, it became apparent that a comprehensive treatise on grounding was warranted, but it would necessitate the collective wisdom of national and international experts. Recognizing the complexity of grounding issues, Moni concluded his existing research was insufficient to address this challenge independently. He reached out to the distinguished Dr. Norbert Doerry, whose qualifications in shipboard grounding were unparalleled. Dr. Doerry has led or been in leadership roles of multiple IEEE standards working groups. He has also published extensively on ship design and ship power system design. Dr. Doerry wholeheartedly grasped the importance of this endeavor and graciously accepted the offer to collaborate. To ensure a well-rounded perspective, Moni also engaged the IEC expertise of Dr. John Prousalidis from Greece. Dr. Prousalidis significant experience with cold ironing (shore power) and European practices proved invaluable.

The twenty-month journey to create this book on shipboard grounding was born from the recognition of a critical gap in the field and the collective effort of renowned experts in the domain. The authors met online almost every week during this journey; we learned much during these focused discussions on shipboard grounding. In particular, we discovered the need to clearly distinguish between power system grounding, common-mode grounding, and protective earthing; this discrimination led to the book focusing chapters on these three types of grounding.

This book is designed to be a valuable resource for academicians and students in the classroom; it offers a wealth of drawings, analytical presentations, data, guidance, models, simulations, and photos. For shipboard power system design engineers, it serves as a comprehensive design guide.

We sincerely hope that this work will contribute significantly to the understanding and practice of shipboard power system grounding and electrical safety will enhance significantly.

1

Introduction

1.1 General

Shipboard power distribution systems have existed since the late nineteenth century. Only until recently, both commercial and naval ships have predominantly employed ungrounded power systems with nominal system voltages less than 1 kV. Ungrounded systems, equivalent to isolated systems (IEC), have a desirable feature in that the unintentional grounding of one conductor results in low ground fault currents and thus enables continued operation with a single ground fault. The ship's engineers can wait for a favorable time to find and clear the ground fault. For many years, ungrounded power systems served the maritime industry well.

However, over the past decades, the total electric load on many types of ships has risen, first due to the addition of heat loads that were previously served by steam prior to the adoption of diesel and gas turbine engines, and second due to the introduction of integrated electric propulsion in the form of an integrated power system (IPS) in the 1990s. In response to this growing load, ships started to employ increasingly higher power generation systems with nominal system voltages greater than 1 kV up to 13.8 kV. For these higher voltages, employing an ungrounded system is not recommended due to voltage stresses on system insulation and due to potential safety concerns. Instead, the use of a high-resistance grounded (HRG) system has become prevalent for distribution systems with nominal system voltages above 1 kV and more recently has been employed in some systems with a nominal system voltage as low as 440 V. The desire to be able to easily integrate commercial equipment designed for shore-based facilities has even resulted in some solidly grounded secondary distribution systems.

Additionally, power electronic conversion equipment, such as variable frequency drives (VFDs), has become prevalent in shipboard systems; VFDs enable

Design of Shipboard Power System Grounding/Earthing, First Edition. Norbert Doerry, Mohammed M. Islam, and John Prousalidis.
© 2025 The Institute of Electrical and Electronics Engineers, Inc.
Published 2025 by John Wiley & Sons, Inc.

motors and motor loads to operate at higher efficiencies, reduce inrush current, and increase displacement power factor. On the other hand, the integration of VFDs requires significant effort to ensure harmonic distortion and common mode (CM) currents and voltages are properly controlled. Controlling CM currents and voltages requires additional grounding considerations.

1.2 Grounding and Earthing Definitions

In terrestrial systems, the term "ground" (USA) or "earth" (IEC) refers to the voltage potential of the soil at a particular location and is used as a reference potential for measuring the voltage of other conductors. In some terrestrial power systems, the soil itself (at ground voltage potential) may be used as one of the conductors in the power circuit.

In shipboard systems, the term "ground" or "earth" refers to the voltage potential of the ocean. Since most ships have metallic hulls and structure in direct contact with the ocean, the hull and structure are also said to be at "ground" or "earth" potential. However, because of safety and corrosion concerns, the ship's hull and structure are not normally used as one of the conductors in the power circuit.

Bonding is the act of deliberately connecting exposed metal parts that are not designed to carry electrical currents under normal operation to the hull of the ship via a low-impedance path. Bonding is primarily performed as a safety measure to prevent electrical shock to personnel caused by capacitively coupled voltages on the exposed metal parts, inductively coupled currents in exposed metal parts, or insulation failures. The impacts of electric currents on humans are discussed in Appendix C.

The connection of exposed metal parts to the chassis or frame of a piece of equipment is indicated on circuit diagrams by IEC Symbol 5020 (Figure 1.1f). A terminal, such as one connected to the chassis, that is intended to be connected to ground for the purpose of implementing bonding is called a protective earth (PE) terminal and is represented by IEC Symbol 5019 (Figure 1.1c). This book will additionally use the symbol depicted in Figure 1.1d to represent the connection of

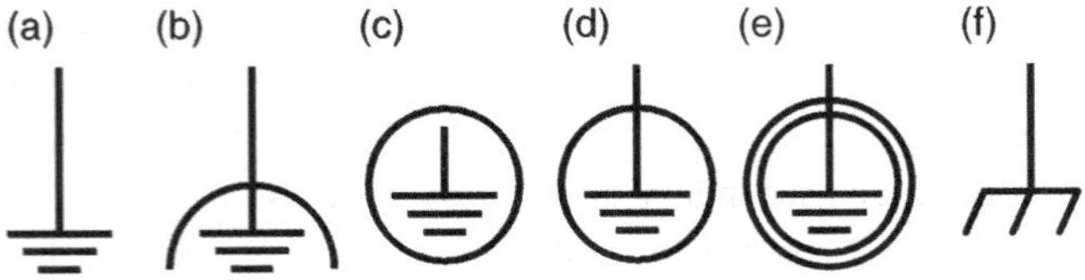

Figure 1.1 Ground symbols: (a) IEC Symbol 5017, (b) IEC Symbol 5018, (c) IEC Symbol 5019, (d) Protective earth, (e) CM ground, and (f) IEC Symbol 5020.

the terminal intended for bonding (IEC Symbol 5019) to the ship's hull. This symbol is used to represent a protective earth, and its incorporation into a design is called protective earthing.

The term "equipotential" means that two conductors have equal voltages with respect to a reference voltage. Two conductors that are electrically connected with a low impedance are equipotential. "Equipotential bonding" and "equipotential grounding" are equivalent to "bonding"; the conductors are at the same voltage as the hull of the ship.

Power distribution system grounding (or earthing (IEC)) is the act of deliberately inserting a solid connection or an impedance between a conductor or neutral of a power system and the hull of the ship. Grounding is usually done to limit conductor voltages with respect to the ship's hull, to provide a path for fault current in the case of a ground fault, and to provide a path for CM currents under normal operations. A ground fault is an unintentional electrical connection between a power system conductor and the ship's hull. A ground fault can be a "solid" ground fault with little resistance, or a fault, such as an arc fault, with a higher resistance.

One may also encounter IEC Symbol 5018 (Noiseless earth), depicted as Figure 1.1b; it is intended to represent a special grounding system for a particular application to minimize CM disturbance or noise on a grounding conductor. These special grounding systems are not covered by this book.

This book distinguishes between power distribution system grounding, protective earthing, and CM grounding. For this book, power distribution system grounding addresses ground currents at frequencies less than three times the fundamental frequency, while CM grounding addresses frequencies at or above three times the fundamental frequency. The two types of grounding do interact; this interaction must be accounted for in the design of each. IEC Symbol 5017 (Figure 1.1a) is used to indicate power system grounding. This book additionally uses the symbol depicted in Figure 1.1e to denote an intentional CM ground. Parasitic connections to ground participate in protective earthing (bonding), power system grounding circuits, and CM circuits; by convention, parasitic component connections to ground are displayed as IEC Symbol 5017. In summary, of the six ground symbols depicted in Figure 1.1, this book will use only the three depicted in Figure 1.2.

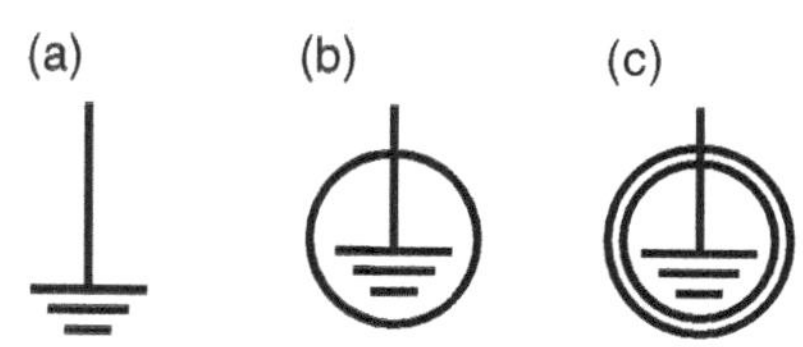

Figure 1.2 Ground symbols used in this book: (a) system ground, (b) protective earth, and (c) common mode ground.

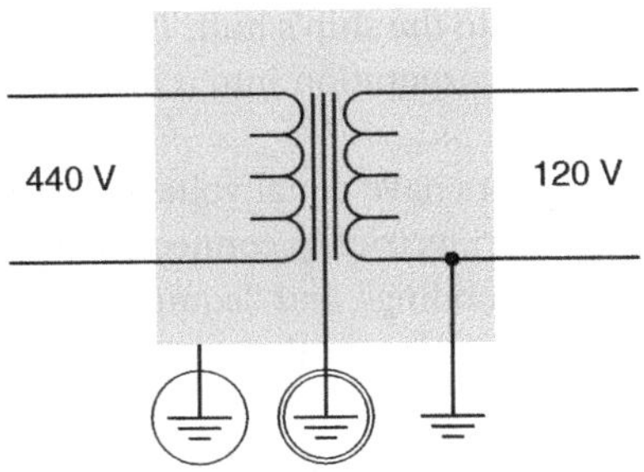

Figure 1.3 Use of ground symbols.

Figure 1.3 depicts the use of the three different ground symbols in the case of a single-phase transformer. The protective earth grounding (left) connects the exposed metallic structure and components of the transformer enclosure (not intended to carry current under normal operation) to ground through bonding. The CM ground (middle) is used to connect the transformer shield between the primary and secondary windings to ground; this shield is used to prevent capacitive coupling of CM currents between the primary and secondary windings. Finally, one of the conductors of the secondary winding is grounded (right) to form a system ground for a solidly grounded distribution system.

Electric current from the power system that, under normal conditions, flows through conductors that are not intended to carry power system current, such as conductors associated with protective earthing, is called objectionable current. Objectionable currents may arise when inappropriate connections are made between the power system and protective earthing conductors.

As defined by IEEE Std 45.1-2023 (2023), the nominal system voltage is "the designated voltage for a power system used as a reference value for establishing other power quality measures. For direct current, single-phase AC, and three-phase AC systems, the nominal system voltage is measured line-to-line." For AC systems, the nominal system voltage is expressed as a root-mean-square (rms) of the fundamental frequency component of the voltage waveform.

1.3 Common Mode Terminology

The neutral voltage of a set of conductors with respect to a reference voltage potential (typically ground) at any instant in time is the average value of the instantaneous conductor voltages with respect to the reference voltage. The neutral voltage is also called a CM voltage. Strictly speaking, the neutral voltage is a calculated quantity from the voltage measurements of multiple conductors and may not correspond to the voltage of any one conductor.

For a three-phase system, the neutral voltage with respect to ground can be expressed as

$$v_{ng} = \frac{1}{3}\left(v_{ag} + v_{bg} + v_{cg}\right) \tag{1.1}$$

$$
\begin{aligned}
v_{ag} &= v_{an} + v_{ng} \\
v_{bg} &= v_{bn} + v_{ng} \\
v_{cg} &= v_{cn} + v_{ng}
\end{aligned}
\tag{1.2}
$$

where

v_{ng}, neutral-to-ground voltage
v_{ag}, v_{bg}, v_{cg}, phase-to-ground voltage
v_{an}, v_{bn}, v_{cn}, phase-to-neutral voltage

Through substitution

$$v_{ng} = \frac{1}{3}\left(v_{an} + v_{ng} + v_{bn} + v_{ng} + v_{cn} + v_{ng}\right)$$

$$v_{an} + v_{bn} + v_{cn} = 0$$

The sum of the phase voltages with respect to the neutral is identically zero.

The instantaneous sum of the conductor currents in a set of conductors is called the CM current. In a balanced system, the CM current of a set of conductors is equal to zero; the sum of the currents in one direction is equal to the sum of the currents in the opposite direction. CM currents often, but not always, have a return path to a CM source via ground connections.

CM current is unintended; the intended differential mode (DM) currents of a set of conductors add to zero when measured in the same direction.

Power electronics, such as VFDs, and circuit asymmetry can be significant CM voltage sources in shipboard power systems. CM circuits include line impedances, parasitic capacitances to ground, and CM grounding systems. In some cases, inductive coupling with bonded conductors can also lead to CM voltages and currents.

A neutral conductor is intended to have a voltage potential equal to the neutral voltage of a set of conductors. However, due to voltage distortion of the phase voltages, the voltage potential of the neutral conductor may vary somewhat from the neutral voltage of an associated set of conductors.

1.4 Types of Power Distribution Systems

IEEE 45.1-2023 identifies the following types of buses to categorize shipboard power systems:

- Primary bus: distribution systems with a nominal system voltage greater than 1 kV.
- Distribution bus: distribution systems with a nominal system voltage between 400 V and 1 kV.

- Secondary bus: distribution systems with a nominal system voltage no greater than 400 V.
- Special bus: circuits for unique purposes such as medical use, control system power, or special control.

This book adapts the IEEE 45.1-2023 terminology to refer to power distribution systems:

- Primary distribution system: a power distribution system with an IEEE Std 45.1-2023 primary bus.
- Low-voltage distribution system: a power distribution system with an IEEE 45.1-2023 distribution bus.
- Low-voltage secondary distribution system: a power distribution system with an IEEE 45.1-2023 secondary bus.
- Low-voltage special circuits: a power system circuit with an IEEE 45.1-2023 special bus.

With the exception of high-voltage shore connections, the terms high voltage and medium voltage are not generally used in this book. For shipboard power systems, the terms high voltage and medium voltage usually refer to nominal system voltages above 1 kV; but they are not consistently defined by the various regulatory bodies, classification societies, and standards organizations. Instead, to avoid confusion, this book refers to primary distribution systems or refers to specific voltage levels.

Figure 1.4 depicts the relationships among the four types of power distribution systems. Not all ships will have all of the distribution systems; smaller mechanical

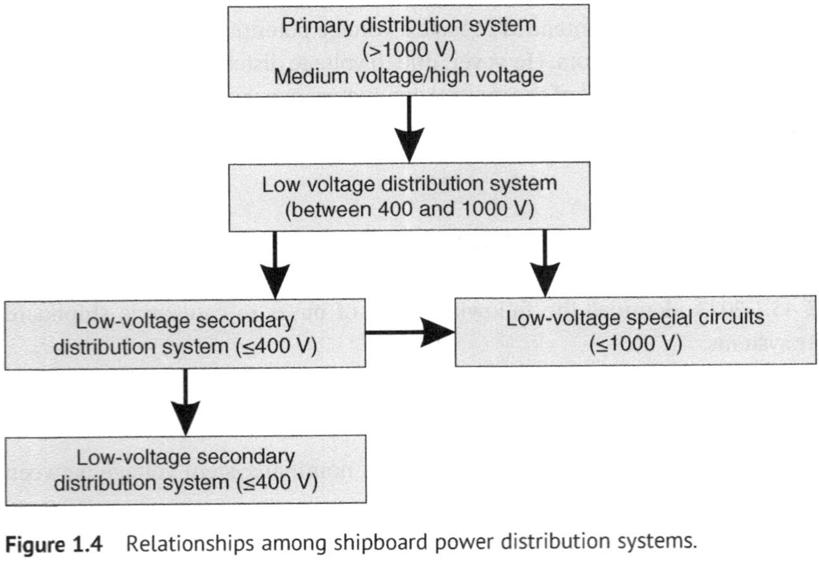

Figure 1.4 Relationships among shipboard power distribution systems.

drive ships typically do not have a primary distribution system. Although low-voltage special circuits may have a nominal system voltage up to 1 kV, most will have a nominal system voltage no more than 400 V. Some low-voltage secondary distribution systems may be derived from other low-voltage secondary distribution systems, typically with a higher nominal system voltage. Low-voltage special circuits may be derived from either low-voltage distribution systems or low-voltage secondary distribution systems.

1.5 Types of Power System Grounding Systems

A power system grounding (or earthing) system is composed of the equipment used to insert a solid connection or an impedance between a conductor or neutral of a power system and the hull of the ship. The impedance, if employed, is typically designed to manage steady-state ground fault currents at the fundamental frequency. The impedance will also impact the transient response during fault initiation and fault clearing.

A solidly grounded system incorporates a very low impedance (approaching 0 Ω) between a conductor or neutral of a power system and the hull of a ship. A ground fault typically results in large fault currents that activate protective devices to isolate the fault. An advantage of a solidly grounded system is that the insulation systems in cables and equipment need only be designed to accommodate the line-to-ground voltage.

An ungrounded system does not have a grounding system. The power system is only electrically connected to the ship's hull via parasitic resistances, capacitances, and inductances. Parasitic capacitances typically dominate for AC systems and parasitic resistances typically dominate for dc systems. A ground fault will result in relatively low amounts of fault current that flows through the ground fault and the parasitic impedances. Continued operation without clearing the first ground fault is usually possible. Generally, the crew should localize and isolate the ground fault as soon as operational conditions permit; a second ground fault on a different phase will result in large fault currents and protective devices tripping. The insulation systems in cables and equipment must be designed to accommodate the full line-to-line voltage between each line and ground, in addition to a transient voltage that occurs due to a resonance between the line inductance and parasitic capacitances to ground.

An HRG system incorporates a grounding (earthing) resistor between the power system (typically a neutral conductor) and the ship's hull. The grounding resistor has a resistance high enough to preclude fault protection circuits from activating on overcurrent due to a ground fault. Like an ungrounded system, an HRG system allows continued operation with one line-to-ground fault. By inserting a resistance

Table 1.1 Recommended grounding methods.

Application	Ungrounded	High-resistance grounded	Solidly grounded
AC primary distribution system	No	Yes	No
AC low-voltage distribution system	Yes	Yes	No
AC low-voltage secondary distribution system	No	No	Yes
AC low-voltage special circuits	See Chapter 9	See Chapter 9	See Chapter 9
DC primary distribution system	No	Yes	No
DC low-voltage distribution system	Yes	Yes	No
DC low-voltage secondary distribution system	Yes	No	Yes
DC low-voltage special circuits	See Chapter 9	See Chapter 9	See Chapter 9

into the ground fault circuit, when a ground fault clears, the charge trapped in the parasitic line-to-ground capacitance has a way to dissipate. The line-to-ground voltage of the conductors quickly returns back to normal; the voltage stress on insulation systems is thereby reduced. HRG systems also facilitate automated detection of a ground fault; the voltage across the grounding resistor is a simple indicator of a ground fault.

Recommendations for the application of each grounding type to the different power distribution system types are shown in Table 1.1 and discussed in detail in Chapter 9.

Equipment may incorporate filter networks that include connections to ground for the purpose of managing electromagnetic interference (EMI) and CM voltages and currents. These filter networks include EMI filters and are typically focused on controlling CM currents and voltages at frequencies substantially higher than the fundamental frequency. EMI filters are one type of a broader set of CM grounding systems.

Finally, equipment enclosures and other accessible conducting components of equipment (not directly connected to the power system) are bonded together and then connected to earth via a protective earth (protective ground) connection. The ground pin on a 120 V appliance plug is an example of protective earthing.

1.6 Modeling and Simulation

This book provides physics-based models of shipboard power cables and other equipment. These models, however, are by necessity simplified; they do not account for certain types of nonlinearities and environmental factors. The cable models, for example, do not account for frequency dependencies of the inductances, to include the impact of the skin effect at high frequencies. Wherever possible, direct measurement of model parameters should be made on physical equipment and components. In general, design margins should be employed to account for the uncertainty associated with model simplifications and parameter estimation. System modeling should include parameter sweeping to ensure the final design is robust to model errors.

The simulations in this book use LTspice; LTspice is a circuit simulation software program available for free (as of 2023) from Analog Devices (https://www.analog.com/en/design-center/design-tools-and-calculators/ltspice-simulator.html). LTspice is derived from the Simulation Program with Integrated Circuit Emphasis (SPICE), which was developed by Laurence Nagel at the Electronics Research Laboratory of the University of California, Berkeley, and released in 1973.

The models presented in this book can also be implemented in other simulation tools. The MATLAB/Simulink environment from MathWorks is another simulation environment often used in the analysis of shipboard power systems. The techniques presented in this book also apply to modeling within MATLAB/Simulink.

1.7 Book Overview

While there are other books that discuss shipboard power system design, the discussion on grounding systems is usually brief and incomplete. This book is unique in that it comprehensively covers the issues associated with shipboard power system grounding and hull currents.

IEEE Std. 3003.1-2019 (2019) provides recommended practices for system grounding of industrial and commercial power systems. This book covers many of the same topics, but focuses on the unique characteristics of shipboard power systems and their impact on grounding system design. This book also discusses a number of topics that are unique to ships.

In a similar vein, in the United States, the National Electrical Code (NFPA 702023 2023) provides grounding requirements for terrestrial applications and floating buildings where the ground means the earth. It specifically does not cover ships where the ground means the water the ship is floating in (and, by extension, the metal hull or ground plate).

This book does not extensively cover EMI/EMC issues such as the coupling between signal cables and power cables, or the coupling between signal cables. Other resources, such as NAVSEA S9407-AB-HBK-0102010 (2010), should be consulted for recommended shielding practices and cable separation recommendations.

Chapter 2, System Grounding: Shipboard Ungrounded AC Systems. Applicable to AC power systems with a nominal system voltage no greater than 1 kV, the chapter provides guidance for modeling cables and other components of an ungrounded system and shows the potential hazards due to intermittent ground faults. Other special cases, such as stuck breaker contacts and ground faults on rectifier outputs, are examined. The chapter also discusses ground fault detection and localization methods, including the impact of an ungrounded system on ground fault circuit interrupter (GFCI) operation.

Chapter 3, System Grounding: Shipboard HRG AC Low-Voltage Distribution Systems. Chapter 3 covers HRG for distribution systems with a nominal system voltage no greater than 1 kV. The damping effect of the HRG on voltage transients due to intermittent ground faults is demonstrated. The multiple ways of implementing an HRG are discussed: neutral grounding resistor (NGR), zigzag transformer, wye-delta transformer, and wye-broken delta transformer. The issue of grounding transformer saturation due to DC currents is covered, as well as ground fault detection and localization.

Chapter 4, System Grounding: Shipboard Solidly Grounded AC Systems. Chapter 4 covers solidly grounded AC secondary distribution systems. Examples are provided to demonstrate why a delta-wye transformer should be used instead of a wye-wye transformer for supplying power to a solidly grounded AC system. The chapter also discusses the need to limit EMI filter capacitance to avoid GFCI nuisance tripping.

Chapter 5, System Grounding: Shipboard HRG AC Primary Distribution Systems. Chapter 5 covers HRG for AC distribution systems with a nominal system's voltage greater than 1 kV. The advantages of using a wye-broken delta grounding transformer configuration over the other configurations are explained. For transient analysis, more detailed models of cables, to include conductor and cable shields, are presented. Methods to predict parasitic capacitances for many different types of power system apparatus are provided. In addition to fault detection and localization, cable terminations for primary distribution system cables are discussed. The characteristics of shielded insulated bus pipe (IBP) are described.

Chapter 6, System Grounding, Shipboard Ungrounded DC Systems. Chapter 6 covers ungrounded DC low-voltage distribution systems and low-voltage secondary distribution systems. Models for unshielded two-conductor and four-conductor dc cables are provided. CM current performance of 12-pulse rectifiers during a ground fault is presented. The chapter also includes a discussion of ground fault detection and localization.

Chapter 7, System Grounding: Shipboard HRG DC Systems. Chapter 7 covers HRG for DC primary distribution systems and DC low-voltage distribution systems. A basis for determining the HRG resistance is presented. The chapter also provides a model for shielded four-conductor DC cables, as well as methods for ground fault detection and localization.

Chapter 8, System Grounding: Shipboard Solidly Grounded DC Systems. Chapter 8 covers solidly grounded DC low-voltage and low-voltage secondary distribution systems. Ground fault detection and localization are discussed, as well as the performance of GFCI and similar devices in a low-voltage secondary distribution system.

Chapter 9, Designing Shipboard Power System Grounding/Earthing Systems. Chapter 9 provides recommendations for the types of grounding systems that should be used with each of the distribution system types. It also provides insight on the impact of not isolating power systems with large voltage differences. The chapter presents a series of ship power system one-line diagrams highlighting the grounding system design to include all of the grounding systems described in Chapters 2 through 8. These one-line diagrams include:

- Shipboard power system for mechanical propulsion with low ship service load.
- Shipboard power system for mechanical propulsion with low ship service load and grounding bus.
- Shipboard power system for integrated power systems (IPS) with grounding transformers.
- Shipboard power system for IPS with generator neutral grounding resistors.
- Shipboard power system for IPS with grounding bus.
- Low-voltage AC zonal distribution system.
- Zonal AC primary distribution system with HRG in generator switchboards.
- Zonal AC primary distribution system with HRG in zonal switchboards.
- Commercial ship DC distribution system – integrated converters.
- Commercial ship DC distribution system – separate converters.
- Zonal DC primary distribution system.

Chapter 10, Power Conversion Equipment Grounding. Chapter 10 discusses the grounding implications of different power conversion options, to include: three-phase transformers, isolated power conversion equipment, and non-isolated power conversion equipment. Means for controlling CM voltages and currents in power conversion equipment are also presented. The chapter provides a number of examples of power conversion equipment configurations to explain how system grounding, bonding/protective earthing, and CM grounding apply to power conversion equipment.

Chapter 11, Shore Power (Cold Ironing) Connection Grounding. Chapter 11 covers shore power (cold ironing) connection grounding. A description of low-voltage shore power connections and high-voltage shore power connections is provided. The chapter includes details on equipotential grounding of the ship's hull with the shore-side ground, as well as a method for monitoring the equipotential bonding.

Chapter 12, Vehicle Connections Grounding. Chapter 12 covers grounding issues associated with embarked vehicles. It includes discharging of static electricity from hovering helicopters, power system grounding, and equipotential bonding of embarked vehicles.

Chapter 13, Common Mode Grounding: Impact of Common Mode Currents and Voltages on Grounding System. Chapter 13 covers the impact of CM currents and voltages on grounding systems. The chapter begins with a presentation of CM fundamentals. It continues with a discussion on the relationship of CM currents and voltages on EMI and philosophies and means for controlling CM currents and voltages. The chapter includes the theory of how CM chokes and CM shunts work.

Chapter 14, Protective Earthing: Bonding. Chapter 14 provides design considerations for implementing equipotential bonding of equipment to the ship's hull. A short discussion on testing of effective bonding is also included.

Chapter 15, Current Related Corrosion. Chapter 15 covers current-related corrosion, primarily on the ship's hull. The chapter provides an introduction to galvanic corrosion theory, including a discussion on how to use polarization curves to understand galvanic corrosion. The impact of galvanic corrosion on the ship's hull and other shipboard components and equipment is discussed. A short description is included of bearing corrosion due to CM currents flowing from a rotor, through the bearing, and into the equipment foundation. The chapter presents methods to counter current-related corrosion to include shaft brushes, sacrificial anodes, and impressed current cathodic protection.

Chapter 16, Lightning Protection Systems. Chapter 16 provides a description of lightning protection systems, along with design considerations to include: air terminals, down conductors, low impedance ground connection, and surge protection.

Chapter 17, Grounding Systems for Nonmetallic Hull Ships. Chapter 17 covers design considerations for power system grounding of nonmetallic hull ships.

Appendix A, Glossary. Appendix A provides a glossary of terms used in this book.

Appendix B, Acronyms and Abbreviations. Appendix B lists acronyms and abbreviations.

Appendix C, Impact of Electric Current on Humans. Appendix C provides information on the impacts of electric current on humans.

The aim of this book is to provide the reader with the background and insight necessary to analyze and design grounding systems for shipboard power systems.

1.8 Legal Notice

While this book provides many recommendations, class rules and regulations should always be consulted and followed. A thorough analysis should be conducted to ensure recommendations in this book are appropriate for any given application.

The examples, data, and other information provided in this document are believed by the authors to be correct. However, the examples, data, and other information should be independently verified and tested before being used in any design, particularly those where incorrect operation could result in equipment damage or injury to personnel.

We make no warranties, express or implied, that the examples, data, and other information provided in this book are free of error, that they are consistent with class rules, regulations, or other industry standards, or that they are appropriate for meeting the requirements of any application. The authors and publisher expressly disclaim the implied warranties of merchantability and of fitness for any particular purpose. The authors and publisher also disclaim all liability for direct, indirect, incidental, or consequential damages that result from any use of the examples, data, or other information in this book.

References

IEEE Std 3003.1-2019 (2019). *IEEE Recommended Practice for System Grounding of Industrial and Commercial Power Systems*. New York: Institute of Electrical and Electronics Engineers. Available from http://ieeexplore.ieee.org.

IEEE Std 45.1-2023 (2023). *IEEE Recommended Practice for Electrical Installations on Shipboard – Design*. New York: Institute of Electrical and Electronics Engineers. Available from http://ieeexplore.ieee.org.

NAVSEA S9407-AB-HBK-010 Revision 2 Change 2 (2010). *Handbook of Shipboard Electromagnetic Shielding Practices*. Washington, DC: US Navy, Naval Sea Systems Command.

NFPA 70 2023 Edition (2023). *National Electrical Code*. Quincy, MA: National Fire Protection Association. Available from www.nfpa.org.

2

System Grounding: Shipboard Ungrounded AC Systems (No Greater than 1 kV)

2.1 Characteristics

Three-phase AC power distribution systems with a nominal system voltage less than 1 kV have been prevalent in ship designs since the mid-twentieth century, and most have employed an ungrounded system. In IEC documents, these systems are designated IT (Isolation Terre). As depicted in Figure 2.1, ungrounded AC systems apply to all the different types of power systems except primary distribution systems; in some cases, other grounding methods are preferred.

As depicted in Figure 2.2, ungrounded power systems do not have equipment identified as grounding/earthing system equipment. There are, however, other high-impedance electrical connections (not shown) between the power system conductors and the ship's hull (ground/earth) to include

- Parasitic capacitance of cables and other equipment.
- Instrumentation such as ground fault detectors.
- Electromagnetic interference (EMI) filter capacitors.

Ungrounded AC systems are effectively capacitively grounded systems.

In the event of a ground fault (a short between a power system conductor and the ship's hull), current flows from the un-faulted conductors through the high-magnitude parasitic line-to-ground impedance (typically parasitic capacitances) and back through the faulted conductor at the site of the ground fault, as depicted in Figure 2.3. Since the parasitic line-to-ground impedance is distributed, the flow of fault current from the un-faulted conductors to ground occurs throughout the power system. For the faulted conductor, however, the fault current also enters at the site of the fault.

Design of Shipboard Power System Grounding/Earthing, First Edition. Norbert Doerry, Mohammed M. Islam, and John Prousalidis.
© 2025 The Institute of Electrical and Electronics Engineers, Inc.
Published 2025 by John Wiley & Sons, Inc.

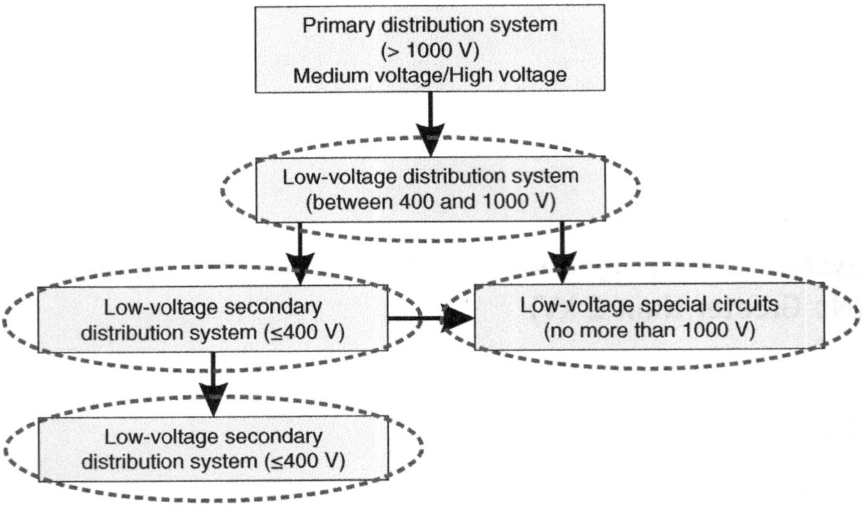

Figure 2.1 Applicability of ungrounded AC power systems.

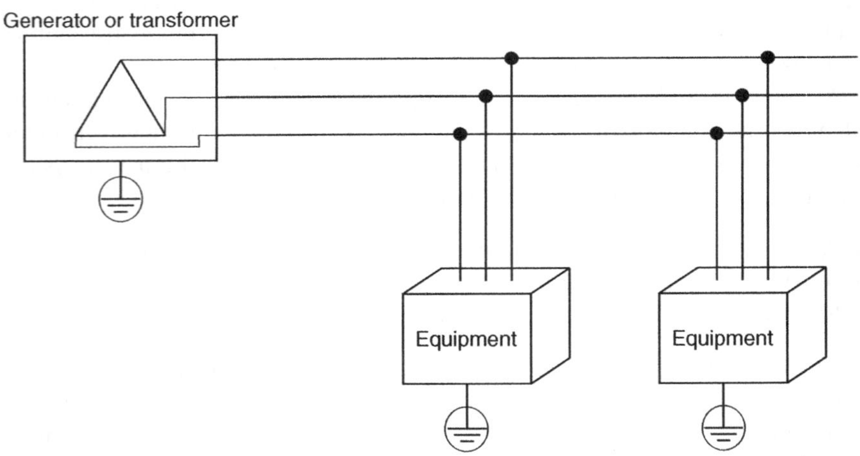

Figure 2.2 Ungrounded three-phase AC system with delta generator or transformer.

The steady-state ground fault current (at the fundamental frequency), expressed as a complex number, for each of the un-faulted phases is given by

$$i_{fault\_phase} = j\,\omega\,C_P\,V_{line-line}$$

$$j = \sqrt{-1}$$

$$\omega = 2\pi f$$

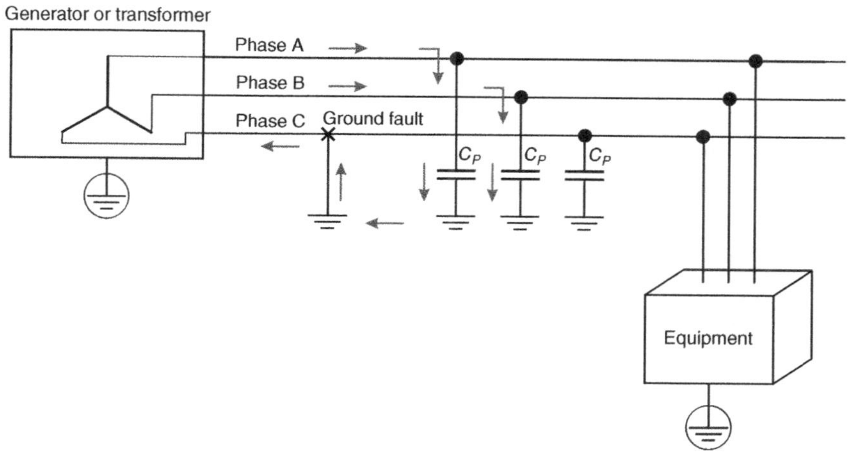

Figure 2.3 Path of ground fault current in an ungrounded system (wye source).

where

$V_{line\text{-}line}$, line-to-line complex voltage (V_{rms}).
ω, electrical frequency (rad/s).
f, electrical frequency (Hz).
C_p, line-to-ground (parasitic + EMI filter) capacitance (F).

The total steady-state ground fault current (A_{rms}) is the sum of the two currents from the un-faulted conductors. These currents are phase-shifted 60 electrical degrees.

$$i_{fault} = j\omega C_P V_{line-line} + j\omega C_P V_{line-line}\left(\cos\frac{\pi}{3} + j\sin\frac{\pi}{3}\right)$$

$$i_{fault} = -\frac{\sqrt{3}}{2}\omega C_P V_{line-line} + j\frac{3}{2}\omega C_P V_{line-line}$$

$$\left|i_{fault}\right| = \sqrt{3}\omega C_P V_{line-line} \tag{2.1}$$

The total steady-state ground fault current is equal in magnitude to the ground fault current supplied by each of the un-faulted conductors multiplied by $\sqrt{3}$.

For many ships, the total steady-state ground fault current in an ungrounded system will be on the order of amps. Ships with a relatively small distribution system and fewer EMI filters will have a lower steady-state ground fault current, while ships with a more expansive distribution system and many EMI filters will have higher steady-state ground fault currents.

One of the advantages of an ungrounded system is that, with the relatively small steady-state ground fault current, normal operation of the power system can continue when one phase has a ground fault. The ground fault should be cleared quickly, however, because a second ground fault on another phase will result in a line-to-line short circuit. Until this line-to-line short circuit is cleared, the extremely high short-circuit current will pass through the ship's structure, cable trays, and other metallic items between the locations of the two ground faults. This short-circuit current can cause a considerable amount of damage or could be hazardous to personnel.

In ungrounded systems, single-phase loads can be either line-to-line or line-to-neutral, as depicted in Figure 2.4. Single-phase load currents should be balanced as much as possible on the three phases. Unbalanced currents can lead to localized neutral-to-ground voltages, which can result in ground currents flowing through the line-to-ground capacitances.

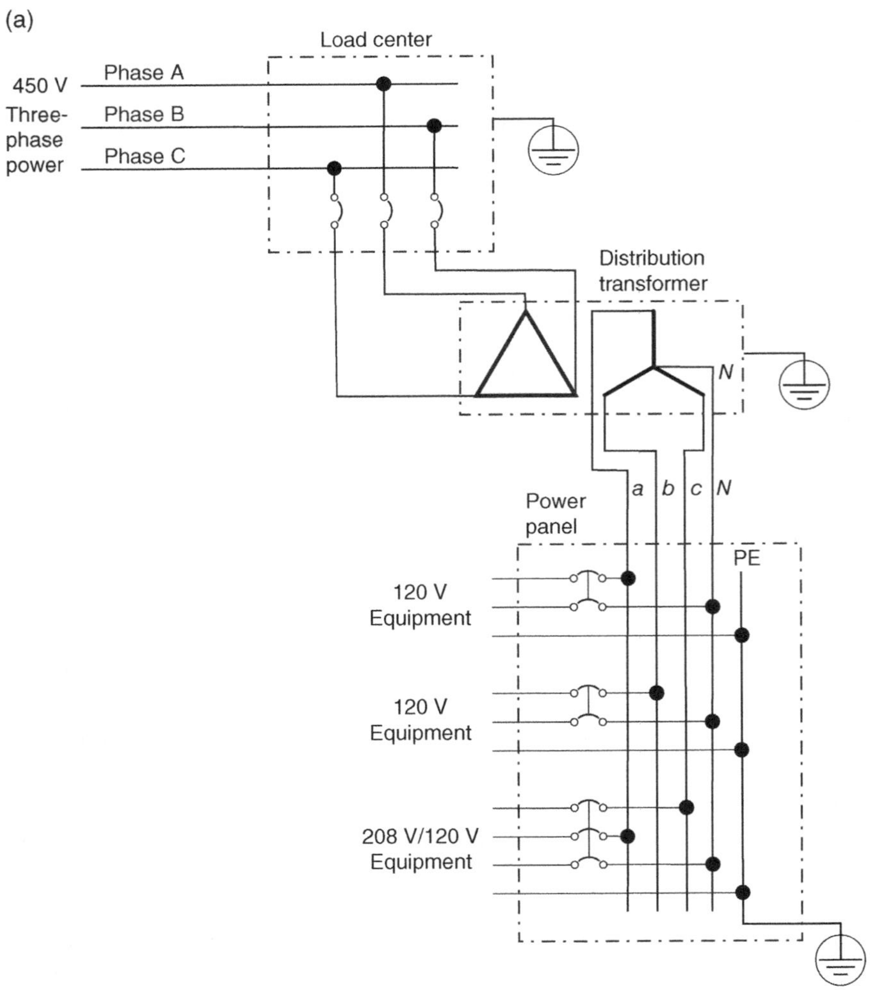

Figure 2.4 Supplying single-phase loads with (a) a three-phase four-wire transformer and (b) a single-phase transformer.

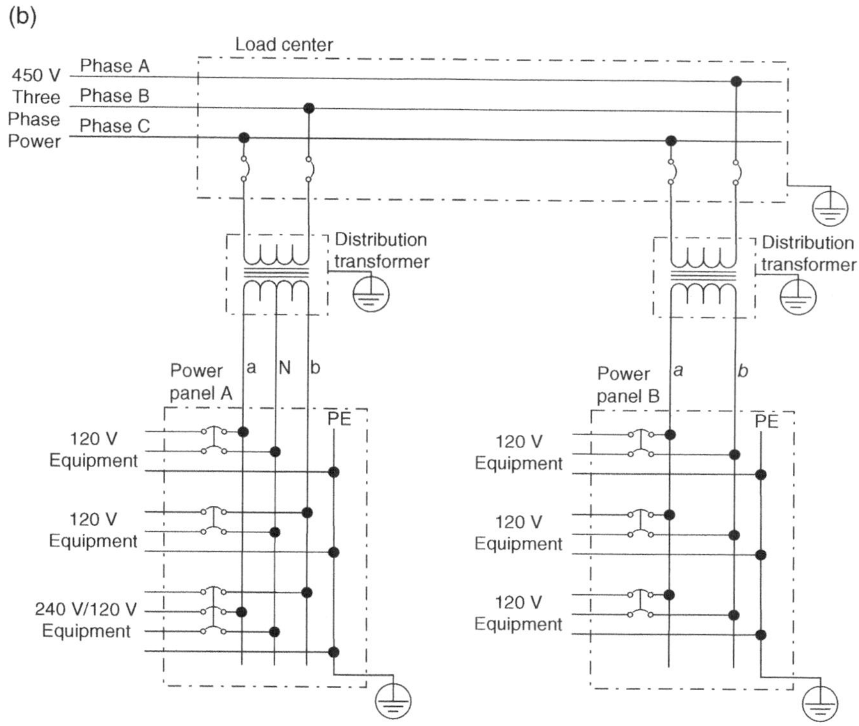

Figure 2.4 (Continued)

Since none of the conductors are guaranteed to be at ground (hull) potential, for personnel safety, single-phase loads should employ two-pole switches and circuit breakers, and three-phase loads should employ three-pole switches and circuit breakers. For secondary distribution systems with a potential steady-state ground fault current of 5 mA or more, all receptacle outlets located in potentially wet locations (such as bathrooms, kitchens, exterior decks, topside, well decks, helicopter hangars, etc.) should be served with a ground fault circuit interrupter (GFCI) or equivalent. In ungrounded systems, GFCIs that interrupt all power conductors should be used. Class society rules and national regulations should be consulted for minimum requirements.

2.2 Modeling Shipboard Ungrounded AC Low-Voltage Distribution Systems

Determining the value of C_p during the design of a shipboard system is difficult. The capacitance value provided in cable specifications is usually the conductor-to-conductor capacitance (C_{ll}), not the conductor-to-ground capacitance (C_p). Since

AC cables for systems with a nominal voltage no greater than 1 kV are typically not shielded, the conductor-to-ground capacitance is a function of the geometry of the cable installation within the cable tray and ship's structure. Generally, the power system designer does not attempt to calculate the contribution of cables to C_p and instead concentrates on maximizing the conductor-to-ground impedance of instrumentation equipment and EMI filters. MIL-STD-1399 Section 300 Part 1 (2018), for example, has a limit of 0.1 μF for the contribution of each EMI filter to C_p.

To potentially identify design issues with an ungrounded system, it may prove beneficial to conduct a transient analysis of the power system subjected to ground faults. Sections 2.2.1 through 2.2.6 provide guidance for developing suitable models.

2.2.1 Low-Voltage AC Cable Model

Figure 2.5 depicts the geometry of a low-voltage three-phase AC unshielded cable. Figure 2.6 depicts a model of the three-phase cable that includes the mutual inductance of the conductors. Modeling the mutual inductance (M) is very important because the inductance (L_a) included in cable data sheets assumes that the three phases are balanced. The balanced inductance, L_a, is calculated using Eq. (2.2).

$$L_a = L_S - M \tag{2.2}$$

Hence, the inductance cited in the data sheets can be considerably smaller than both the self-inductance (L_S) and the mutual inductance (M). During unbalanced conditions, Eq. (2.2) does not hold, and the inductances that govern the transient response can be considerably greater.

The model depicted in Figure 2.6 is a modification of a pi model; the resistance is split into three parts instead of being lumped together with the inductance. This is done to prevent the direct shorting of the line-to-ground capacitance when a ground fault is applied to the end of a cable segment; a direct short would result in an unrealistic current spike in the simulation. Modeling an inductor at the ends of the cable segment should be avoided because doing so can result in unrealistic voltage spikes when a ground fault is removed from the connection of one cable segment to another; the currents through the end inductors of the two cable segments may not be equal when the fault is removed.

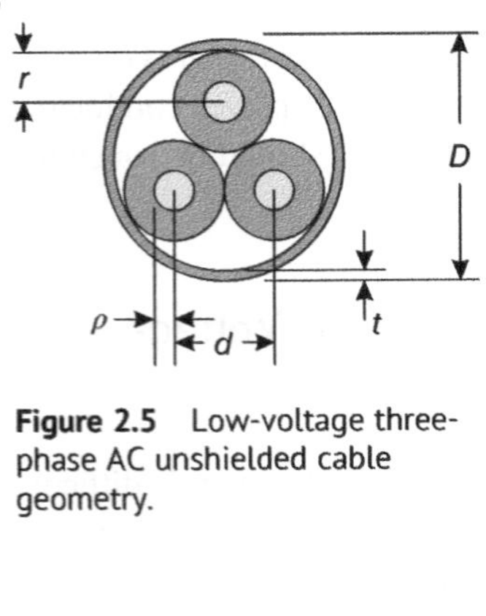

Figure 2.5 Low-voltage three-phase AC unshielded cable geometry.

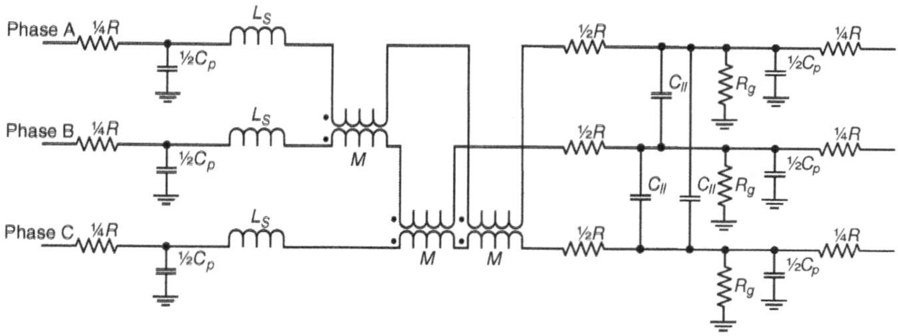

Figure 2.6 Dynamic model of a low-voltage three-phase AC unshielded cable for transient analysis.

The self-inductance (μH) of a round wire of radius ρ m and length l m is given by equation (7) of Grover (2009):

$$L_S = 0.2l\left[\ln\left(\frac{2l}{\rho}\right) - \frac{3}{4}\right]$$

The mutual inductance (μH) with respect to a conductor at a distance d m away is given by equation (1) of Grover (2009):

$$M = 0.2l\left[\ln\left(\frac{l}{d\sqrt{2}} + \sqrt{1 + \frac{l^2}{2d^2}}\right) - \sqrt{1 + \frac{2d^2}{l^2}} + \frac{d\sqrt{2}}{l}\right]$$

If the length l is much greater than the conductor separation d, then:

$$L_S = 0.2l\left[\ln(2) + \ln(l) - \ln(\rho) - \frac{3}{4}\right] \tag{2.3}$$

$$M \approx 0.2l\left[\ln\left(\frac{l\sqrt{2}}{d}\right) - 1\right]$$

$$M \approx 0.2l\left[\ln\left(\sqrt{2}\right) + \ln(l) - \ln(d) - 1\right] \tag{2.4}$$

Neglecting the impact of resistance and parasitic capacitance, the equation to calculate the voltage between the ends of the conductor for phase a of the cable (v_{L_a}) is thus:

$$v_{L_a} = L_S\frac{di_a}{dt} + M\frac{di_b}{dt} + M\frac{di_c}{dt}$$

$$v_{L_a} = L_S\frac{di_a}{dt} + M\frac{di_b + i_c}{dt} \tag{2.5}$$

where

v_{L_a} is the voltage across both the self-inductance and mutual inductance of phase a

i_a, i_b, and i_c are the phase currents.

For normal balanced operation, the sum of the currents is zero.

$$i_a + i_b + i_c = 0$$

$$i_b + i_c = -i_a$$

The voltage across the self-inductance and mutual inductance for phase a is

$$v_{L_a} = (L_S - M)\frac{di_a}{dt}$$

$$L_S - M = 0.2l\left[\ln(2) + \ln(l) - \ln(\rho) - \frac{3}{4} - \ln\left(\sqrt{2}\right) - \ln(l) + \ln(d) + 1\right]$$

$$L_a = L_S - M = 0.2l\left[\ln\left(\frac{d}{\rho}\right) + 0.597\right]\ \mu\text{H} \tag{2.6}$$

Equation (2.6) assumes the cable in unarmored. An armored cable will have a higher inductance. The work of Summers (1948) suggests the inductance of armored cable will be on the order of 15% higher than for unarmored cable. Physical testing or electromagnetic modeling of the cable should be performed if a more accurate value is required. Although the use of armored cable was once common onboard ships, most shipboard cable now is unarmored due to the high weight of the steel armor.

For unarmored cable, L_a is the value for the balanced inductance that is typically provided in cable data sheets. The conductor radius ρ can be determined from data sheets based on the gauge or cross-sectional area of the conductor. Three useful references for determining the conductor radius are IEEE Std 1580-2021 (2021), MIL-HDBK-299 (1991), and ASTM B8-11 (2017). With values for L_a and ρ, the distance between conductors (d) can be calculated (it is not normally provided in the data sheets) using Eq. (2.7).

$$d = \rho e^{\left(\frac{L_a}{0.2} - 0.597\right)} \tag{2.7}$$

The units for L_a are μH per meter, and the units for d are the same as for ρ (suggest using meters for consistency).

Alternatively, d can be approximated by

$$d = 2r = 2\rho + 2t_{insulation}$$

Where the thickness of the insulation ($t_{insulation}$) may be found in cable datasheets or approximated from data in IEEE Std 1580-2021 (2021).

With an estimate for the length of the cable (l), L_S and M (μH) can be estimated using Eqs. (2.3) and (2.4).

The resistance R is usually provided in the cable data sheets. IEEE Std 1580-2021 (2021), MIL-HDBK-299 (1991), and ASTM B8-11 (2017) may also be consulted for values. The resistance is provided for a given temperature. Use Eq. (2.8) from MIL-HDBK-299 (1991) to adjust the resistance of copper conductors from one temperature to another temperature.

$$R_T = R_{T0}(1 + \alpha_R(T - T_0)) \tag{2.8}$$

where

R_T, resistance at the desired temperature T (°C)
R_{T0}, resistance at temperature T_0 (°C)
α_R, temperature coefficient (1/°C).

For $T_0 = 20°C$, $\alpha_R = 0.00393$ (1/°C).
For $T_0 = 25°C$, $\alpha_R = 0.00385$ (1/°C).

Three-phase AC low-voltage distribution systems typically use unshielded three-conductor cables. Since the capacitance depends on the geometries of the cable conductors and the grounded conductors, calculating the line-to-ground capacitance is therefore very complex, but necessary to determine the required value of the grounding resistance. The ground conductors are typically a combination of the ship structure and the elements of the cable trays or cable hangers. Ideally, a finite element analysis program would be used to develop estimates of the line-to-ground capacitances. Of course, early in the design process, the details necessary to produce a finite element model likely do not exist.

One can approximate the line-to-ground capacitance of an unshielded cable by treating the capacitance as two capacitances in series: the conductor to the cable outer sheath capacitance (conductor-to-sheath capacitance) and the capacitance from the outer sheath to the structure (sheath-to-ground capacitance).

$$\frac{l}{C_p} = \frac{1}{C_1} + \frac{1}{C_2} \tag{2.9}$$

where

C_1, conductor-to-sheath capacitance (F)
C_2, sheath-to-ground capacitance (F)
l, length of the cable (m).

The capacitance per unit length of two coaxial cylinders is given by

$$C_{c\_axial} = l\frac{2\pi\varepsilon_0\kappa}{\ln(r_2/r_1)}$$

where

r_1, radius of the inner cylinder (m)
r_2, radius of outer cylinder (m)
ε_0, electric permittivity of free space
$\varepsilon_0 = 8.85$ pF/m
κ, dielectric constant.

While a three-conductor cable is not coaxial, we can produce a rough approximate value for C_1 by estimating values for r_1 and r_2 using Eqs. (2.10) and (2.11).

$$r_2 = \sqrt{\frac{rD}{2}} \tag{2.10}$$

$$r_1 = \rho \tag{2.11}$$

where

r, outer radius of the conductor insulation

D, outer diameter of the cable sheath

ρ, outer radius of the conductor.

C_1 can be estimated using Eq. (2.12).

$$C_1 \approx l \frac{2\pi\varepsilon_0 \kappa}{\ln\left(\frac{\sqrt{\frac{rD}{2}}}{\rho}\right)} \tag{2.12}$$

Nominal dielectric constants (κ) of some insulators are

Material	Dielectric constant (κ)
Ethylene propylene rubber (EPR)	2.24 (can vary significantly)
Nylon	4.00
Polyethylene (cross-linked) (XLPE)	2.30
Polypropylene	2.24
Silicone rubber	2.6
Polyvinyl chloride	2.7

C_2 may be estimated by applying Eq. (2.13) for an infinite-length straight conductor parallel to a conducting plane in air.

$$C_2 \approx l \frac{2\pi\varepsilon_0}{\operatorname{acosh}\left(\frac{2h}{D}\right)} \tag{2.13}$$

where

h, representative distance (m) of the center of the cable to the grounded conducting plane.

The value of h should be based on the type of cable hangers employed. If details on the cable hangers are not known, a value of 0.20 m may be used until better information is obtained. One may vary this value to determine the sensitivity of the transient response to this value.

The value of C_p obtained by using Eqs. (2.9)–(2.13) should be considered very crude. The actual value can easily vary from 80% to 150% of the calculated value when the length of cable is known with certainty.

One can approximate the capacitance between the conductors using Eq. (2.14).

$$C_{ll} \approx l \frac{\pi \varepsilon_0 \kappa}{\operatorname{acosh}\left(\frac{d}{2\rho}\right)} \tag{2.14}$$

2.2.2 Bus Duct

Bus duct consists of current-carrying busbars in an enclosure made of either aluminum or steel. The configuration of the busbars, the number of busbars per phase, and the configuration of the enclosure vary widely among manufacturers and can vary for different current ratings for a specific manufacturer.

In high-current applications, bus duct may be more economical to use and easier to arrange compared to multiple paralleled cables. The maximum ampacity of three-conductor cable is typically less than 400 A for cables that have a reasonable bend radius. Bus duct is available in ratings in excess of 6000 A, and with the proper fittings, it has an effective "bend radius" considerably less than that for cable.

The metal enclosure is normally physically attached to the ship's structure; hence, the enclosure is bonded to ground.

Some of the electrical properties of bus duct are normally provided by the manufacturer. These properties include resistance and reactance at line frequency under balanced operation. Capacitances and mutual inductances are not typically provided and must be calculated or measured. Piatek et al. (2013) provide methods for calculating these properties for arbitrary bus duct configurations. Alternatively, finite element methods can be employed.

2.2.3 EMI Filters

In some ungrounded AC systems, the line-to-ground capacitance of the EMI filters can be the dominant source of line-to-ground capacitance. If possible, one should obtain the EMI filter schematic or at least the value of the line-to-ground capacitance from the manufacturer. If it is not possible, once should consider using 0.1 μF for the line-to-ground capacitance for each EMI filter. This value is the maximum permissible according to MIL-STD-1399 Section 300 Part 1 (2018).

2.2.4 Generators and Motors

Generator and motor data sheets typically do not provide the line-to-ground capacitance. For AC low-voltage distribution systems, salient pole generators and motors with a rated speed of between 900 and 1800 rpm can be assumed to have a line-to-ground capacitance between 0.02 and 0.10 μF for rated power less than 3000 kVA. The capacitance tends to be higher for lower rpm and higher kVA ratings.

Low-voltage AC turbine generators with a rated speed of 3600 rpm can be assumed to have a line-to-ground capacitance of roughly 0.01 μF per rated MVA.

The actual line-to-ground capacitance depends on machine design, and the manufacturer should be consulted if values more accurate than those provided here are needed. These assumptions are extrapolated from data provided by Westinghouse Electric Corporation Central Station Engineers (1964) for higher voltage machines.

2.2.5 Transformers

As with generators and motors, data sheets for transformers typically do not provide the line-to-ground capacitance. For AC low-voltage distribution systems, the line-to-ground capacitance can be assumed to be within the range of 0.04–0.09 μF per rated MVA. Transformers with a grounded transformer shield will have a higher line-to-ground capacitance. The manufacturer should be consulted if more accurate values are required.

2.2.6 System Simulation

Figure 2.7a depicts a simple model of an ungrounded power system used for simulation in LTspice. The circuit parameters are representative but are not associated with any particular ship design. Figure 2.7b depicts how the elements of the model align with the physical components of the power system. Figure 2.8 shows the impact of a ground fault on phase *c* to all of the phase voltages. Figure 2.9 depicts the ground fault current. In general, the response consists of voltage and current spikes, then voltage and current transients for several milliseconds to tens of milliseconds, followed by a steady-state fault condition. The spikes and transients reflect the interaction of the cable inductances, resistances, and capacitances with the sudden shift in circuit topology. The magnitude of the transient voltages can be more than double the nominal system voltage. In some designs, line-to-ground surge protective devices (SPDs) are used to limit voltage spikes; the maximum continuous operating voltage of the SPD should be greater than the line-to-line voltage. The current spikes can have a magnitude of hundreds of amps or more.

After the fault clears, the neutral voltage will generally have a DC offset equal to the negative value of the line-to-neutral voltage of the phase that was faulted when the fault cleared. This DC offset can persist for a long time because the neutral-to-ground resistance is extremely high, and therefore the charge on the parasitic capacitances cannot be quickly dissipated.

One of the risks associated with an ungrounded system is the transient response to an intermittent ground fault. Figure 2.10 depicts a situation where a system with

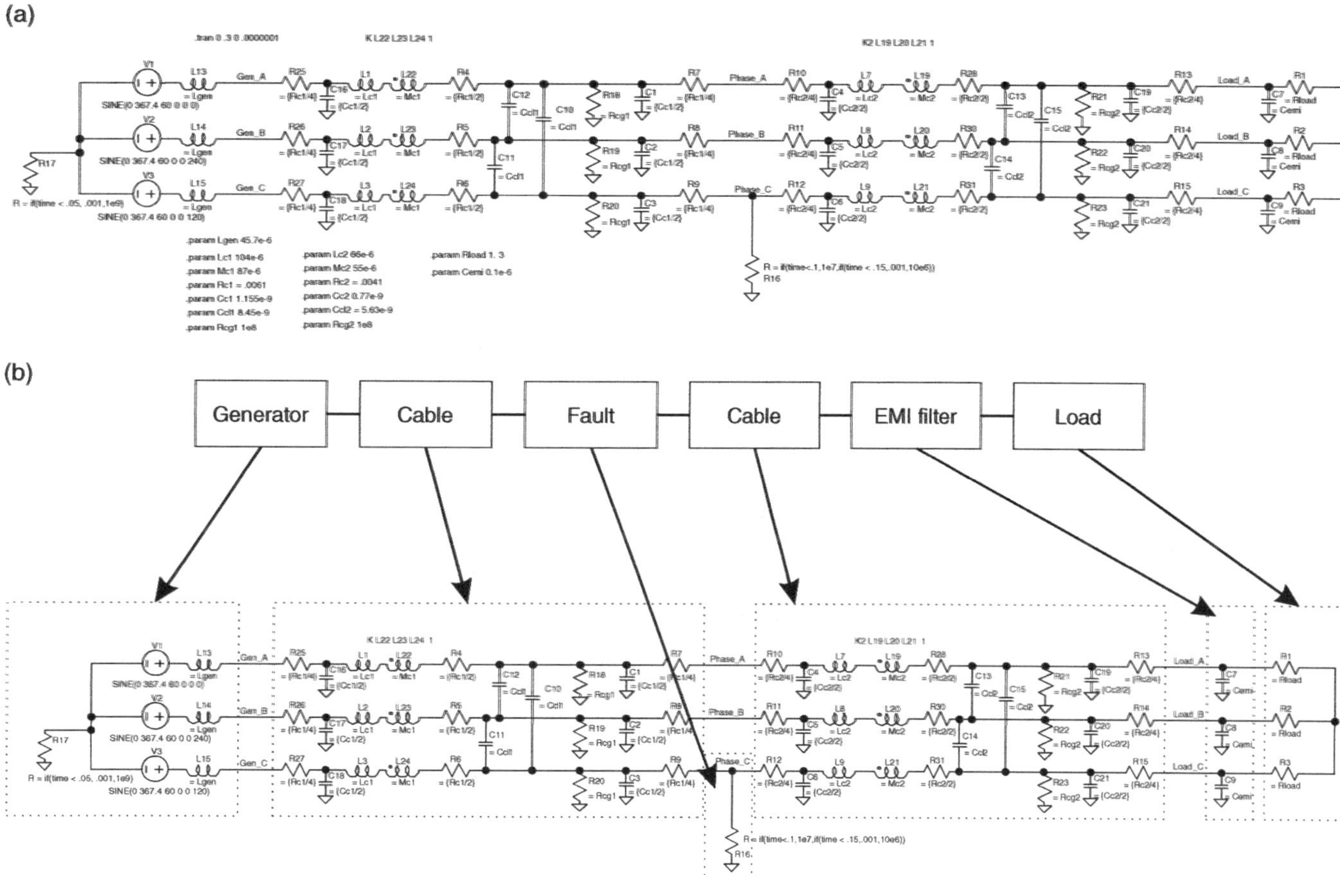

Figure 2.7 (a) LTspice simple model of an ungrounded system with ground fault; (b) identification of model boundaries for system components.

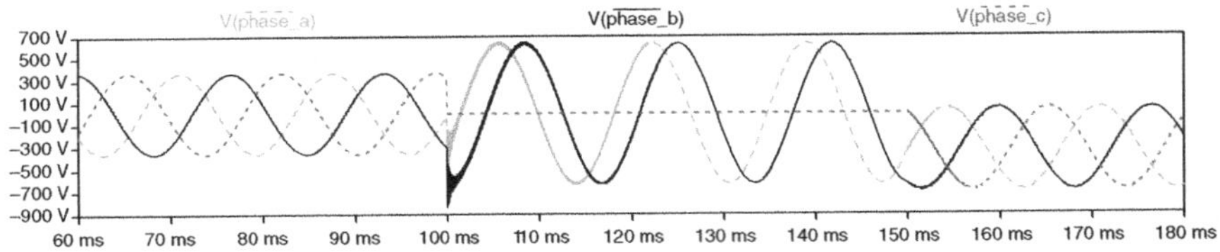

Figure 2.8 Line-to-ground phase voltages in an ungrounded system with ground fault (phase *c*).

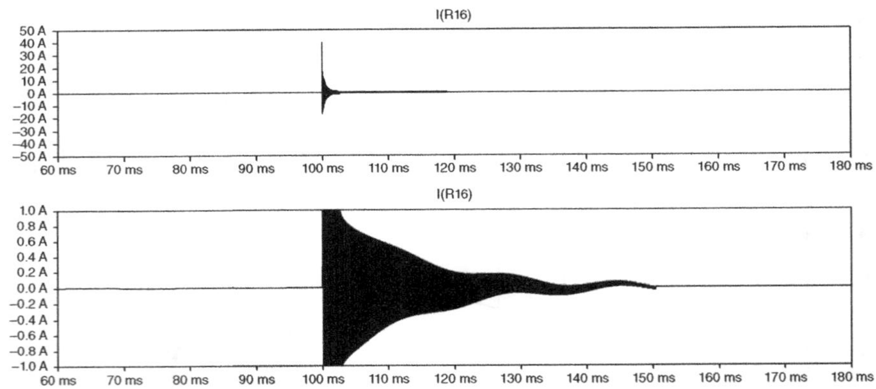

Figure 2.9 Current through ground fault in an ungrounded system.

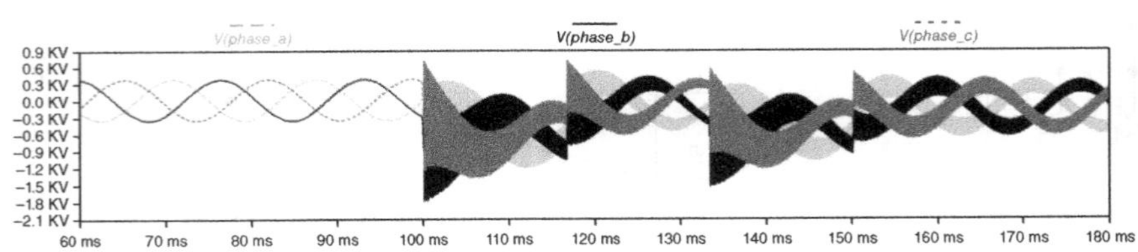

Figure 2.10 Impact of intermittent ground fault on an ungrounded system.

substantial EMI filter capacitance to ground (3 μF) experiences a ground fault that is present for roughly 1 ms, then off for roughly15.6 ms, repeated four times. After the fourth ground fault period, the ground fault is removed. As can be seen in Figure 2.10, the line-to-ground voltages experience large voltage spikes and high-frequency transients, with peak values potentially exceeding five times the line-to-ground voltage associated with the nominal system voltage. The magnitudes of the transients depend on the instantaneous capacitor voltages and inductor currents at the time the fault initiates and clears. Line-to-ground SPDs are usually selected to suppress spikes; however, the relatively long transients can result in SPD failures if the SPD is not appropriately rated. Because of the intermittent nature of the fault, ground fault detection equipment may have difficulty detecting this condition, possibly resulting in the system operating in this condition for a prolonged period of time. This possible condition is one reason ungrounded systems are not recommended for terrestrial power systems in IEEE Std 3003.1-2019 (2019).

Under normal, un-faulted conditions, the parasitic capacitance may provide a high-impedance connection between the power system conductors and the metallic components of equipment that are not intended to conduct current. To prevent a potentially hazardous voltage difference between these metallic components and the ship's structure and hull, they should be bonded via a low-impedance path to the ship's hull. These connections are designated as equipment grounding or "protective earth"; they are indicated by the protective earth symbol in Figure 2.2. The subject of bonding/protective earthing is discussed in greater detail in Chapter 14.

Figure 2.11 depicts the same circuit with a simple model of an arc fault. The arc fault susceptibility window is initiated at 100 ms and ends at 200 ms. During the arc fault susceptibility window, a fault is initiated if the line-to-ground voltage on phase *c* exceeds 340 V and extinguishes when the current experiences a zero crossing. As expected, and as depicted in Figure 2.12, during the arc fault susceptibility window, the phase *c* voltage to ground is no greater than 340 V. The other two phases, however, can approach 1 kV line-to-ground; this is more than 2.5 times the normal peak voltage. At 200 ms, when the susceptibility window ends, the three-phase voltages have a DC offset that slowly decays to zero.

When performing transient analysis of a ground fault, the importance of modeling the mutual inductance and self-inductance of the cable cannot be overemphasized. Figure 2.13 depicts the transient response of the same circuit used in Figure 2.10, with the exception that the self-inductance L_s and mutual inductance M are replaced with an inductance with the balanced inductance, L_a. The response is significantly less severe when the impact of the mutual inductance is not accounted for.

2.3 Ground Fault Detection

Detecting a ground fault in an ungrounded system is straightforward. One measures each of the line-to-ground voltages; when one of the voltages is near zero, then one has a ground fault. A traditional ground fault detection system is depicted in Figure 2.14. Three ground detection lamps are connected in wye with the neutral solidly grounded, often through a push-button switch. If the push button is

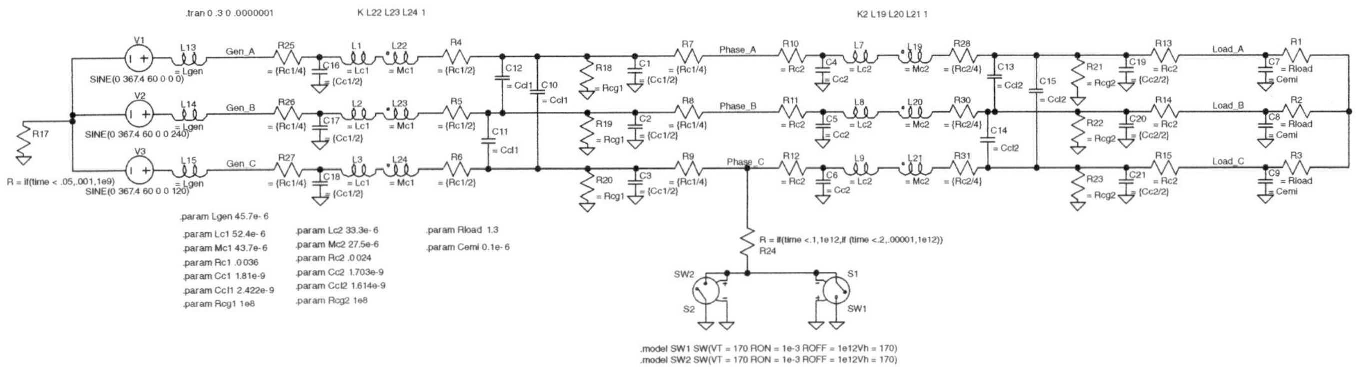

Figure 2.11 LTspice model to determine the impact of arc ground fault on an ungrounded system.

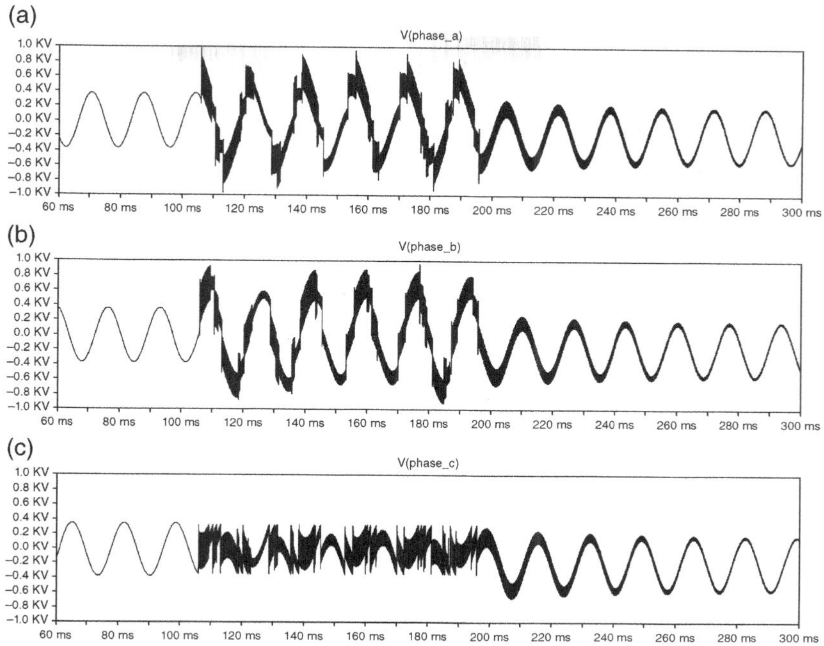

Figure 2.12 Impact of arc ground fault on an ungrounded system: (a) phase *a*; (b) phase *b*; (c) phase *c*.

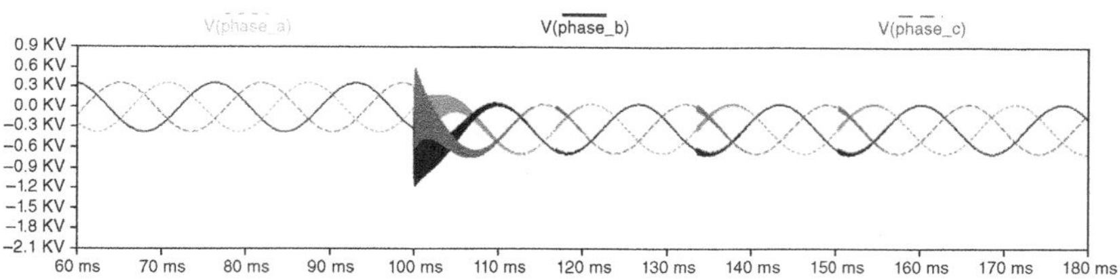

Figure 2.13 Impact of not modeling the cable with a mutual inductance.

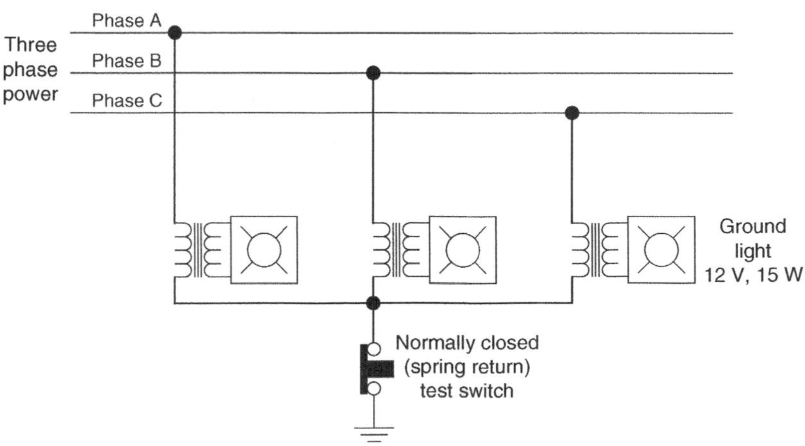

Figure 2.14 Lamp-based ground detection system.

normally closed, then normally (un-faulted) all three lights will be on, but will be dim because they will be connected to the line-to-neutral voltage. If a ground fault is present, one of the lights will be extinguished, and the other two will be bright. The light that is not on will correspond to the faulted phase, while the other two lights experience the full line-to-line voltage. If one presses the normally closed push button, then all three lights should be dimly lit; this can be used to ensure that one of the light bulbs is not burned out.

Some ground fault detectors use a normally open push-button switch. One must press the push-button switch to see if there is a ground fault.

Often, for voltages higher than 240 V_{rms} line-to-line, the light bulbs are connected to the circuit via a transformer to allow the use of standard light bulbs. In some models, light-emitting diodes (LEDs) are used instead of incandescent lamps. Figure 2.15 depicts a typical implementation of ground fault detection lights.

When a three-phase AC system is used to power a rectifier, and the fault occurs on the DC output of the rectifier, the ground fault manifests itself as a sequential ground fault on each of the AC phases. A model of such a system is depicted in Figure 2.16; a ground fault is applied on the negative output of the rectifier at

Figure 2.15 Ground fault detection panel display. *Source:* Mohammed M. Islam.

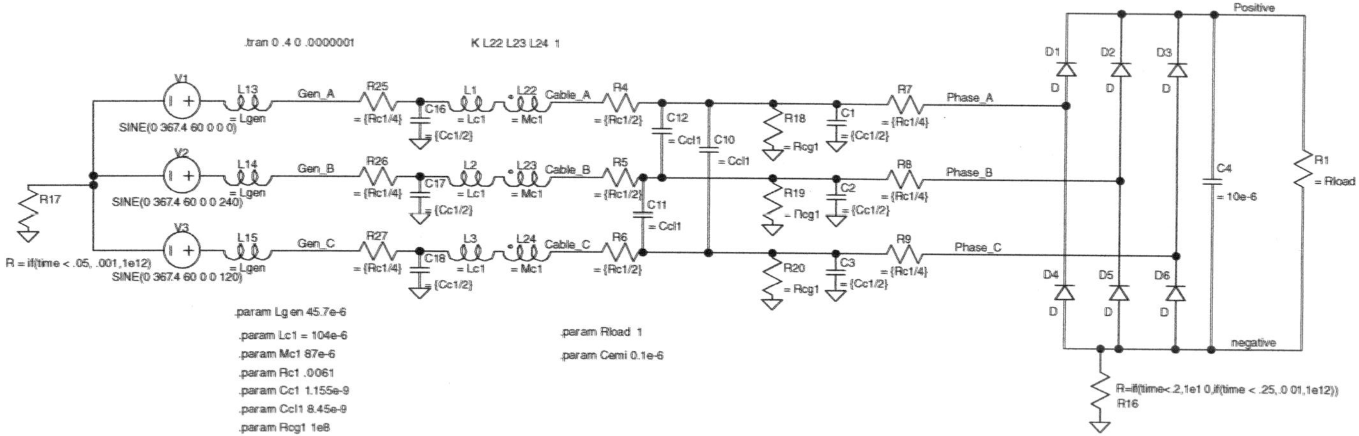

Figure 2.16 LTspice model of an ungrounded three-phase AC system feeding a passive rectifier.

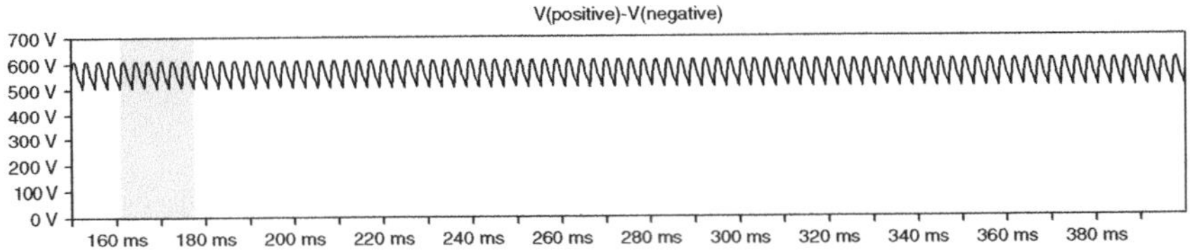

Figure 2.17 Rectifier output voltage (shaded area corresponds to one cycle).

200 ms and removed at 250 ms. Figure 2.17 depicts the rectifier output voltage; it is unaffected by the ground fault and exhibits the characteristic six pulses during a single cycle. As expected for a six-pulse rectifier, the AC line current as depicted in Figure 2.18 is heavily distorted. However, the AC line current is not impacted by the ground fault. The heavily distorted AC line current results in a heavily distorted rectifier input AC line-to-line voltage, as depicted in Figure 2.19; the line-to-line voltage is also not impacted by the ground fault. The line-to-ground voltage is shown in Figure 2.20. During the ground fault, the three light bulbs in Figure 2.14 ground detection circuit would all be the same brightness; one could not discern a ground fault from them. After the ground fault clears, the DC offset voltage on the line-to-ground capacitance will slowly decay through the line-to-ground resistance. The line-to-ground resistance can be extremely high, even higher than the 100 MΩ resistance modeled here.

Another interesting situation is when a bus-tie breaker connecting two online generators is open, but one of the contacts for one of the phases is stuck closed. As depicted in Figure 2.21 for a stuck contact on phase *c*, the ground lights will appear to show a "rolling ground fault" as the phase lights sequentially become dimmer and brighter at a frequency equal to the difference in frequencies of the two generators. Closing the bus-tie breaker will eliminate the indication of a ground fault as the two systems will be in synchronism.

Insulation monitoring devices are also used to detect ground faults. These devices inject either a voltage or current signal between the power system conductors and ground and evaluate the conductor insulation resistance by measuring the current or voltage impact of the injected signal. If the insulation resistance falls beneath a preset value, usually on the order of 50 kΩ, the operator is notified of the low insulation resistance. A panel meter such as the one depicted in Figure 2.22 is typically provided to inform the operator of the insulation resistance.

For many insulation monitoring devices, only one should be installed per system to avoid interference among multiple units. This includes ground detection lamps since their resistance to ground may be interpreted as low insulation resistance.

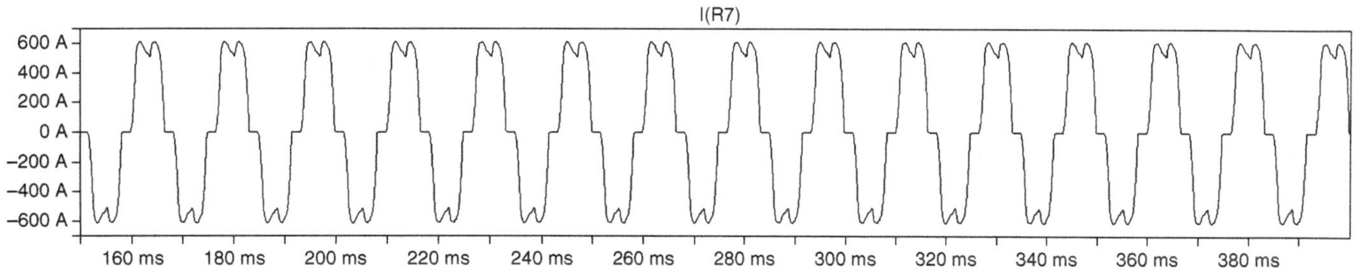

Figure 2.18 Rectifier AC phase current.

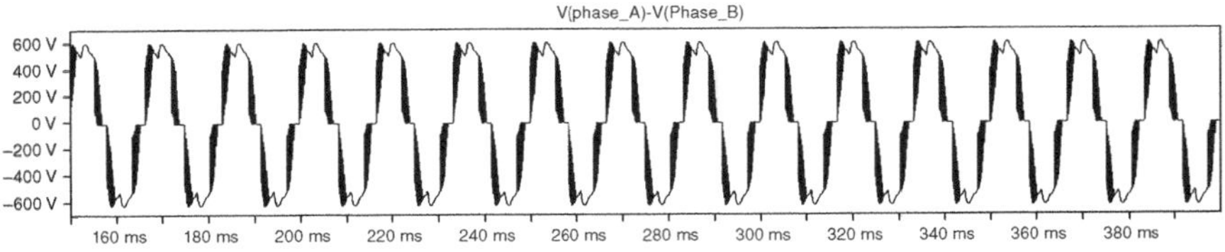

Figure 2.19 Rectifier AC line-to-line voltage.

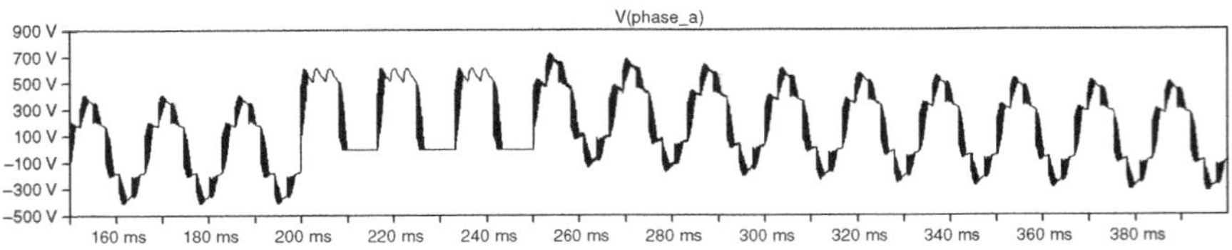

Figure 2.20 Line-to-ground voltage of AC system feeding a passive rectifier with ground fault on the negative conductor.

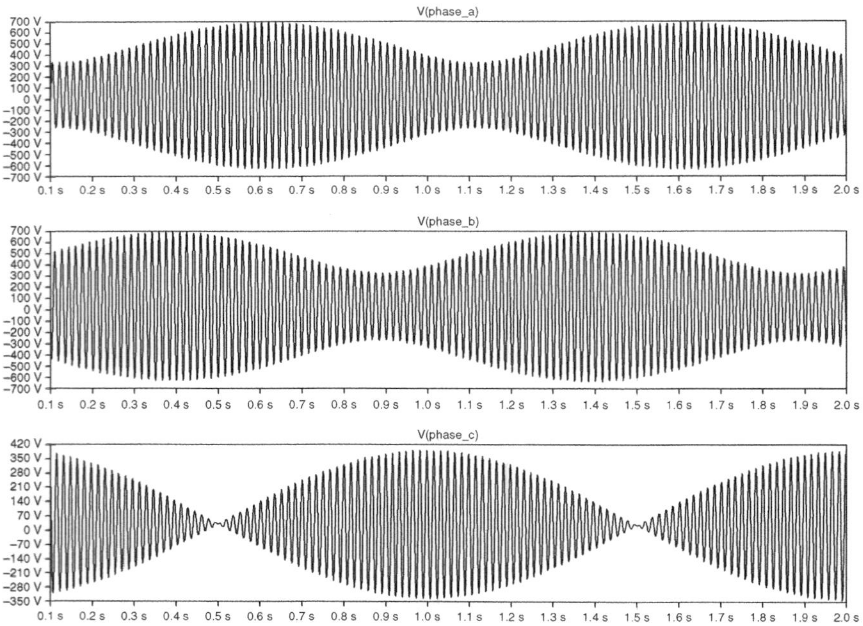

Figure 2.21 Line-to-ground phase voltages with stuck bus-tie breaker contact.

Figure 2.22 Insulation monitoring device panel meter. *Source:* Mohammed M. Islam.

2.4 Ground Fault Localization

Although it is straightforward to identify that a ground fault is present in an ungrounded system, finding the ground fault can be challenging. The traditional approach in a radial distribution system is to sequentially isolate feeder cables by turning off the associated circuit breakers. If the ground fault disappears, then the fault will be located to the last feeder turned off. That feeder is turned back on, and breakers on downstream panels are then sequentially turned off until the section of the power system with the ground fault is identified. A manual inspection of the cable and equipment within this section is performed to identify the cause of the fault.

This process is very time-consuming and disruptive to ship operations. Multiple ground faults on the same phase can be particularly challenging to locate if a systematic process is not employed.

An alternate approach that is not as disruptive, but also will not be successful in all cases, is to use a clamp-on current meter (Figure 2.23) around each unshielded three-phase feeder cable leaving a load center or power panel. The window area of

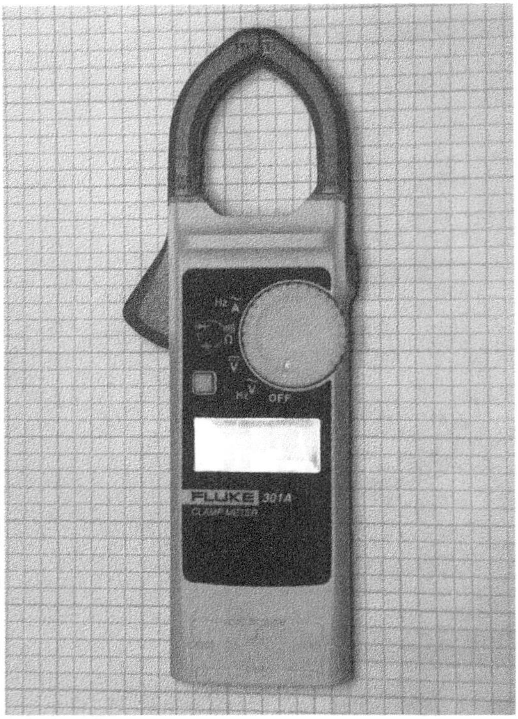

Figure 2.23 Clamp-on current meter. *Source:* Norbert Doerry (author 1).

the clamp-on current meter must be large enough to fully enclose the cable. By enclosing all three conductors, the common mode (CM) current is measured. The feeder cable with a ground fault will almost always have the largest CM current. Figure 2.24 depicts a schematic of a simple system with one generator feeder and three load feeders. Figure 2.25 demonstrates that the faulted feeder cable has the largest CM current.

In practice, it may be difficult to gain sufficient access to feeder cables to clamp on the current probe. Installing cables with sufficient room to enable clamping on current probes can greatly assist the process of localizing which feeder has a ground fault. This method will also not work well with cables that have shields or drain lines that are grounded on both ends; these additional conductors will carry a portion of the return CM current and thus will reduce the magnitude of the measured CM current.

With unshielded cable, it may be possible to locate where on a faulted feeder the ground fault is located because the CM current is likely to have a discrete change in magnitude between measurements on either side of the ground fault. Everywhere else, the change in CM current will be smooth due to the distributed nature of the

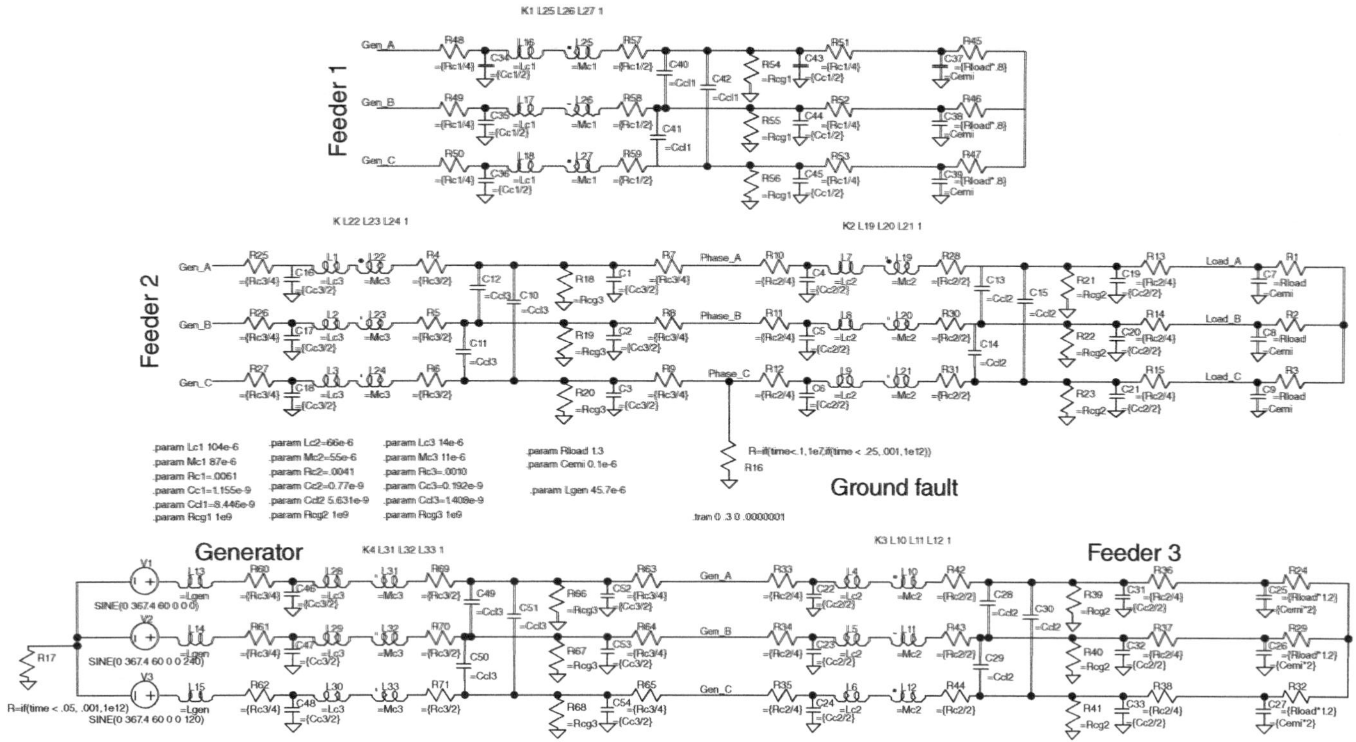

Figure 2.24 LTspice model of a system with one generator feeder and three load feeder cables.

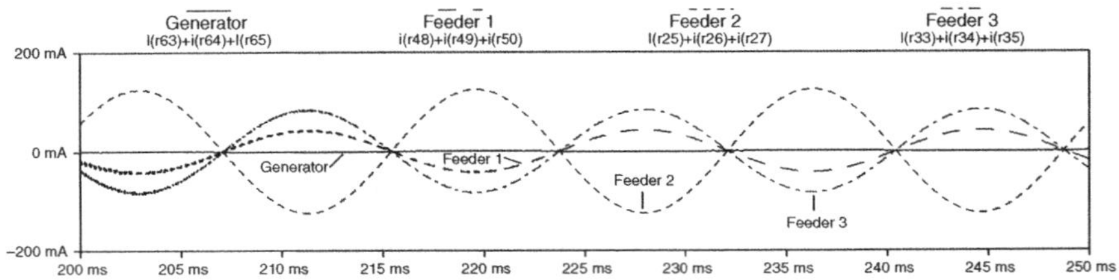

Figure 2.25 CM current at each feeder cable (feeder 2 is faulted).

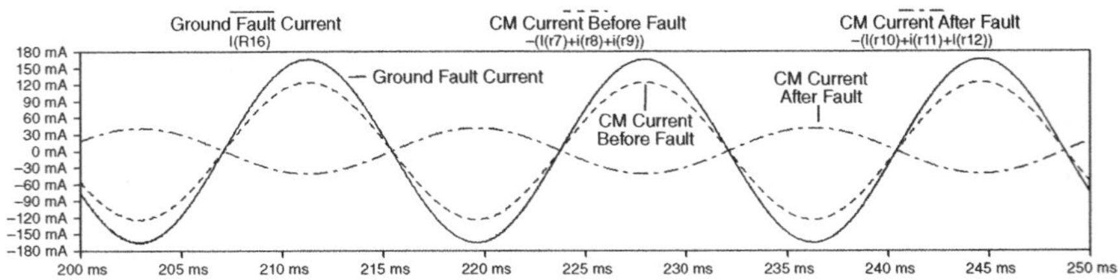

Figure 2.26 CM current of feeder before and after the fault location.

parasitic capacitance. This method does not work however, if the ground fault is in user equipment, within a load center/power panel/junction box where multiple cables are connected together, or if the ground fault happens to be in a location where the contribution of CM current from either side is roughly the same in magnitude. For the schematic depicted in Figure 2.24, Figure 2.26 demonstrates how the CM current changes immediately before and after the fault on feeder 1. Note that the CM currents are about 180° shifted in phase on either side of the ground fault; both sides of the cable feed into the ground fault.

Multiple ground faults on the same phase may require a combination of the traditional and alternate approach to successfully identify the faulted feeder cables.

Some insulation monitoring devices have the capability of identifying which feeder is faulted by integrating measurements from permanently installed current transformers (CTs) around all the conductors (but not the shields) of each of the feeders. With the capability to measure both the phase and magnitude of the CM current, the insulation monitoring devices have an improved capability in identifying the faulted feeder cable. The injection of a high-frequency (as compared to the fundamental frequency of the power system) signal between the power system and ground can also facilitate detection by the CTs of the faulted feeder. Furthermore, handheld devices can be employed to detect the high-frequency signal at short ranges to better localize the ground fault on the feeder cable.

The properties of the transient voltage that is observed when the ground fault initially happens can also be used to identify where the ground fault is located.

Extracting features of these transients to identify the fault location is the subject of ongoing research and may sometime in the future be reflected in commercially available projects.

2.5 Electrical Insulation Impacts

Because ungrounded systems can operate for extended periods of time with a ground fault, the insulation between conductors and ground should be rated for the full line-to-line voltage (instead of line-to-neutral voltage). The rated voltage for the insulation should be at least $\sqrt{3}$ times the nominal system voltage to account for the steady-state line-to-ground voltage during the fault and for the potential offset DC offset after the fault clears. The design of the insulation system should also account for the line-to-ground voltage spikes and transients that occur when a ground fault is initiated and cleared.

If simulations show the potential for high transient voltages upon clearing of faults, consideration should be given to using insulation with rated voltages higher than $\sqrt{3}$ times the nominal system voltage. Alternatively one could consider using a high-resistance ground system.

2.6 One-Line Diagram Symbology

Schematic symbols, such as those shown in Figure 2.14, are used in detailed circuit schematics. In one-line diagrams, ground detection systems, ground detection lamps, and insulation monitoring devices are often represented by the symbols shown in Figure 2.27. The top connection to the symbols represents a connection to the three phases, while the bottom connection represents a connection to the ship's hull.

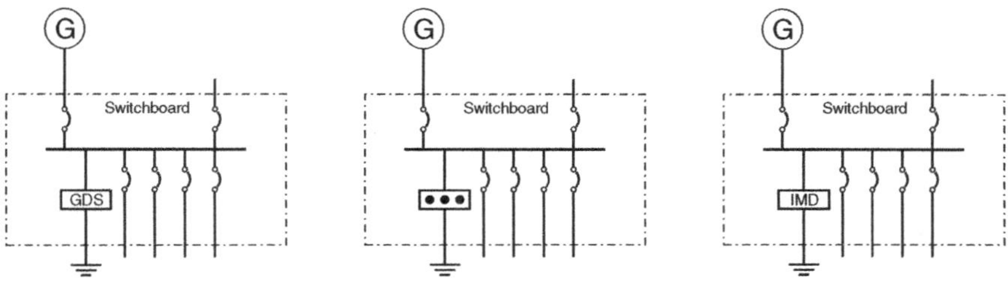

Figure 2.27 One-line diagram symbols for ground detection system, ground detection lamps, and insulation monitoring device.

2.7 Case Studies

2.7.1 Modeling a Low-Voltage AC Cable

Problem statement: determine the elements of a cable model for a 65 m cable bundle of five paralleled three-phase unshielded 400 mcm cable for a 440 V nominal system voltage system.

The following information is obtained from available sources:

R	0.031 Ω/1000 ft	MIL-HDBK-299 (1991)
$t_{insulation}$	0.050 in	MIL-DTL-24643/16
r_{sheath}	1.1015 in	MIL-DTL-24643/16
Insulation	Silicone rubber	MIL-DTL-24643/16
k	2.6	Section 2.2.1
2ρ	0.742 in	MIL-DTL-24643 (2011)

Convert to metric:

R	$0.0001017 \times l\ \Omega$
$r_{insulation}$	0.01069 m
r_{sheath}	0.02800 m
ρ	0.009423 m
l	65 m

Calculate inductances and capacitances:

$$L_S = 0.2l\left[\ln(2) + \ln(l) - \ln(\rho) - \frac{3}{4}\right]$$

$$L_S = 0.2l\left[\ln(2) + \ln(65) - \ln(0.009423) - \frac{3}{4}\right] = 1.756l\ \mu\text{H}$$

$$M \approx 0.2l\left[\ln\left(\sqrt{2}\right) + \ln(l) - \ln(d) - 1\right]$$

$$d = 2r_{insulation} = 2 \times 0.01069 = 0.02138\ \text{m}$$

$$M \approx 0.2l\left[\ln\left(\sqrt{2}\right) + \ln(65) - \ln(0.02138) - 1\right] = 1.473l\ \mu\text{H}$$

$$C_1 \approx l\frac{2\pi\varepsilon_0\kappa}{\ln\left(\frac{\sqrt{r_{insulation}r_{sheath}}}{\rho}\right)}$$

$$C_1 \approx l\frac{2\pi \times 8.85 \times 2.6}{\ln\left(\frac{\sqrt{0.01069 \times 0.02800}}{0.009423}\right)} = 237.9l\ \text{pF}$$

$$C_2 \approx l \frac{2\pi \times 8.85}{\operatorname{acosh}\left(\dfrac{0.20}{0.02800}\right)} = 20.9l\ \text{pF}$$

$$\frac{l}{C_p} = \frac{1}{C_1} + \frac{1}{C_2}$$

$$\frac{l}{C_p} = \frac{1}{237.9l} + \frac{1}{20.9l}$$

$$C_p = 19.2l\ \text{pF}$$

$$C_{ll} \approx l \frac{\pi \varepsilon_0 \kappa}{\operatorname{acosh}\left(\dfrac{d}{2\rho}\right)}$$

$$C_{ll} \approx l \frac{\pi \times 8.85 \times 2.6}{\operatorname{acosh}\left(\dfrac{0.02138}{2 \times 0.009423}\right)} = 140.9l\ \text{pF}$$

Calculate the parameters per cable (l equals length of cable):

$$L_{S_{cable}} \approx 65\ \text{m} \times 1.756\ \mu\text{H/m} = 114\ \mu\text{H}$$

$$M_{cable} \approx 65\ \text{m} \times 1.473\ \mu\text{H/m} = 95.7\ \mu\text{H}$$

$$R_{cable} = 65\ \text{m} \times 0.0001017\ \Omega/\text{m} = 0.00661\ \Omega$$

$$C_{p\_cable} \approx 65\ \text{m} \times 19.2\ \text{pF/m} = 1248\ \text{pF}$$

$$C_{ll} \approx 65\ \text{m} \times 140.9\ \text{pF/m} = 9159\ \text{pF}$$

Calculate the parameters per cable bundle (multiply capacitance of single cable by number of cables, and divide resistance and inductance of a single cable by number of cables):

$$L_{S_{bundle}} \approx \frac{114\ \mu\text{H}}{5} = 22.8\ \mu\text{H}$$

$$M_{bundle} \approx \frac{95.7\ \mu\text{H}}{5} = 19.1\ \mu\text{H}$$

$$R_{bundle} = \frac{0.00661\ \Omega}{5} = 0.00132\ \Omega$$

$$C_{p\_bundle} \approx 5 \times 1248\ \text{pF} = 0.006240\ \mu\text{F}$$

$$C_{ll\_undle} \approx 5 \times 9159\ \text{pF} = 0.0458\ \mu\text{F}$$

2.7.2 Modeling a Cable Line-to-Line Capacitance for GFCI Calculations

Problem statement: for a secondary 120 V AC 60 Hz single-phase power distribution system powered by a single-phase transformer and distributed to multiple loads via a power panel, determine whether a GFCI will have sufficient ground fault current to trip.

In the secondary distribution system depicted in Figure 2.28, the feeder cables to each load are three-conductor 14 AWG cables; two conductors are used for power, and the remaining conductor is used for a protective earth. The ground fault current on one phase originates from the un-faulted phase voltage in the power panel which is connected to the secondary of the transformer, travels through

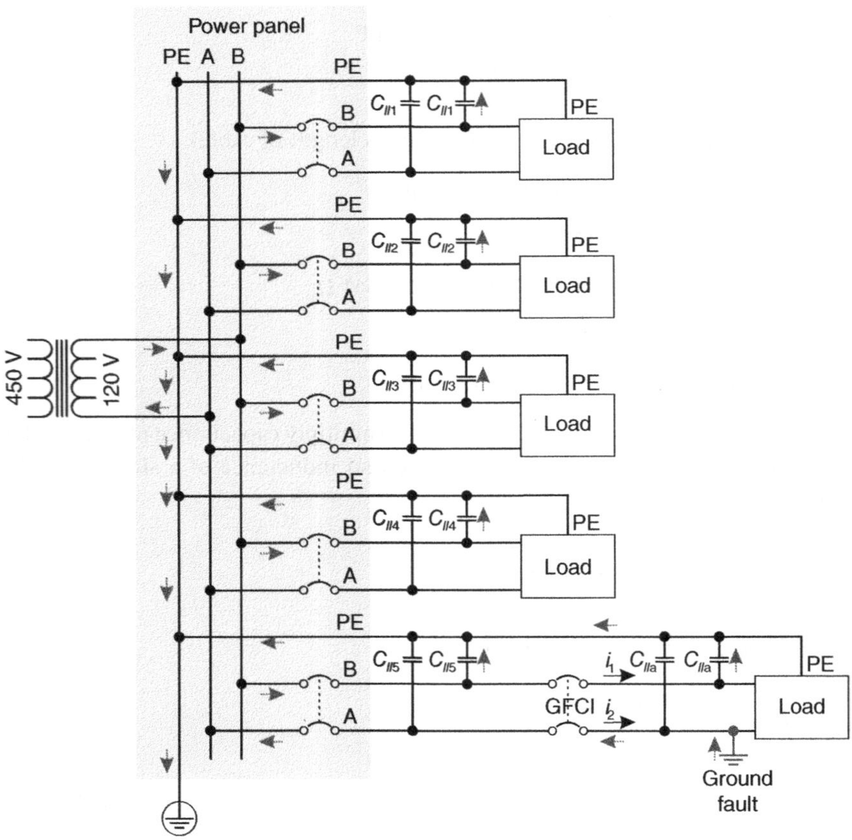

Figure 2.28 Secondary distribution system (120 V single-phase) with GFCI and ground fault.

the line-to-line capacitance of the cables (or through line-to-ground EMI capacitors if present) to the protective earth connection. In the power panel, the fault current flows through the protective earth connection to the ship's hull; from the ship's hull, the current flows through the ground fault, then back to the faulted phase voltage in the power panel which is also connected to the secondary of the transformer.

The GFCI will trip if the current magnitudes of i_1 and i_2 in Figure 2.28 differ by more than 5 mA. The fault current is the source of the difference in magnitude.

The line-to-line capacitance between the un-faulted conductor and the protective earth conductor on the transformer side of the GFCI limits the amount of fault current. To provide enough fault current (5 mA) for the GFCI to trip requires the line-to-ground impedance on the source side of the GFCI to be no more than:

$$X = \frac{120\ \text{V}}{5\ \text{mA}} = 24\ \text{k}\Omega$$

Assuming all of the impedance is due to line-to-ground capacitance, the minimum capacitance between the power conductor and the protective earth conductor on the transformer side of the GFCI is thus:

$$C_{min} = \frac{1}{X2\pi f} = \frac{1}{24 \times 10^3 \times 2 \times \pi \times 60} = 0.11\ \mu\text{F}$$

C_{min} is a lower bound, if the load current is largely resistive, then the current of i_2 will be largely resistive and the current of i_1 will be a combination of the resistive current and the capacitive current. For the difference in magnitude of the two currents to be greater than 5 mA, the capacitive current must be larger than 5 mA. The exact value will depend on the load current. If the load is not drawing any current, than C_{min} is sufficient to trip the GFCI.

The question now becomes: how many meters of cable will be required on the transformer side of the GFCI to achieve the minimum capacitance?

Data for a representative cable (LSTSGU-4) includes

Conductor diameter: 0.0726 in	ASTM-B8-11 (2017)
Insulation thickness: 0.018 in	MIL-DTL-24643/16 (2009)
Assume $k = 2.3$ for thermoset insulation	MIL-DTL-246436/16 (2009) 2.3 is representative of XLPE

From this data, the line-to-line capacitance (which in this case is also line-to-ground capacitance since one of the conductors is grounded) is given by

$$C_{ll} \approx l \frac{\pi \varepsilon_0 \kappa}{\operatorname{acosh}\left(\frac{d}{2\rho}\right)}$$

$$\rho = \frac{0.0726}{2} \text{ in} = 0.0363 \text{ in}$$

$$d = 2 \times 0.018 + 0.0726 \text{ in} = 0.1086 \text{ in}$$

$$C_{ll} \approx l \frac{\pi(8.85)\,2.3}{\operatorname{acosh}\left(\frac{0.1086}{0.0726}\right)} \text{ pF} = 66.7 \times l \text{ pF}$$

The number of meters of cable required to achieve C_{min} (without any EMI filters) is

$$l = \frac{0.11 \times 10^{-6}}{66.7 \times 10^{-12}} = 1649 \text{ m}$$

It is unlikely that a power panel with several dozen feeder cables would have the total length of feeder cables exceeding 1649 m. However, the line-to-ground capacitance of EMI filters is included in C_{min}, EMI filters for 120 V applications typically have a line-to-ground capacitance between about 0.001 and 0.033 μF; depending on the design of the EMI filters and the number of loads that have them, C_{min} may be achieved with EMI filters alone.

For a power panel with only a couple of EMI filters, the line-to-ground capacitance is likely less than C_{min}; for circuits connected to such a power panel, the installation of a GFCI does not provide any additional protection.

The line-to-ground capacitance on the load side of a GFCI should also be limited to well below 0.11 μF to avoid tripping due to a ground fault on another feeder not protected by a GFCI, as depicted in Figure 2.29. The feeder cables themselves are not likely to result in a line-to-ground capacitance of 0.11 μF; the presence of multiple EMI filters on the same GFCI protected circuit is likely required. If the line-to-ground capacitance is larger than 0.11 μF, then the line-to-ground current will be greater than 5 mA; if the ground fault current is sufficiently large such that i_1 in Figure 2.29 is more than 5 mA greater in magnitude than i_2, then the GFCI will trip. The GFCI will trip even though the ground fault is not on the feeder it is meant to protect. Many hours could be wasted looking for a ground fault on the GFCI-protected circuit when the ground fault is actually on a circuit not protected by a GFCI.

If a secondary distribution panel is powered by a three-phase transformer with a delta secondary, as depicted in Figure 2.30, the currents through the parasitic (and EMI) capacitors of the un-faulted phases are added as vectors, 120 electrical

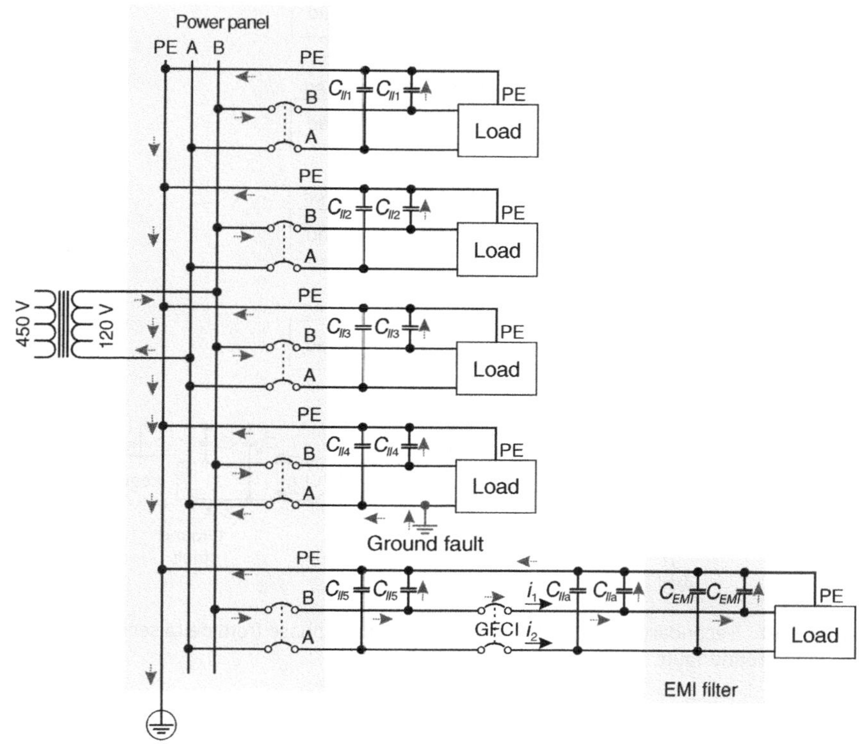

Figure 2.29 Secondary distribution system (120 V single-phase) with GFCI and ground fault with large EMI filters.

degrees apart. This can result in the required meters of cabling increasing (approaching doubling), depending on the length of cable connected to each of the un-faulted phases. A circuit analysis or simulation should be performed to ensure sufficient ground fault current will be present to trip the GFCI. Circuit analysis should also be performed if the secondary distribution is powered by a delta-wye transformer, and loads are applied line-to-neutral. As shown in Figure 2.31, if the neutral side of the load feeder has a ground fault, then the secondary distribution system is essentially solidly grounded; if the feeder cables are relatively balanced, then little fault current will flow; the GFCI is unlikely to trip. If a phase conductor has a ground fault as shown in Figure 2.32, then the phase currents of the various parasitic (and EMI) capacitors to the faulted phase (phase a in this example) must be summed.

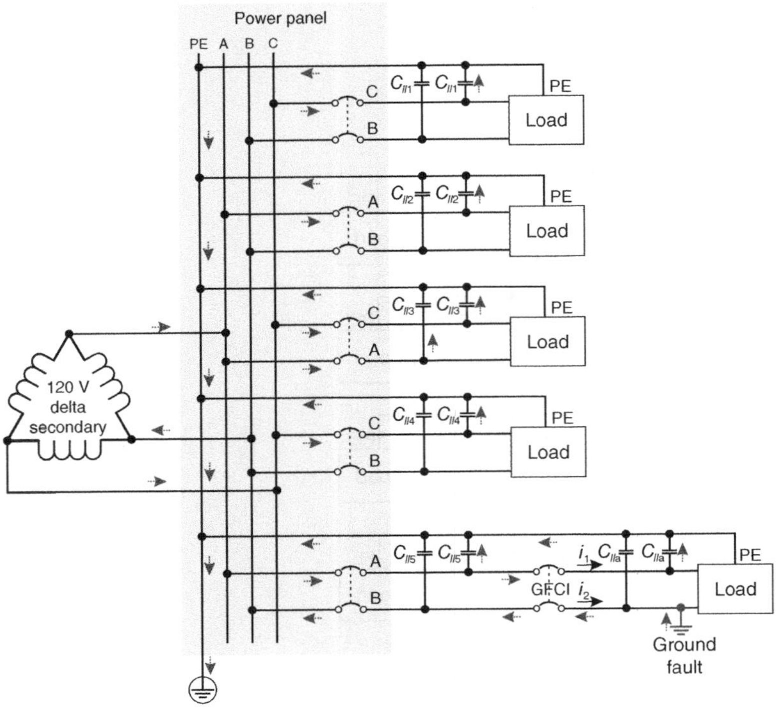

Figure 2.30 Secondary distribution system (120 V three-phase from delta secondary) with GFCI and ground fault.

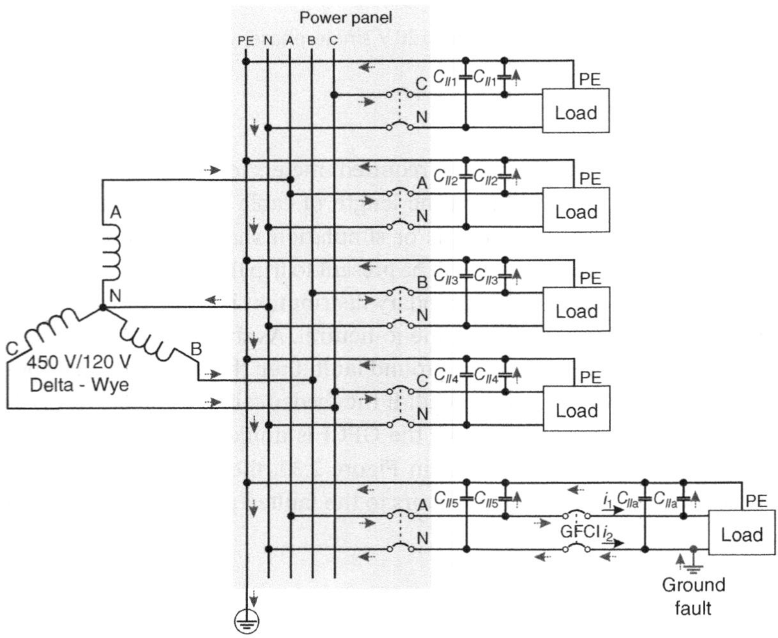

Figure 2.31 Secondary distribution system (120 V three-phase from wye secondary) with GFCI and ground fault on neutral.

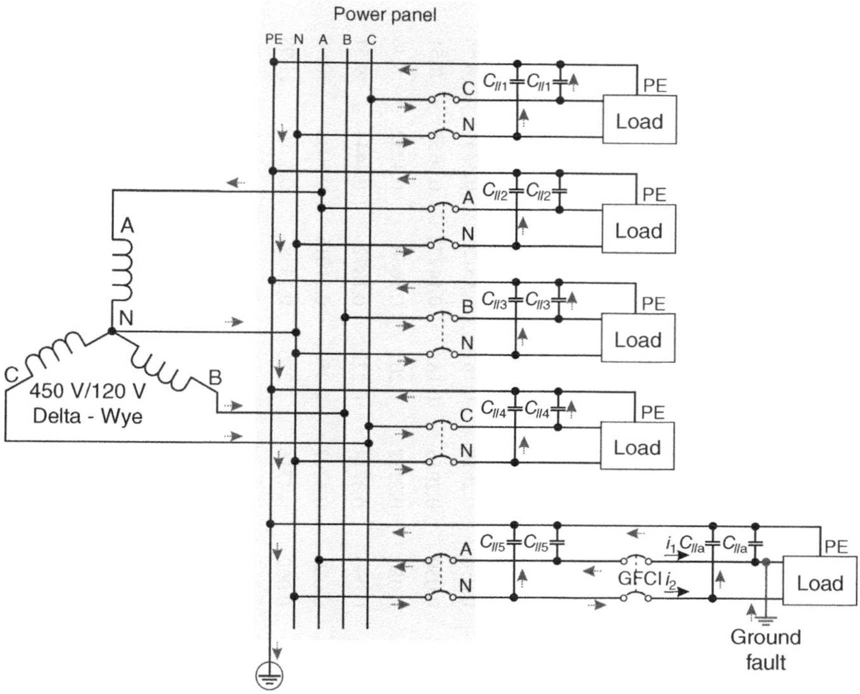

Figure 2.32 Secondary distribution system (120 V three-phase from wye secondary) with GFCI and ground fault on phase *a*.

2.8 Reference Cable Data

LSTSGU cable is common onboard US Navy ships where unarmored, watertight, low-smoke, halogen-free three-conductor cable with a rating of 1 kV is required. Tables 2.1 and 2.2 provide dimensional data and resistance data from existing standards and specifications. They also present the results of calculations based on the equations in Section 2.2.1. LSTSGU cable uses a thermoset insulation; for these calculations, the dielectric constant for XLPE (which is a thermoset insulation) was used. The resistance values are presented for 25°C; these values should be adjusted as described in Section 2.2.1 if the resistance at another temperature is desired. The calculation of the parasitic capacitance assumes the distance of the cable from the ground plane is 0.20 m. The self and mutual inductances are calculated at three conductor lengths: 10, 25, and 50 m.

Table 2.1 Cable dimensions for LSTSGU cable.

Cable	AWG	D (in)	$r-\rho$ (in)	t (in)	2ρ (in)	r (in)	ρ (in)	d (in)	2ρ (mm)
LSTSGU-3	16 (Class B)	0.411	0.018	0.030	0.0576	0.047	0.0288	0.0936	1.46
LSTSGU-4	14 (Class B)	0.449	0.018	0.030	0.0726	0.054	0.0363	0.1086	1.84
LSTSGU-9	10 (Class B)	0.575	0.015	0.040	0.1130	0.072	0.0565	0.143	2.87
LSTSGU-14	9 (Class B)	0.718	0.018	0.040	0.1260	0.081	0.063	0.162	3.20
LSTSGU-23	7 (Class B)	0.812	0.018	0.050	0.1590	0.098	0.0795	0.195	4.04
LSTSGU-30	30(19) Navy	0.852	0.030	0.050	0.2020	0.131	0.101	0.262	5.13
LSTSGU-40	4 (Class C)	0.900	0.030	0.050	0.2345	0.147	0.11725	0.2945	5.96
LSTSGU-50	3 (Class C)	0.969	0.035	0.050	0.2630	0.167	0.1315	0.333	6.68
LSTSGU-60	2 (Class C)	1.060	0.035	0.050	0.2955	0.183	0.14775	0.3655	7.51
LSTSGU-75	1 (Class C)	1.134	0.035	0.050	0.3332	0.202	0.1666	0.4032	8.46
LSTSGU-100	0 (Class D)	1.266	0.035	0.060	0.3744	0.222	0.1872	0.4444	9.51
LSTSGU-125	00 (Class D)	1.408	0.040	0.060	0.4203	0.250	0.21015	0.5003	10.68
LSTSGU-150	000 (Class D)	1.515	0.040	0.060	0.4716	0.276	0.2358	0.5516	11.98
LSTSGU-200	0000 (Class D)	1.669	0.050	0.060	0.5301	0.315	0.26505	0.6301	13.46
LSTSGU-250	250 MCM (Class C)	1.794	0.050	0.060	0.5764	0.338	0.2882	0.6764	14.64
LSTSGU-300	300 MCM (Class D)	1.957	0.050	0.075	0.6314	0.366	0.3157	0.7314	16.04
LSTSGU-350	350 MCM (Class D)	2.073	0.050	0.075	0.6820	0.391	0.341	0.782	17.32
LSTSGU-400	400(127) Navy	2.203	0.050	0.075	0.7420	0.421	0.371	0.842	18.85
Reference see Table 2.2	1	1	1	1	2 and 3	Calc	Calc	Calc	Calc

Table 2.2 Cable electrical data for LSTSGU cable.

	$l = 10$ m		$l = 25$ m		$l = 50$ m						
Cable	L_s (μH)	M (μH)	L_s (μH)	M (μH)	L_s (μH)	M (μH)	L_a (μH/m)	R (Ω/1000 ft)	R (Ω/m)	C_{ll} (pF/m)	C_p (pF/m)
LSTSGU-3	18.9	15.4	51.9	43.0	110.8	93.0	0.355	4.300	0.014108	60	11.4
LSTSGU-4	18.5	15.1	50.8	42.3	108.4	91.5	0.338	2.680	0.008793	67	11.7
LSTSGU-9	17.6	14.5	48.5	40.9	104.0	88.8	0.305	1.080	0.003543	90	12.6
LSTSGU-14	17.4	14.3	48.0	40.3	102.9	87.5	0.308	0.859	0.002818	87	13.2
LSTSGU-23	16.9	13.9	46.8	39.4	100.6	85.7	0.299	0.543	0.001781	97	13.7
LSTSGU-30	16.4	13.3	45.6	37.9	98.2	82.7	0.310	0.358	0.001175	85	14.0
LSTSGU-40	16.1	13.1	44.9	37.3	96.7	81.5	0.304	0.290	0.000951	91	14.3
LSTSGU-50	15.9	12.8	44.3	36.7	95.6	80.3	0.305	0.210	0.000689	90	14.6
LSTSGU-60	15.7	12.7	43.7	36.2	94.4	79.4	0.300	0.190	0.000623	95	15.0
LSTSGU-75	15.4	12.5	43.1	35.7	93.2	78.4	0.296	0.134	0.000440	100	15.3
LSTSGU-100	15.2	12.3	42.6	35.2	92.0	77.4	0.292	0.106	0.000348	106	15.8
LSTSGU-125	15.0	12.0	42.0	34.7	90.9	76.2	0.293	0.089	0.000291	105	16.3
LSTSGU-150	14.7	11.8	41.4	34.2	89.7	75.3	0.289	0.067	0.000221	111	16.7
LSTSGU-200	14.5	11.6	40.8	33.5	88.6	73.9	0.293	0.053	0.000174	106	17.3
LSTSGU-250	14.3	11.4	40.4	33.1	87.7	73.2	0.290	0.044	0.000146	110	17.7
LSTSGU-300	14.1	11.3	39.9	32.8	86.8	72.4	0.287	0.038	0.000124	115	18.2
LSTSGU-350	14.0	11.1	39.6	32.4	86.0	71.8	0.285	0.032	0.000104	119	18.6
LSTSGU-400	13.8	11.0	39.1	32.1	85.2	71.0	0.283	0.027	0.000090	125	19.0
Reference	Calc	Calc	Calc	Calc	Calc	Calc	Calc	1	Calc	Calc	Calc

(Continued)

Table 2.2 (Continued)

Cable	$l = 10$ m		$l = 25$ m		$l = 50$ m						
	L_s (μH)	M (μH)	L_s (μH)	M (μH)	L_s (μH)	M (μH)	L_a (μH/m)	R (Ω/1000 ft)	R (Ω/m)	C_{ll} (pF/m)	C_p (pF/m)
References											
1	MIL-DTL-24643/16 (2009)										
2	MIL-DTL-24643 (2011)										
3	ASTM B8-11 (2017)										
Calc	Calculated from equations in Section 2.2.1										
Notes											
1	The dielectric constant is assumed to be 2.3										
2	Resistance at 25°C										
3	h assumed as 0.2 m (distance to ground plane)										

References

ASTM B8-11 (2017). *Standard Specification for Concentric-Lay-Stranded Copper Conductors, Hard, Medium-Hard, or Soft*. Philadelphia, PA: ASTM International. Available from http://webstore.ansi.org.

Grover, Frederick W. (2009). *Inductance Calculations*. Mineola, NY: Dover Publications, Inc.

IEEE Std 1580-2021 (2021). *IEEE Recommended Practice for Marine Cable for Use on Shipboard and Fixed or Floating Facilities*. New York: Institute of Electrical and Electronics Engineers. Available from http://ieeexplore.ieee.org.

IEEE Std 3003.1-2019 (2019). *IEEE Recommended Practice for System Grounding of Industrial and Commercial Power Systems*. New York: Institute of Electrical and Electronics Engineers. Available from http://ieeexplore.ieee.org.

MIL-DTL-24643 (2011). *Detail Specification Cables, Electric, Low Smoke Halogen-Free, For Shipboard Use General Specification for*. Washington, DC: Department of Defense. Available from http://quicksearch.dla.mil.

MIL-HDBK-299 (1991). *Cable Comparison Handbook, Data Pertaining to Electric Shipboard Cable*. Washington, DC: Department of Defense. Available from http://quicksearch.dla.mil.

MIL-STD-1399 Section 300, Part 1 (2018). *Low Voltage Electric Power, Alternating Current*. Washington, DC: Department of Defense. Available from http://quicksearch.dla.mil.

MIL-DTL-24643/16 (2009). *Detail Specification Sheet Cable, Electrical, −20°C to +105°C, 1000 Volts, Type LSTSGU*. Washington, DC: Department of Defense. Available from http://quicksearch.dla.mil.

Piatek, Z., Baron, B., Jablonski, P. et al. (2013). Numerical Method of Computing Impedances in Shielded and Unshielded Three-Phase Rectangular Busbar Systems. *Progress In Electromagnetics Research B* 51: 135–156. https://doi.org/10.2528/PIERB13030205.

Summers, Samuel D. (1948). Effective Resistance and Inductance of 3-Conductor Shipboard Power Cables. *Transactions of the American Institute of Electrical Engineers* 67 (2): 1345–1350. https://doi.org/10.1109/T-AIEE.1948.5059827.

Westinghouse Electric Corporation Central Station Engineers (1964). *Electrical Transmission and Distribution Reference Book*. East Pittsburgh PA: Westinghouse Electric Corporation.

3

System Grounding: Shipboard HRG AC Low-Voltage Distribution Systems (No Greater than 1 kV)

3.1 Characteristics

As described in Chapter 2, one drawback of the ungrounded AC system is the possibility of significant line-to-ground transient overvoltages due to an intermittent ground fault, as well as a sustained DC neutral offset when the fault clears. Adding a resistance from the power system neutral to ground dampens the transient voltage and provides a means for the DC neutral offset voltage to decay to zero when the ground fault is cleared. The resistance value should be low enough to shorten the time of the overvoltage, but high enough so that fault protection components do not react to the ground fault current. This latter property distinguishes a high-resistance grounded (HRG) system from a low-resistance grounded (LRG) system. An HRG system enables continued operation with a single ground fault because the fault protection components are not sensitive to the low levels of ground fault current. An LRG system will selectively trip circuit breakers, but at a reduced level of fault current compared to a solidly grounded system. LRG systems are typically used when ground faults in a solidly grounded system would result in fault currents that exceed the interrupting capability of the circuit breakers or would cause excessive damage. Because of the inability to continue operation with a single ground fault, LRG systems are not typically used in shipboard systems.

As shown in Figure 3.1, HRG is generally not used on AC secondary distribution systems, which normally supply loads with either 120 or 240 V single-phase power; these systems are typically either ungrounded or solidly grounded. AC secondary distribution systems usually power only a small portion of the ship; the risk of intermittent ground faults on an ungrounded secondary distribution system is low. The added complexity of incorporating the HRG system is usually not warranted; however, there is no technical reason to rule out HRG for AC secondary distribution systems.

Design of Shipboard Power System Grounding/Earthing, First Edition. Norbert Doerry, Mohammed M. Islam, and John Prousalidis.
© 2025 The Institute of Electrical and Electronics Engineers, Inc.
Published 2025 by John Wiley & Sons, Inc.

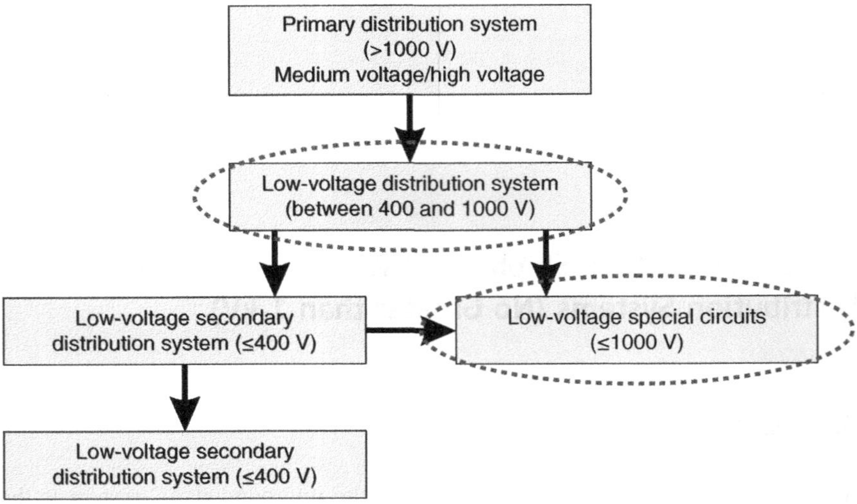

Figure 3.1 Applicability of high-resistance grounded AC low-voltage distribution systems.

On the other hand, AC low-voltage distribution systems operating between 400 and 1000 V can benefit from HRG; they span much larger portions of the ship and serve many more loads. The risk of an intermittent ground fault is much higher. Depending on their characteristics, low-voltage special circuits may also benefit from HRG.

Figure 3.2 depicts how a grounding circuit is used to create the HRG system ground for an AC system powered from a delta winding set from either a generator or transformer. The grounding circuit typically consists of a grounding transformer, a grounding resistor, and instrumentation, as shown in Figure 3.3. Figure 3.4 depicts the use of a neutral grounding resistor (NGR) with a wye-connected winding set from either a generator or transformer. An NGR with associated instrumentation is shown in Figure 3.5.

As depicted in Figure 3.6, in the event of a ground fault, current flows from the un-faulted conductors through the high-magnitude line-to-ground impedance (typically parasitic capacitances) and back through the faulted conductor at the site of the ground fault. Ground fault current also flows from the power system neutral through the ground fault and the back through the faulted conductor. Since the line-to-ground impedance is distributed, the flow of fault current from the un-faulted conductors to ground occurs throughout the power system.

The steady-state ground fault current (at the fundamental frequency) expressed as a complex number, for each of the un-faulted phases, is given by

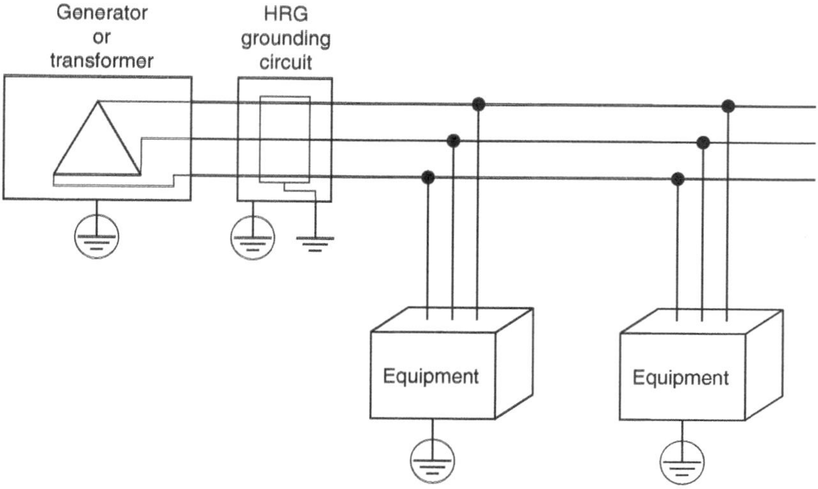

Figure 3.2 HRG three-phase AC system with delta generator or transformer.

Figure 3.3 HRG with grounding transformer and grounding resistor (grounding circuit) and instrumentation. *Source:* Courtesy of I-Gard.

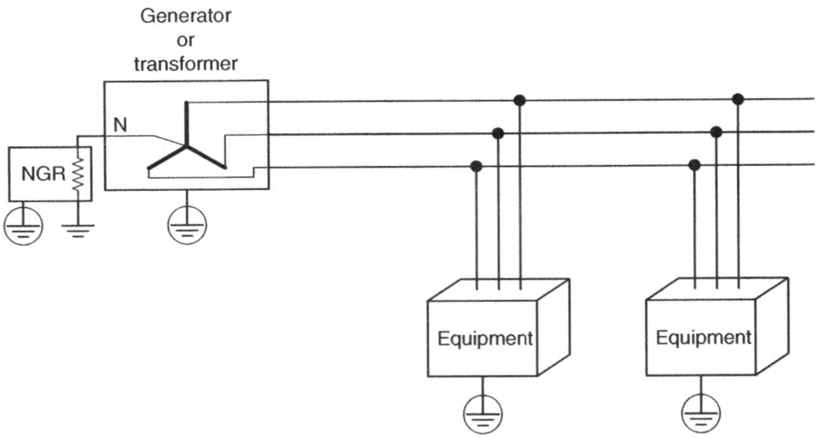

Figure 3.4 Three-phase AC system with wye generator or transformer and NGR.

Figure 3.5 NGR with instrumentation. *Source:* Courtesy of I-Gard.

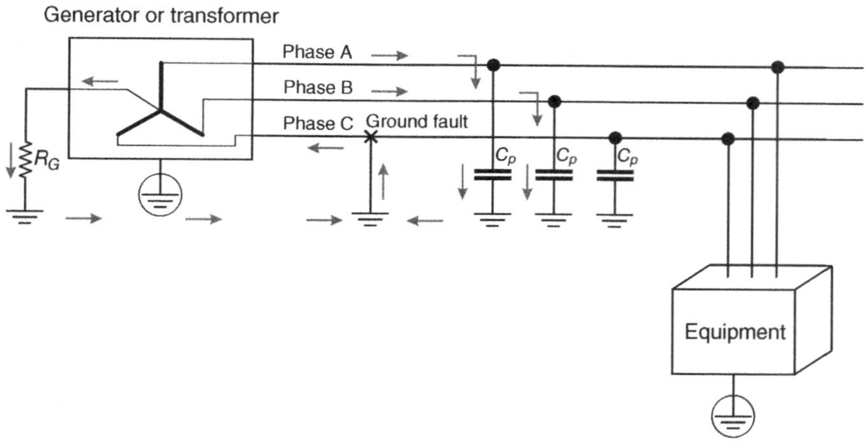

Figure 3.6 Path of ground fault current in a HRG system using an NGR.

$$i_{fault\_u\_phase} = j\omega C_P V_{line-line} \tag{3.1}$$

The steady-state ground fault current (at the fundamental frequency) expressed as a complex number, for the grounding resistor is given by

$$i_{fault\_f\_phase} = \frac{V_{line-line}}{\sqrt{3}R_G} \tag{3.2}$$

$$j = \sqrt{-1}$$

$$\omega = 2\pi f$$

where

$V_{line-line}$, line-to-line voltage (V_{rms})
ω, electrical frequency (rad/s)
f, electrical frequency (Hz)
C_p, line-to-ground capacitance (F)
R_G, grounding resistor resistance (Ω).

The total steady-state ground fault current (A_{rms}) is the sum of the two currents from the un-faulted conductors and the current through the grounding resistor:

$$i_{fault} = j\omega C_P V_{line-line}\left(\cos\frac{\pi}{6} + j\sin\frac{\pi}{6}\right) + j\omega C_P V_{line-line}\left(\cos\frac{\pi}{6} - j\sin\frac{\pi}{6}\right) + \frac{V_{line-line}}{\sqrt{3}R_G}$$

$$i_{fault} = \frac{V_{line-line}}{\sqrt{3}}\left(\frac{1}{R_G} + j3\omega C_P\right) \tag{3.3}$$

$$|i_{fault}| = \frac{V_{line-line}}{\sqrt{3}}\sqrt{\left(\frac{1}{R_G}\right)^2 + (3\omega C_P)^2} \tag{3.4}$$

If a grounding resistance is chosen to be nearly equal to the impedance of the paralleled line-to-ground parasitic capacitances, then:

$$R_G \approx \frac{1}{3\omega C_p} \tag{3.5}$$

$$3R_G C_p \approx \frac{1}{\omega} \tag{3.6}$$

Now $3R_GC_p$ is the time constant of the exponential decay of the neutral voltage; from Eq. (3.6), the time constant is about equal to $\frac{1}{\omega}$, where $\frac{2\pi}{\omega}$ is the period of the fault current waveform. Hence, there will be over six time constants (2π) of decay within one period of the waveform; any DC offset on the system neutral will have decayed to essentially zero within one period.

Figure 3.7 depicts a simple model of an HRG power system used for simulation in LTspice. The circuit parameters are representative, but are not associated with any particular ship design. Figure 3.8 shows the impact of a ground fault on phase *c* to the phase voltages. Figure 3.9 depicts the ground fault current, and Figure 3.10 depicts the current through the grounding resistor.

In general, the response consists of transient voltage and fault current spikes for the first millisecond or so, followed by a steady-state fault condition. The transients reflect the interaction of the cable inductances, resistances, and capacitances to the sudden shift in circuit topology. The magnitude of the transient voltages can be more than double the nominal system voltage. In some designs, line-to-ground SPDs are used to limit these voltage spikes. The fault current spikes can have a magnitude of 10s of amps. Compared to the ungrounded system, the voltage and current spikes are similar because the grounding resistor value is still considerably greater than the cable resistances. The steady-state fault current is larger than in an ungrounded system because of the additional fault current through the grounding resistor. When the fault is removed, the phase voltages quickly return to their pre-fault waveforms and the grounding resistor current drops to zero. The DC offset of the neutral voltage present in an ungrounded system when the fault clears is not present in an HRG system. As shown in Figure 3.11, the line-to-line voltages only show a transient at the initiation of the ground fault; otherwise, the line-to-line voltages are not impacted by the ground fault.

If the EMI capacitance is greatly increased to 3 μF, the grounding resistor must be accordingly reduced. With the new grounding resistor, the line-to-ground voltages during the ground fault are nearly identical to the previous case, as shown in Figure 3.12. While the ground fault current through the grounding resistor has

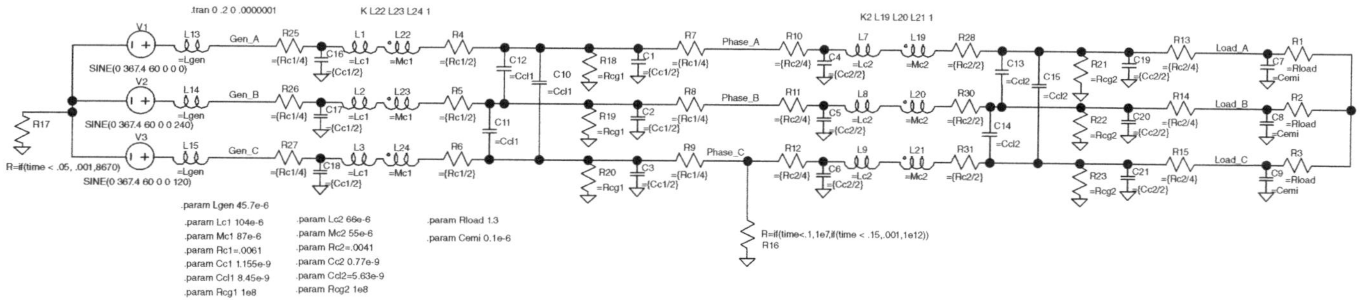

Figure 3.7 LTspice model of an HRG system using NGR with ground fault.

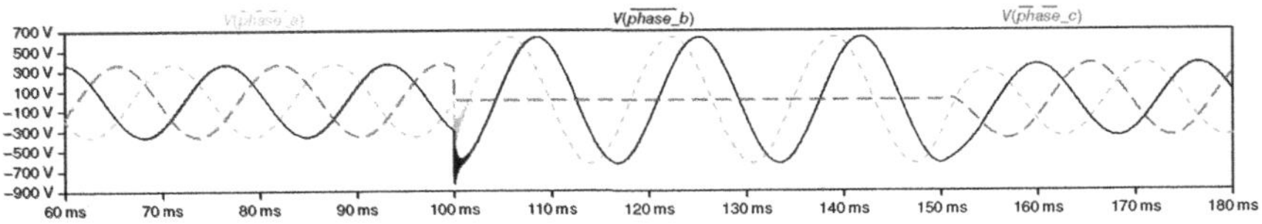

Figure 3.8 Line-to-ground phase voltages in an HRG system using NGR with ground fault.

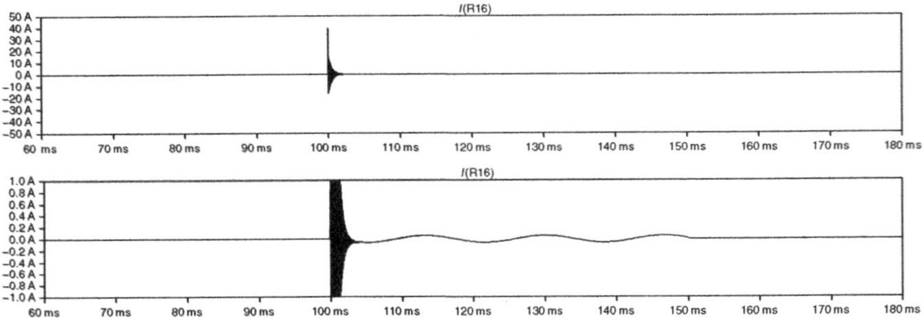

Figure 3.9 Current through the ground fault in an HRG system using NGR.

Figure 3.10 Current through the grounding resistor in an HRG system using NGR with ground fault.

Figure 3.11 Line-to-line voltages during the ground fault in an HRG system using NGR.

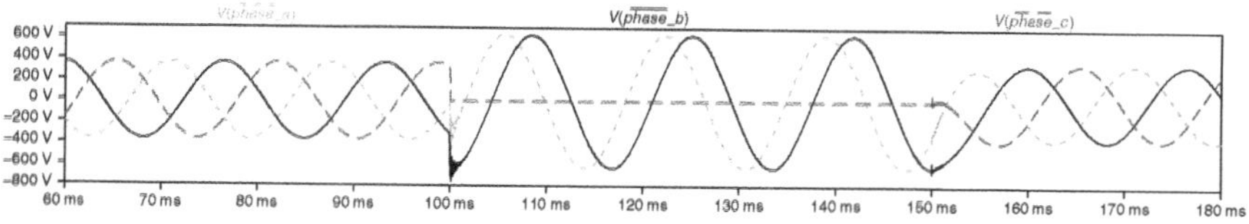

Figure 3.12 Line-to-ground voltages during ground fault with high EMI capacitance in an HRG system with NGR.

nearly the same shape as the previous case, as shown in Figure 3.13, the current magnitude is increased by a factor roughly equal to the ratio of the EMI capacitance; in this case, the EMI capacitance is the major contributor to the line-to-ground capacitance.

One big advantage of a high-resistance ground system compared to an ungrounded system is the response to an intermittent ground fault. Figure 3.14 shows the damping of the transient voltages and the neutral voltage returning to ground potential within one cycle of the removal of the ground fault. The fault is applied for 1 ms every cycle. Voltage spikes still occur, but can be controlled with SPDs. The waveforms in Figure 3.14 are directly comparable to the waveforms in Figure 2.10 for an ungrounded system; the high-resistance ground greatly reduces the transient response.

The impact of an arc fault, where a ground fault is initiated when the line-to-ground fault exceeds an arcing threshold and extinguishes when the ground fault

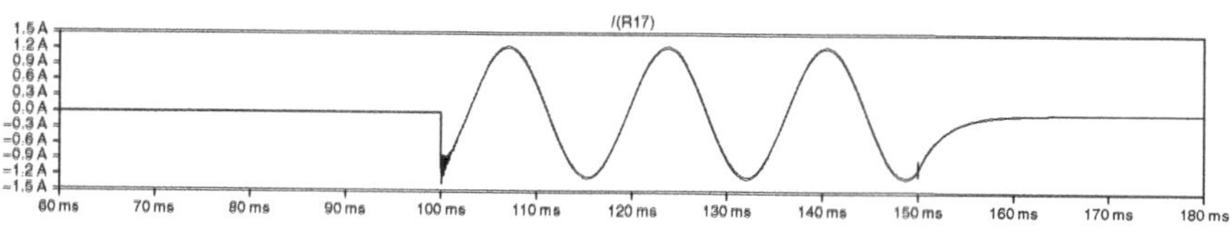

Figure 3.13 Current through the NGR during ground fault with high EMI capacitance in an HRG system.

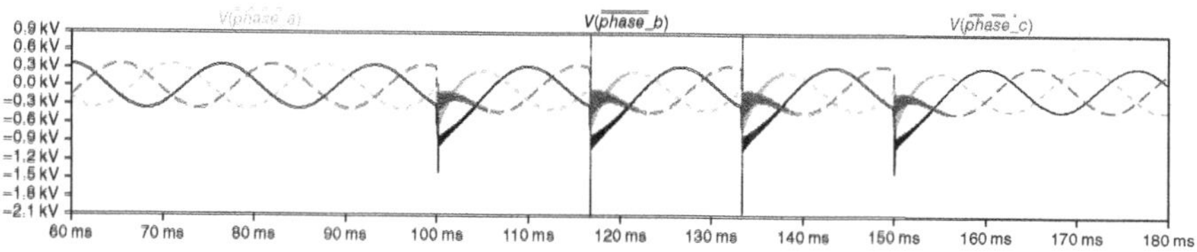

Figure 3.14 Impact of intermittent ground fault on an HRG system with NGR.

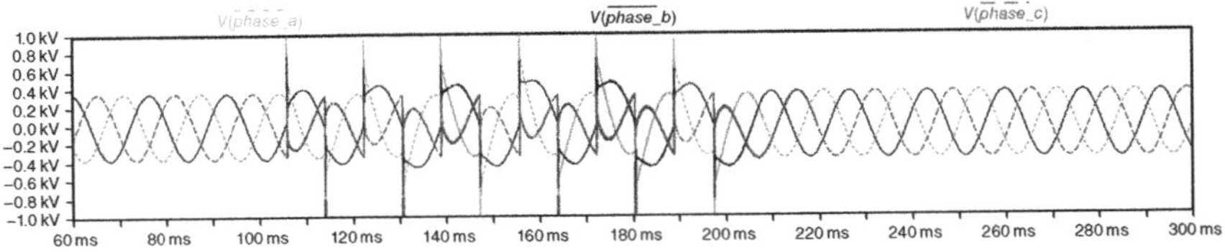

Figure 3.15 Impact of arc fault on a HRG system with NGR.

current passes through zero, is also much less in an HRG system compared to an ungrounded system. Compare Figure 3.15 with Figure 2.12, where the only difference is the insertion of a high-resistance ground. The high-resistance ground quickly dampens the transient ringing; the arc fault is not reinitiated by the line-to-ground voltage ringing. While a voltage spike still occurs every half cycle, the overall response is less severe.

As with ungrounded systems, under normal, un-faulted conditions, parasitic capacitances may provide a high-impedance connection between the power system conductors and the metallic components of equipment that are not intended to conduct current. To prevent a potentially hazardous voltage difference between these metallic components and the ship's structure and hull, they should be bonded via a low-impedance path to the ship's hull. These connections are designated as "protective earth" and indicated by the protective earth symbol in Figure 1.2. The subject of bonding/protective earthing is discussed in more detail in Chapter 14.

As with ungrounded systems, all loads should be line-to-line, and load currents balanced on the three phases as much as possible. Unbalanced currents can lead to localized neutral-to-ground voltages, which can result in ground currents flowing through the line-to-ground capacitances.

3.2 Grounding Circuit

IEEE Std. C57.32-2015 (2015) provides requirements, terminology, and test procedures for grounding resistors and grounding transformers. Specifications for grounding transformers and grounding resistors should invoke this standard.

3.2.1 Grounding Resistor Characteristics

Grounding resistors are typically specified by a rated voltage, rated time, and an initial rated current. They are normally wire-wound power resistors using materials such as nickel-chromium or stainless-steel wire.

For grounding resistors directly connecting a neutral conductor to ground, the rated voltage is the line-to-ground voltage corresponding to the nominal system voltage. In some grounding resistor datasheets, the voltage is specified as the nominal system voltage, assuming that the maximum voltage that the resistor will experience is the nominal system voltage divided by the square root of 3. Some grounding transformer configurations require a different rated voltage.

The initial rated current assumes the grounding resistor is at ambient temperature and corresponds to the magnitude of the calculated line-to-ground fault current through the line-to-ground capacitance. As the resistor heats up due to ground fault current flow, its resistance will increase and the current through it will decrease. Simulations or calculations should be performed to ensure acceptable performance at the highest expected temperature and associated resistance. Grounding resistor manufacturer datasheets should provide sufficient information to estimate the steady-state resistance of the grounding resistor.

Some manufacturers have off-the-shelf designs for low-voltage grounding resistors. Options are usually provided for several levels of initial rated current typically between 2 and 10 A. When selecting such a grounding resistor, the initial rated current required should be rounded up to the next available level for the grounding resistor.

For shipboard applications, one of the primary reasons for using high-resistance grounding is to enable continued operation of the power system with one line-to-ground fault. For this reason, grounding resistors should be designed with either extended-time ratings or steady-state time ratings as defined in IEEE Std. C57.32-2015 (2015).

For low-voltage systems, the grounding resistor during a ground fault can be expected to dissipate heat in the order of several kilowatts to the compartment the grounding resistor is located in. Typically, grounding resistors for low-voltage systems do not use forced air or any other type of dedicated cooling. Therefore, the ventilation system of the space housing the grounding resistor should be provided with sufficient capacity to remove this additional heat load. Sufficient volume should be reserved around the grounding resistor to enable the compartment air to adequately cool the resistor.

If the designer has low confidence in the estimate for line-to-ground capacitance, it may prove beneficial to specify several taps on the grounding resistor to reflect different resistance values. During the commissioning tests of the shipboard power system, the line-to-ground capacitance can be measured and the appropriate taps used in the grounding circuit.

For estimating purposes, the approximate dimensions of a steady-state time rating grounding resistor intended for a 480 V nominal system voltage and 5 A initial rated current are 0.3 m high, 0.8 m long, and 0.4 m wide. For low-voltage applications, these dimensions can be scaled such that the volume is proportional to the rated nominal system voltage multiplied by the initial rated current.

3.2.2 Neutral Ground

If the neutral connection of wye-wound generator or transformer secondary is available for connection, then a grounding resistor can be directly connected between this neutral connection and the hull of the ship. The rated voltage is the line-to-ground voltage corresponding to the nominal system voltage. The initial rated current is the line-to-neutral voltage divided by R_G.

3.2.3 Zigzag Transformer

Figure 3.16 depicts a zigzag transformer configuration used in HRG grounding circuits. Each of the transformers has a turns ratio of 1. As will be shown, if v_a, v_b, and v_c are connected to the phase conductors, v_n will be the voltage of the neutral conductor. The grounding resistor (R_G) is connected between v_n and the hull of the ship.

The equations for the zigzag transformer are

$$v_a - v_n = v_1 - v_2$$
$$v_b - v_n = v_2 - v_3$$
$$v_c - v_n = v_3 - v_1$$

Summing these three equations results in

$$(v_a - v_n) + (v_b - v_n) + (v_c - v_n) = 0$$

Since the neutral voltage is defined as the voltage reference for which the sum of the phase voltages with respect to the voltage reference sum to zero, v_n is by

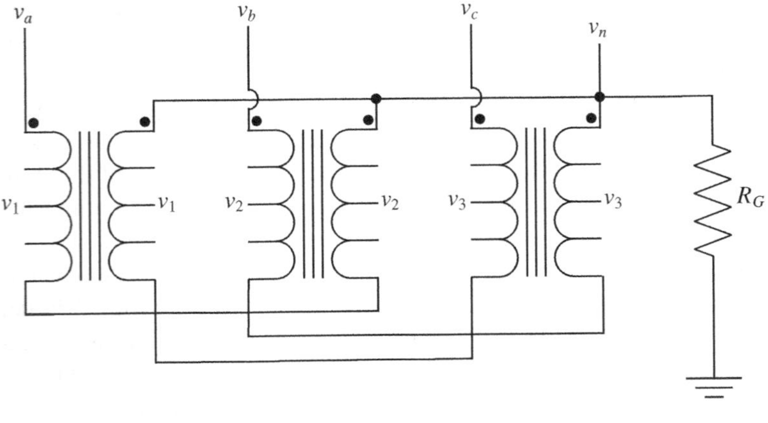

Figure 3.16 Zigzag transformer HRG configuration.

definition the neutral voltage. A grounding circuit employing a zigzag transformer is typically depicted, as shown in Figure 3.17.

Each winding of the zigzag transformer should be rated for 1/3 the ground fault current through the grounding resistor. Because the primary and secondary winding currents of each transformer are identical (including phase), and the winding currents of series connected transformers are identical, all of the winding currents are equal. The insulation on the transformer windings should be rated for the full line-to-line voltage.

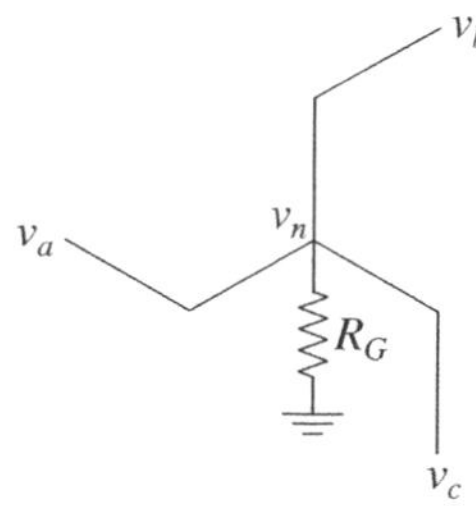

Figure 3.17 Zigzag transformer HRG grounding circuit.

3.2.4 Wye-Delta Transformer

Figure 3.18 depicts a wye-delta transformer configuration used in HRG grounding circuits. Each of the transformers typically has a turns ratio of 1. As will be shown, if v_a, v_b, and v_c are connected to the phase conductors, v_n will be the neutral voltage. The grounding resistor (R_G) will normally be connected between v_n and the hull of the ship.

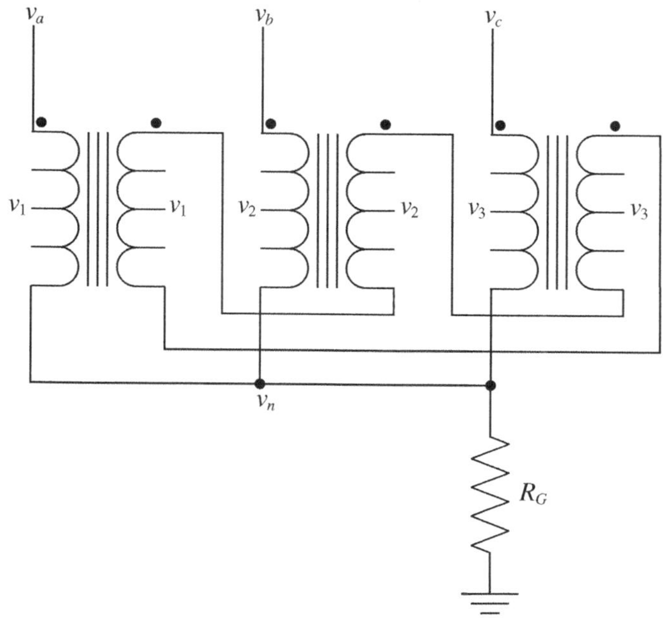

Figure 3.18 Wye-delta transformer HRG configuration.

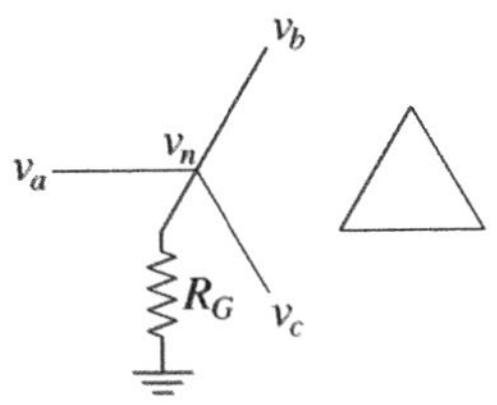

Figure 3.19 Wye-delta transformer HRG grounding circuit.

The wye-delta transformer equations for the wye connection are

$$v_a - v_n = v_1$$
$$v_b - v_n = v_2$$
$$v_c - v_n = v_3$$

The equation for the delta connection is

$$v_1 + v_2 + v_3 = 0$$

Summing the equations on the wye connection results in

$$(v_a - v_n) + (v_b - v_n) + (v_c - v_n) = v_1 + v_2 + v_3 = 0$$

Since the neutral voltage is defined as the voltage reference for which the sum of the phase voltages with respect to the voltage reference sum to zero, v_n is by definition the neutral voltage. A grounding circuit employing a wye-delta transformer is typically depicted, as shown in Figure 3.19.

3.2.5 Wye-Broken Delta Transformer

Figure 3.20 depicts a wye-broken delta transformer configuration used in HRG grounding circuits. Each of the transformers typically has a turns ratio of 1. The wye common point is hard grounded, and the grounding resistor is inserted into the delta secondary.

The equation for the delta connection is

$$v_a + v_b + v_c = v_{RG}$$

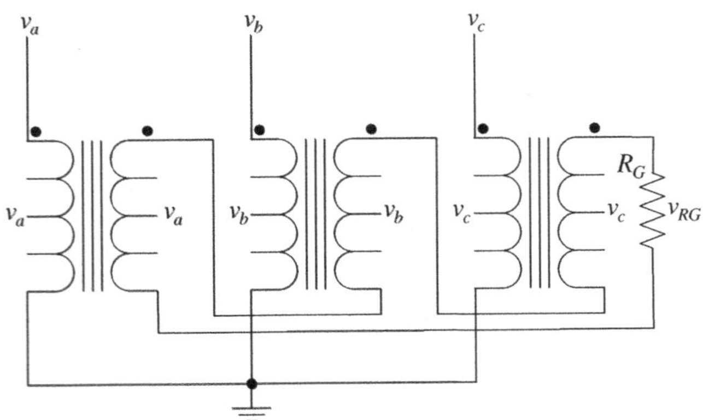

Figure 3.20 Wye-broken delta transformer HRG configuration.

Rewriting this equation:

$$\left(v_a - \frac{1}{3}v_{RG}\right) + \left(v_b - \frac{1}{3}v_{RG}\right) + \left(v_c - \frac{1}{3}v_{RG}\right) = 0$$

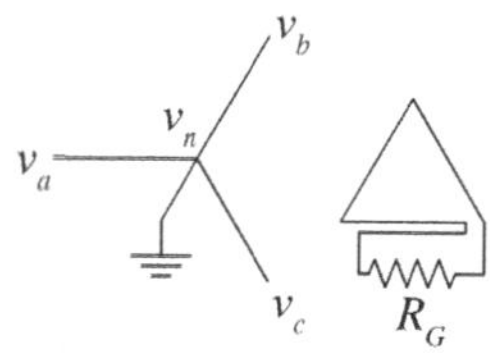

Figure 3.21 Wye-broken delta transformer HRG grounding circuit.

Hence, $\frac{1}{3}v_{RG}$ is the neutral voltage. The current through all of the secondary windings is identical and equal to $\frac{v_{RG}}{R_G}$. Since they are all equal, the current into the ground connection on the primary is $3\frac{v_{RG}}{R_G}$.

A grounding circuit employing a wye-broken delta transformer is typically depicted, as shown in Figure 3.21.

3.2.6 One-Line Diagram Symbology

The schematic symbols shown in Sections 3.2.3 through 3.2.5 are used in detailed circuit schematics. In one-line diagrams, the various grounding transformers, grounding resistors, and ground detection instrumentation comprising a grounding circuit are often represented collectively by the symbols shown in Figure 3.22. Figure 3.22a depicts a grounding transformer with a grounding resistor and ground detection instrumentation; the connection to the switchboard is a three-phase connection. A note should be added to the one-line diagram if a particular type of grounding circuit is required. Figure 3.22b depicts an NGR connected to the neutral conductor of a wye-wound stator of a generator, along with ground detection instrumentation. The NGR symbol can also be used to indicate a connection to the neutral conductor of a transformer wye secondary.

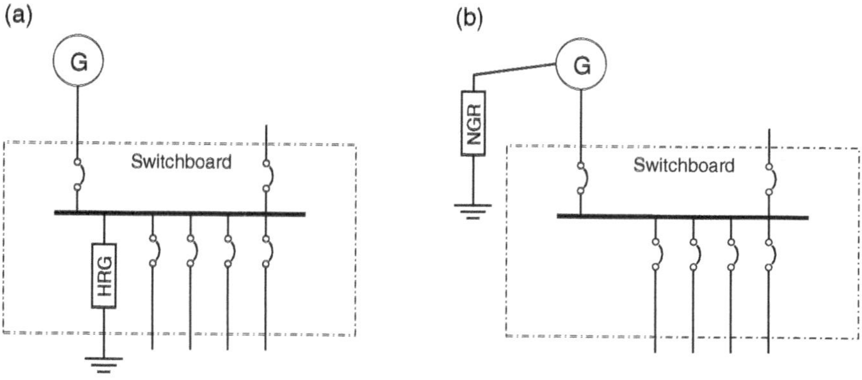

Figure 3.22 (a) HRG circuit on the bus one-line diagram symbol; (b) NGR on the generator neutral one-line diagram symbol.

3.2.7 Grounding Transformer Characteristics

Low-voltage grounding transformers for shipboard applications usually have the following characteristics:

- The rated voltage is the nominal system voltage.
- The rated frequency is the nominal system frequency.
- Because a ground fault may exist for an extended period of time (greater than 10 minutes), the rated continuous neutral current should be greater than the nominal system voltage divided by $\sqrt{3}$ and the ambient temperature grounding resistor resistance. Sufficient margin should be provided to account for system voltage tolerances (typically 5%) and the ambient temperature grounding resistance tolerance (typically 10%). For most applications, a 20% margin should be sufficient.
- Shipboard grounding transformers are normally of the dry type and to resist corrosion, usually employ copper wire for the windings.
- Other characteristics are as described in IEEE Std. C57.32-2015 (2015).

3.2.8 Grounding Transformer Saturation

If rectifiers are connected to an AC low-voltage distribution system that employs grounding transformers, one area of concern is the possibility of grounding transformer saturation due to a DC component of the grounding resistor current if a fault occurs on the DC side of the rectifier. Figure 3.23 depicts an AC system feeding a rectifier load. A ground fault is applied between 100 and 200 ms. Figures 3.24 and 3.25 depict the phase a and line-to-line (ab) voltages. The line-to-line voltages in particular are not sensitive to the ground fault. Figure 3.26 depicts the current through the grounding resistor. This grounding resistor current has a significant DC component.

Transformers are generally not intended to have primary currents with a large DC component. Typically, the maximum desired DC component of the current is on the order of 1% of the rated current. The transformer manufacturer should be consulted to determine the limits for any particular transformer. Ways to mitigate a potential ground fault on the DC side of a rectifier include

- Provide an isolation transformer for the rectifier. This may be desirable to provide wye and delta secondaries as part of a 12-pulse rectifier.
- If the rectifier load is not a critical load, or is a critical load with sufficient redundancy, the breaker for the rectifier input can be tripped when a DC ground fault is detected on the rectifier output. A large common mode DC current is a good indicator of such a ground fault.
- Insert a parallel combination of a blocking capacitor and large resistor between ground and the grounding transformer.

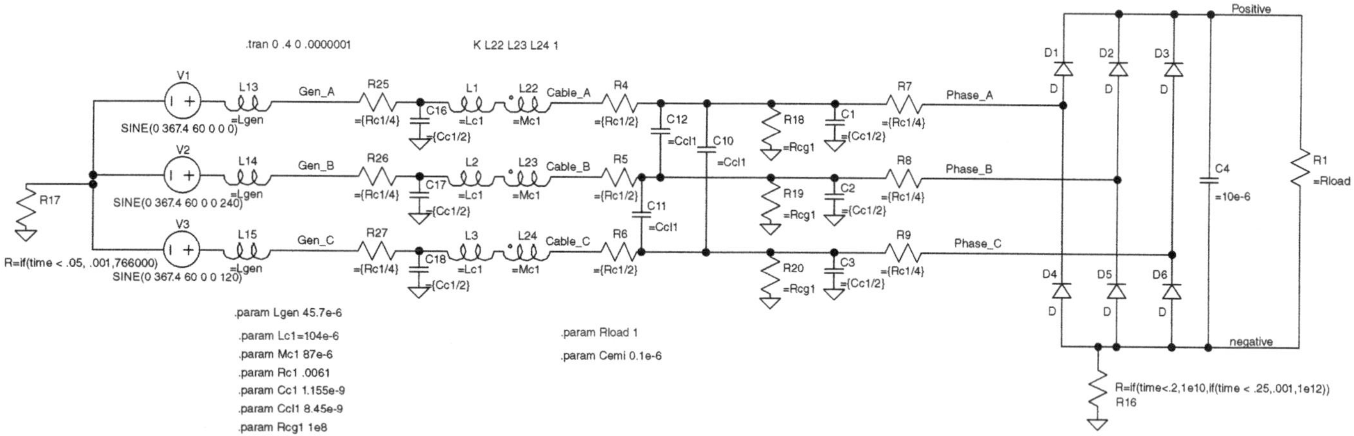

Figure 3.23 LTspice model of an AC HRG power system with rectifier load schematic.

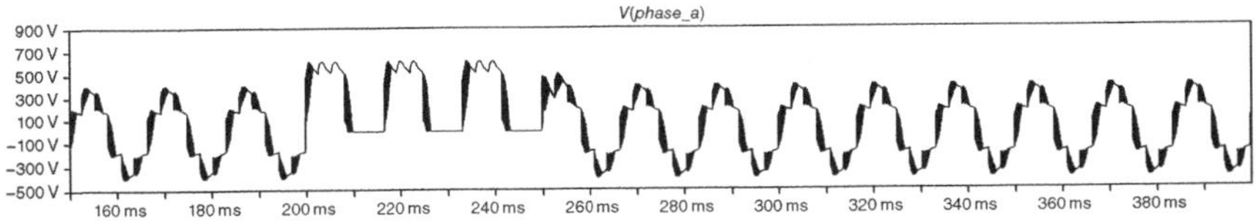

Figure 3.24 AC HRG system with rectifier load phase voltages – ground fault on DC side of rectifier.

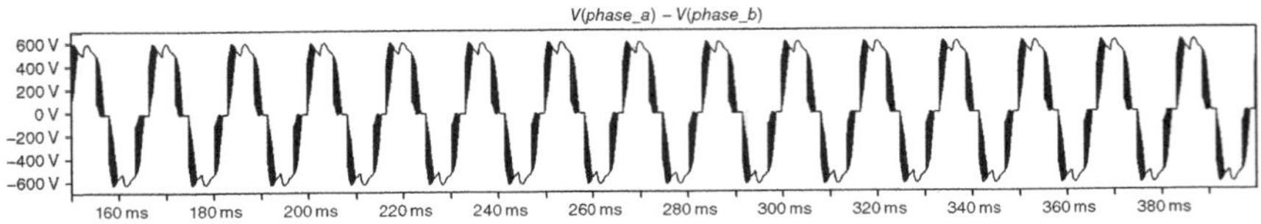

Figure 3.25 AC HRG system with rectifier load line-to-line voltages – ground fault on DC side of rectifier.

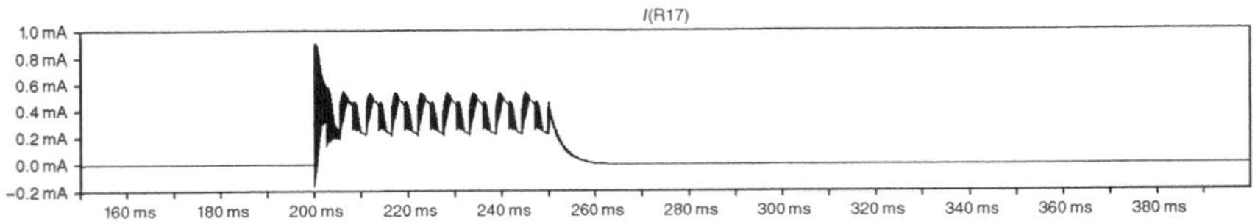

Figure 3.26 AC HRG system with rectifier grounding resistor current – ground fault on DC side of rectifier.

If this last option is employed, the blocking capacitor should have an impedance at the fundamental frequency about 10% of the grounding resistance. The parallel resistor should have a resistance of about 100 times the grounding resistor. The blocking capacitor impedance is chosen so it is small in comparison to the grounding resistor and line-to-ground capacitance impedance. The parallel resistor value significantly reduces the DC fault current to a value that the grounding transformer should tolerate and provides a means to discharge both the line-to-ground capacitance and the blocking capacitor after the ground fault clears. In any case, the grounding circuit should be analyzed to ensure a fault on the DC side of the rectifier will not cause the grounding transformer to overheat and possibly fail. Figure 3.27 depicts the application of the blocking capacitor for the various types of grounding transformer configurations.

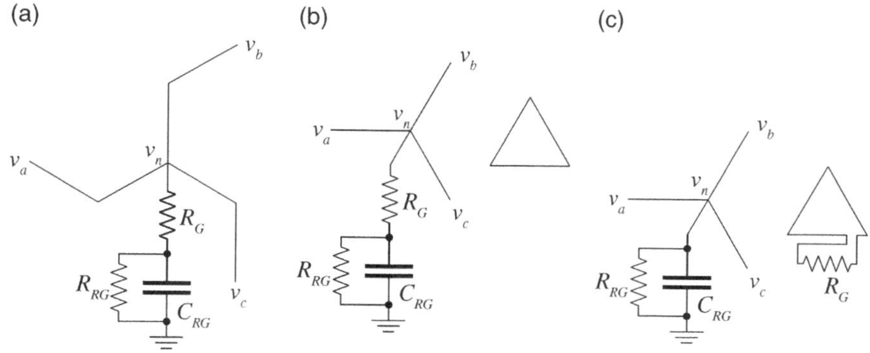

Figure 3.27 Grounding transformers with blocking capacitors (a) zigzag; (b) wye-delta; (c) wye-broken delta.

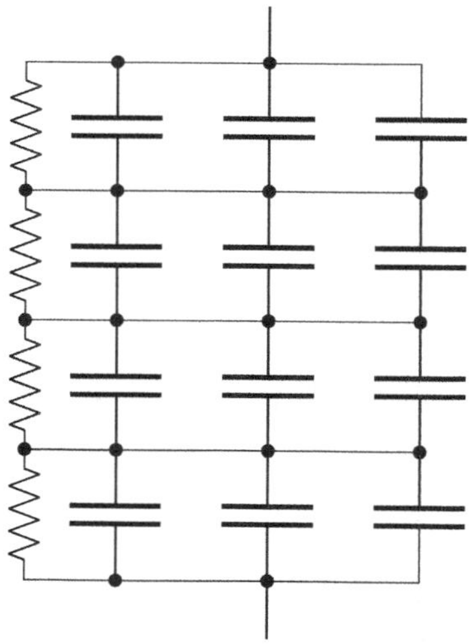

Figure 3.28 Blocking capacitor network.

The blocking capacitor must be bipolar (such as a polypropylene capacitor) and have sufficient voltage rating, capacitance, and ripple current rating. A single capacitor may not be commercially available that meets all of these criteria, especially if typical voltage derating factors of 50%–80% are used to ensure reliability. It may be necessary to assemble a series and parallel network of capacitors to achieve the desired ratings (Figure 3.28). The parallel resistance can be segmented to serve as balancing resistors for the series capacitors.

Figure 3.29 shows the DC component of the grounding resistor (and grounding transformer) decays relatively quickly to zero. While this should prevent the grounding transformer from saturating, a more detailed analysis should be conducted for confirmation. Figure 3.30 shows the blocking capacitor voltage. It rapidly builds until the voltage across the blocking capacitor equals the line-to-ground capacitance voltage. Without the paralleled resistor, the voltage decay after the fault clears in Figure 3.30 would not happen, a large DC voltage would remain on the capacitor, resulting in a DC offset of the neutral voltage. This offset would subject the insulation system with a long-term voltage

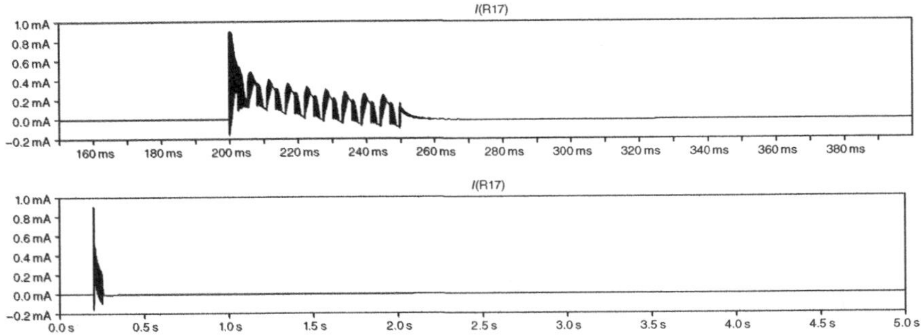

Figure 3.29 Grounding resistor current with blocking capacitor network due to DC ground fault.

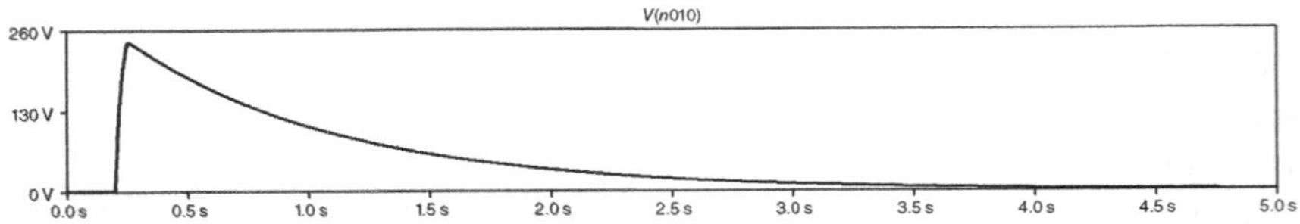

Figure 3.30 Blocking capacitor network voltage due to DC ground fault.

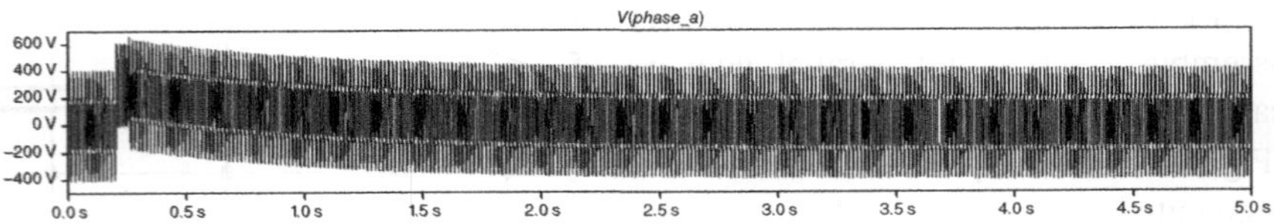

Figure 3.31 Line-to-ground voltages (phase *a*) due to DC ground fault with blocking capacitor network.

stress. The voltage on the blocking capacitor is reflected in the line-to-ground phase voltages, as depicted in Figure 3.31.

3.3 Location of Grounding Circuit

Traditionally, grounding resistors have been co-located with sources of power, typically generators and transformer secondaries. The grounding resistance was determined based on the system configuration with the maximum line-to-ground capacitance. If two or more sources can be paralleled, an automatically controlled grounding switch was inserted in series with each grounding resistor; the control system would ensure only one grounding resistor per set of paralleled sources was in-service at any given time.

The reason typically given for single-point grounding of a power system is to avoid circulating currents. However, for a circulating current to exist between two grounding resistors, a voltage difference between the power system neutrals of the two grounding resistors must also exist. This voltage difference is a CM voltage between the two points of the power system. In most shipboard power systems, the only source for a CM voltage between two switchboards would be due to a CM current interacting with a cable impedance.

The phase voltage differences between two switchboards are given by

$$
\begin{aligned}
v_{a1d} + v_{1cm} - v_{a2d} - v_{2cm} = {} & L_s \frac{d(i_{ad} + i_{ac})}{dt} + M \frac{d(i_{bd} + i_{bc})}{dt} + M \frac{d(i_{cd} + i_{cc})}{dt} \\
& + R(i_{ad} + i_{ac})
\end{aligned}
$$

$$
\begin{aligned}
v_{b1d} + v_{1cm} - v_{b2d} - v_{2cm} = {} & M \frac{d(i_{ad} + i_{ac})}{dt} + L_s \frac{d(i_{bd} + i_{bc})}{dt} + M \frac{d(i_{cd} + i_{cc})}{dt} \\
& + R(i_{bd} + i_{bc})
\end{aligned}
$$

$$
\begin{aligned}
v_{b1d} + v_{1cm} - v_{b2d} - v_{2cm} = {} & M \frac{d(i_{ad} + i_{ac})}{dt} + M \frac{d(i_{bd} + i_{bc})}{dt} + L_s \frac{d(i_{cd} + i_{cc})}{dt} \\
& + R(i_{cd} + i_{cc})
\end{aligned}
$$

The relationships among CM and balanced three-phase voltages and currents can be expressed as

$$
\begin{aligned}
& v_{a1d} + v_{b1d} + v_{c1d} = 0 \\
& v_{a2d} + v_{b2d} + v_{c2d} = 0 \\
& i_{ad} + i_{bd} + i_{cd} = 0 \\
& i_{ac} + i_{bc} + i_{cc} = i_{cm}
\end{aligned}
$$

Summing the three voltage difference equations and substitution results in

$$
\begin{aligned}
& 3v_{1cm} - 3v_{2cm} = L_s \frac{d(i_{cm})}{dt} + 2M \frac{d(i_{cm})}{dt} + Ri_{cm} \\
& v_{1cm} - v_{2cm} = \frac{1}{3}(L_s + 2M) \frac{d(i_{cm})}{dt} + \frac{1}{3} Ri_{cm} \\
& v_{1cm} - v_{2cm} = L_{cm} \frac{d(i_{cm})}{dt} + R_{cm} i_{cm} \\
& L_{cm} = \frac{1}{3}(L_s + 2M)
\end{aligned}
\tag{3.7}
$$

$$R_{cm} = \frac{1}{3}R \tag{3.8}$$

where

v_{a1d}, v_{b1d}, v_{c1d}, phase voltages with respect to neutral at switchboard 1
v_{a2d}, v_{b2d}, v_{c2d}, phase voltages with respect to neutral at switchboard 2
v_{1cm}, CM voltage at switchboard 1 (neutral voltage with respect to ground)
v_{2cm}, CM voltage at switchboard 2 (neutral voltage with respect to ground)
i_{ad}, i_{bd}, i_{cd}, the DM currents through the cable
i_{ac}, i_{bc}, i_{cc}, CM phase current components of the current through the cable
i_{cm}, CM current through the cable
L_s, cable conductor self-inductance
M, cable conductor mutual inductance
L_{cm}, cable CM inductance
R_{cm}, cable CM resistance.

If we use the values from Chapter 2 example, we have

$$L_s = 22.8\ \mu\text{H}$$

$$M = 19.1\ \mu\text{H}$$

$$R = 0.00132\ \Omega$$

Then

$$L_{cm} = \frac{1}{3}(22.8 + 2 \times 19.1)\ \mu\text{H} = 20.3\ \mu\text{H}$$

$$R_{cm} = \frac{1}{3}(0.00132) = 0.00044\ \Omega$$

If we assume the line-to-ground capacitance of each switchboard is 2 μF, then the grounding resistor at each switchboard would be (at 60 Hz):

$$R_{G1} = R_{G2} \approx \frac{1}{3\omega C_p} = \frac{1}{3(2\pi 60)2 \times 10^{-6}} = 442\ \Omega$$

The steady-state circulating current through the grounding resistors would be the CM voltage between the switchboards divided by the sum of the grounding resistors.

$$i_{circulating} = \frac{v_{1cm} - v_{2cm}}{R_{G1} + R_{G2}} = i_{cm}\frac{R_{cm} + j\omega L_{cm}}{R_{G1} + R_{G2}}$$

The magnitude of the circulating current is

$$\left|i_{circulating}\right| = \left|i_{cm}\right|\frac{\sqrt{(R_{cm})^2 + (\omega L_{cm})^2}}{R_{G1} + R_{G2}}$$

At 60 Hz, the relationship for the example is given by

$$|i_{circulating}| = |i_{cm}| \frac{\sqrt{(0.00044)^2 + (2\pi 60 \times 20.3 \times 10^{-6})^2}}{442 + 442} = 8.67 \times 10^{-6}\, i_{cm}$$

The circulating current is a very small fraction of the overall CM current at 60 Hz. Because the resistance is a small fraction of the reactance at 60 Hz and greater, the expression can be simplified for higher frequencies to

$$\frac{|i_{circulating}|}{|i_{cm}|} = f \frac{2\pi L_{cm}}{R_{G1} + R_{G2}}$$

where

f, frequency (Hz).

For this example

$$\frac{|i_{circulating}|}{|i_{cm}|} = f \frac{2\pi L_{cm}}{R_{G1} + R_{G2}} = f \frac{2\pi 20.3 \times 10^{-6}}{442 + 442} = 0.144 \times 10^{-6} \times f$$

Hence, if the CM currents can be kept small, the impact of circulating currents between the grounding resistors should be negligible up to frequencies approaching 1 MHz.

Note that circulating currents will always exist due to the distributed parasitic capacitance to ground interaction with CM voltage differences between different parts of the power system. Because the impedance of the parasitic capacitance at the nominal system frequency is the same order of magnitude as the grounding resistors, at much higher frequencies, the impedance of the parasitic capacitances will be much less than that of the grounding resistors, resulting in the circulating currents due to the parasitic capacitances being much greater than due to the grounding resistors.

This analysis does not support the practice of not allowing multiple grounding resistors in a power system. Because the capacitance to ground in most systems is dominated by load EMI filters and other load characteristics, with contributions by the distribution cabling, employing a grounding circuit at each switchboard based on the line-to-ground capacitance of equipment and cables connected to it can enable a better matching of grounding resistor impedance to the line-to-ground impedance across multiple electrical distribution system plant lineups. Line-to-ground capacitances associated with cables and loads that are only connected via another switchboard when bus-tie breakers are closed are not included.

In this example, a single-point ground would require a resistance half that for each switchboard (221 Ω). While this value would be appropriate in operational lineups where the two switchboards are connected via a closed bus-tie breaker, the resistance value would be half the value desired if the switchboard bus-tie breaker is open.

Allowing multiple grounding resistors also enables avoiding the cost and complexity of installing the grounding resistor disconnect switches and the associated controls. On the other hand, each switchboard would require a grounding transformer.

If a single-point ground is required by regulation or class rules, consideration should be given to implementing a "grounding bus" between all of the switchboards. The grounding resistors (or wye neutral of a wye-broken delta grounding transformer) would connect to this grounding bus. Each switchboard would have a grounding switch to connect the grounding bus to the ship's hull, but only one of the switches would be closed at any one time. If the grounding bus were damaged, other switchboard switches could be closed as needed. The grounding bus should be rated to continuously carry the maximum ground fault current and to carry the maximum line-to-line fault current for the maximum time needed to clear a line-to-line short circuit. Regulations or class rules may specify a minimum conductor size. Each switchboard should measure the voltage or impedance between the grounding bus and the ship's hull to detect and report to the operator if a solid connection between the two is not present. The current through the grounding switch connecting the grounding bus to the ship's hull should also be monitored to detect and report to the operator the presence of a ground fault.

3.4 Ground Fault Detection

Detecting a ground fault in an HRG system is straightforward. As with the ungrounded system, one can measure each of the line-to-ground voltages; when one is near zero, then one has a ground fault. The traditional three-lamp ground fault detection circuit used in ungrounded systems is also applicable in HRG systems.

A ground fault can also be detected by measuring the voltage across and/or the current through the grounding resistor. Under normal operation, the current through the grounding resistor and the voltage across the grounding resistor will be very small. The current will likely be nonzero due to asymmetries in the power system and loads. During a ground fault, the voltage across the grounding resistor will be roughly equal to the line-to-neutral voltage of the source.

Usually, measuring the voltage across the grounding resistor is preferred to measuring the current through the grounding resistor to detect a fault because the voltage measurement will still be valid in the case of a failure within the grounding resistor. In many applications, both current and voltage are measured for redundancy.

Another method is to use three resistors of high resistance connected in wye to derive a neutral point whose voltage is compared to ground. This voltage will be small under normal operation, and nearly equal to the line-to-ground voltage during a ground fault.

Most, if not all, commercially available insulation monitoring devices are not compatible with HRG systems. The insulation value they calculate will be significantly impacted by the grounding resistor.

3.5 Ground Fault Localization

While continued operation of the system is enabled with an HRG system, the ground fault should be localized and cleared as soon as possible to avoid a line-to-line short circuit from a subsequent ground fault on a different phase. Ground fault localization devices are often incorporated as part of the HRG system.

The methods described in Sections 3.5.1 through 3.5.3 apply to feeder cables from a switchboard. If a ground fault is present and none of the feeder cable has a ground fault, then the ground fault lies within a switchboard, a bus tie, a generator cable, or the generator itself. Systematically reconfiguring the power system should enable one to logically deduce the faulted element of the system without securing power to any load. The exact procedure for identifying the faulted element depends on the power system architecture. The general concept is to have one generator online for each switchboard that has at least one generator connected to it. Bus ties are opened to island each generator while still providing power to all loads. One should then be able to determine which island has the ground fault and which islands do not have a ground fault. If the faulted island has more than one switchboard, the system is reconfigured to power them from different islands by closing and opening bus-tie breakers. If the ground fault moves to a different island, then one should be able to determine the faulted switchboard/bus tie. Isolating a bus tie should confirm or rule out that bus tie as a source of the ground fault.

3.5.1 Pulsing System

One common method to localize a ground fault is to use a pulsing system to periodically either short out a section of the grounding resistor or connect a second grounding resistor in parallel with the normal grounding resistor. The period of the pulse is typically between 1 and 3 seconds. When the pulse is "on," the grounding current is increased between 2 and 5 A.

This pulsing current signal can be detected with either a handheld clamp-on ammeter around all three phases, or with permanently installed current transformers around all three phases for every feeder cable. Since most of the current from the grounding resistor will flow through the faulted cable, sensing the pulsing current indicates a faulted cable, and a lack of a significant pulsing current indicates a cable that is not faulted. In this manner, the fault can first be localized to a single feeder cable, and then a handheld clamp-on ammeter can be used to identify where the fault is located on the feeder cable; the pulsing current will disappear when the ground fault is electrically closer to the grounding resistor than the measuring point.

If this method is employed, consideration should be given to providing easy access to the individual cables and room to insert the clamp-on ammeter. These locations should be easily identifiable to the maintenance personnel.

Figure 3.32 is a schematic of a representative three-feeder system powered by a single generator. Feeder 2 has a ground fault on phase *c*, and a pulsing system for ground fault detection halves the grounding resistance for 0.5 seconds each second. Figures 3.33 through 3.35 depict the CM currents in each of the feeder cables. The un-faulted feeder cables (Feeders 1 and 3) have relatively low-magnitude CM currents that are not impacted by the pulsing system. The faulted feeder (Feeder 2) has a relatively larger CM current before the fault (the measurement point is between the grounding resistor and the fault) that nearly doubles when the pulsing system halves the grounding resistance. After the fault (the fault is between grounding resistor and the measurement point), the CM current on the faulted feeder is significantly smaller and does not reflect the pulses from the pulsing system.

3.5.2 Low-Resistance Ground Fault Clearing

In low-resistance ground fault clearing, a grounding resistor with a low resistance and a switch are connected in series between the generator neutral and ground. When the ship is clear of potential danger, the switch is closed, resulting in a ground fault current large enough to trip the appropriately coordinated breaker via its normal I^2t characteristic. The resistor value should be small enough to result in a fault current at least twice the steady-state rating of the largest circuit breaker in the system. The generator, resistor, and switch should be sufficiently rated to handle the fault current for the maximum time it would take to clear the fault.

Connecting the low-resistance grounding resistor and switch to the system neutral via a grounding transformer is not advised because this grounding transformer would also have to be rated to take the ground fault current for the maximum time it would take to clear the fault.

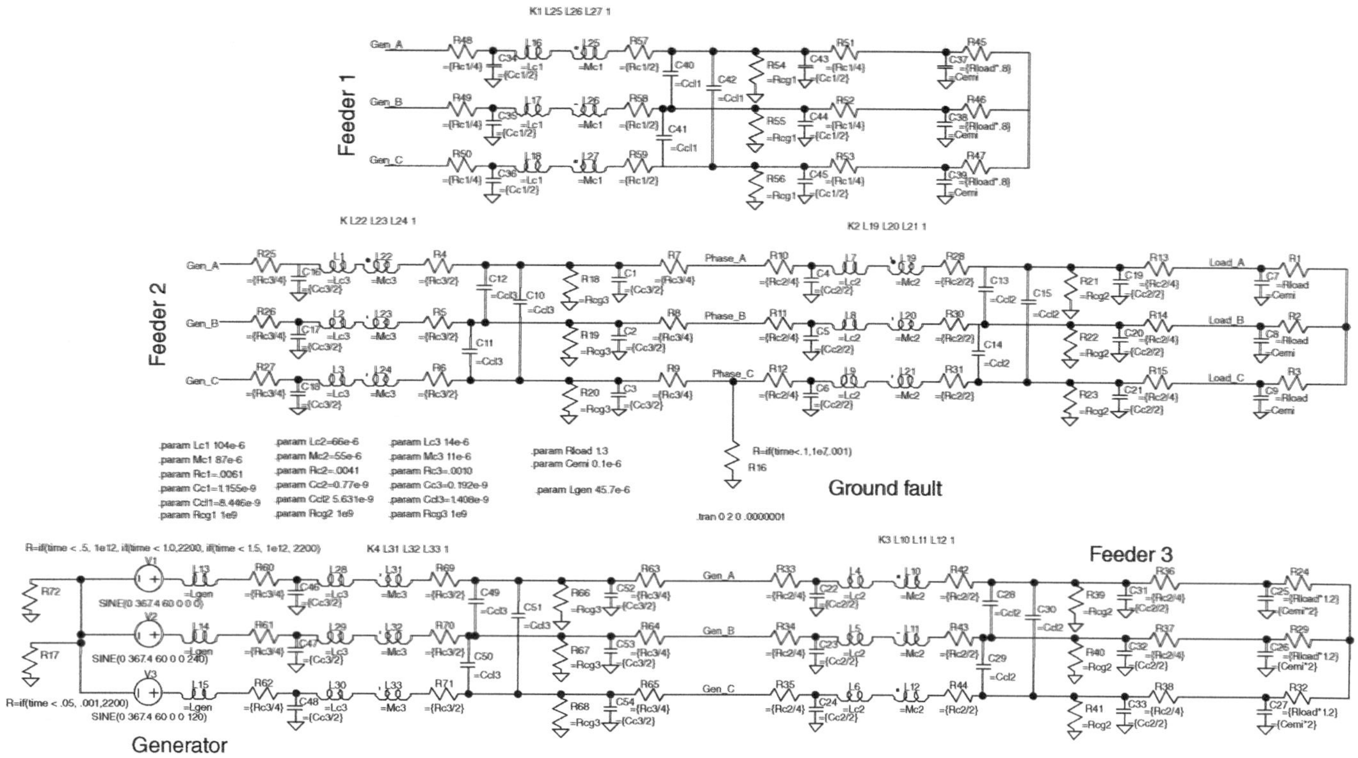

Figure 3.32 LTspice model of three-feeder system with pulsing system for ground fault location.

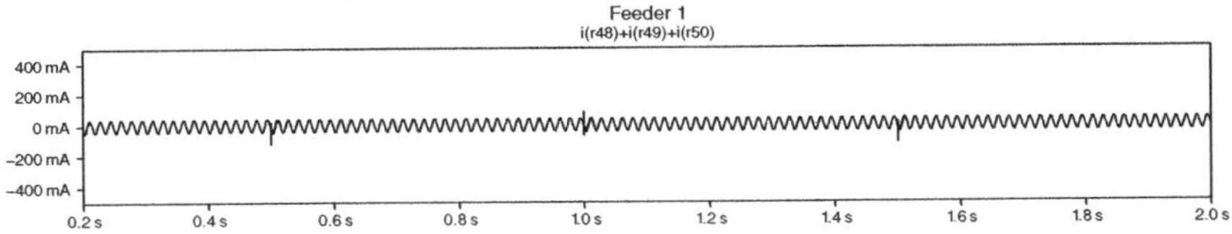

Figure 3.33 CM current on feeder 1.

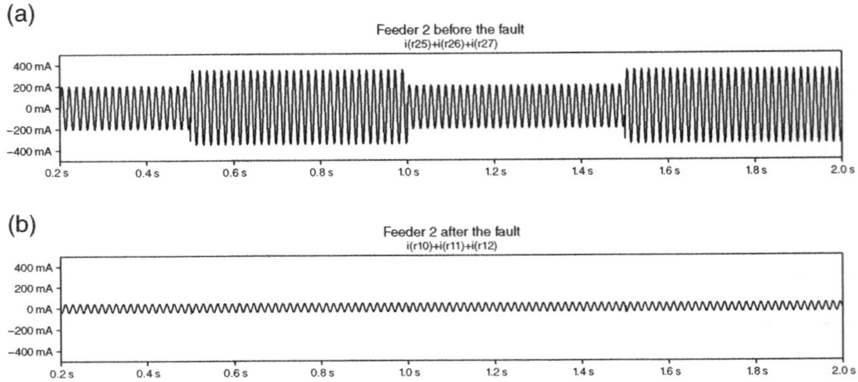

Figure 3.34 CM current on feeder 2 (a) before the ground fault; (b) after the ground fault.

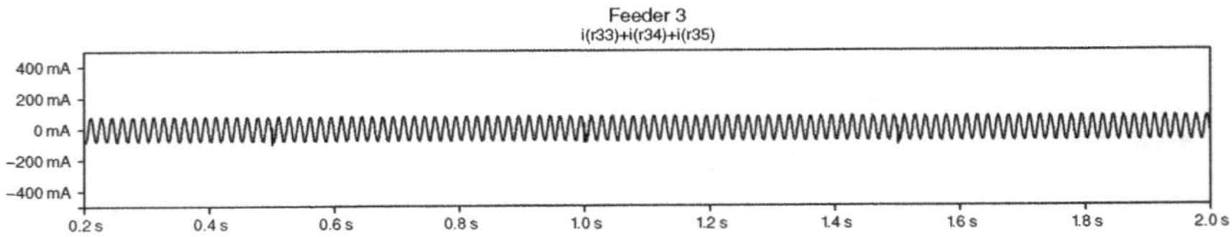

Figure 3.35 CM current on feeder 3.

Because this method localizes and clears the fault at the same time, the operators should prepare for losing any one feeder. Redundant critical equipment powered from different feeders should be brought online prior to switching in the low resistance. In this manner, critical systems are less likely to be lost due to the isolation of the faulted feeder.

3.5.3 Common Mode Current Phase Measurements

In a radial system, one can also determine which feeder has a ground fault on it by comparing the phase of the current through the grounding resistor with the phase of the CM current of each of the feeders. The un-faulted feeders will represent a

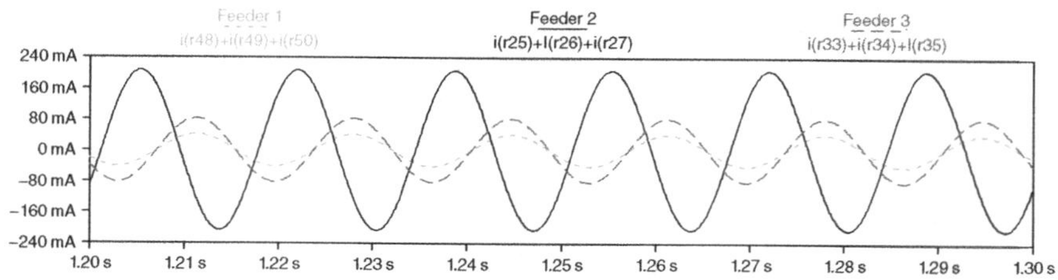

Figure 3.36 CM current through each feeder cable of Figure 3.32.

capacitive, or leading load, and the current waveform through these feeders will lead the current through the faulted feeder, a resistive load, by approximately 90 electrical degrees. Figure 3.36 shows that the un-faulted feeders (Feeder 1 and Feeder 3) have low-magnitude CM currents that lead the current through the faulted feeder (Feeder 2) by about 90 electrical degrees. This difference in phase can be used to determine which feeder is faulted.

While this technique can be used to determine which feeder is faulted, it cannot (without significant investment in current transformers) determine where in the feeder the fault is located. A handheld clamp-on ammeter can be used on the faulted feeder to determine the location of the fault by determining where on the feeder the CM current magnitude drops significantly.

As with the pulsing system, consideration should be given to providing easy access to the individual cables and room to insert the clamp-on ammeter. These locations should be easily identifiable to the maintenance personnel.

3.6 Electrical Insulation Impacts

Because HRG systems can operate for extended periods of time with a ground fault, the insulation between conductors and ground should be rated for the full line-to-line voltage (instead of line-to-neutral voltage). The rated voltage for the insulation should be at least $\sqrt{3}$ times the nominal system voltage to account for the steady-state line-to-ground voltage during the fault. The design of the insulation system should also account for the line-to-ground voltage spikes and transients that occur when a ground fault is initiated and cleared.

3.7 Limiting Ground Fault Current

As recommended by IEEE Std 3003.1-2019 (2019), ground fault current should be limited across all electrical system configurations to less than 10 A to prevent significant damage to faulted equipment. Limiting ground fault current implies limiting the line-to-ground capacitance. In many systems, EMI filters are a significant contributor of line-to-ground capacitance; careful attention to the specification of these filters may be needed to keep the ground fault current under 10 A. The designer has less control over the line-to-ground capacitance of cable, transformers, generators, and motors.

If limiting the ground fault current to less than 10 A across all electrical system configurations is not feasible, the following risk mitigation activities should be considered:

- Provide an automated fault localization system to quickly identify and report to the operator the location of the ground fault.
- Upon detection of a ground fault and localization to a specific feeder, enable automatic or semiautomatic reconfiguration of the power system such that power continuity is provided to critical loads and that sufficient distributed capacitance is removed from the faulted portion of the power system to lower the ground fault current to less than 10 A. For example, this could include opening bus-tie breakers to create two or more independent systems from a normally combined system, or moving a load with an EMI filter having a large line-to-ground capacitance from a normal to an alternate source.

References

IEEE Std 3003.1-2019 (2019). *IEEE Recommended Practice for System Grounding of Industrial and Commercial Power Systems.* New York: Institute of Electrical and Electronics Engineers. Available from http://ieeexplore.ieee.org.

IEEE Std C57.32-2015 (2015). *IEEE Standard for Requirements, Terminology, and Test Procedures for Neutral Grounding Devices.* New York: Institute of Electrical and Electronics Engineers. Available from http://ieeexplore.ieee.org.

4

System Grounding: Shipboard Solidly Grounded AC Systems (No Greater than 400 V)

4.1 Characteristics

Single-phase shipboard loads with a nominal system voltage no greater than 240 V may be provided power via a solidly grounded system to avoid incompatibilities with commercial off-the-shelf equipment. In IEC documents, these systems are designated TN-S (Terre-Neutral). As shown in Figure 4.1, solidly grounded AC systems are usually not employed for primary distribution systems or low-voltage distribution systems. Low-voltage special circuits may employ solid grounding based on the requirements for the low-voltage special circuit.

As depicted in Figure 4.2, single-phase loads are often supplied from one phase of a wye-connected transformer secondary. The term "neutral" is used in single-phase systems to indicate the power conductor that is solidly grounded. While the potential of this conductor is close to the neutral voltage of the three conductors of the three-phase wye-connected system, it is not at the neutral voltage of the two conductors supplying power to the single-phase load. An alternate method of supplying 120 or 240 V is depicted in Figure 4.3 with Power panel A. This option uses a single-phase transformer with a center-tapped secondary. The center tap is solidly grounded and is called the neutral conductor. 120 V loads are connected from one of the other power conductors to the neutral conductor, while 240 V loads are connected to the two non-neutral conductors. If only 240 V loads are supplied, single-phase transformers as depicted in Figure 4.3 with Power panel B may be used.

A protective earth conductor (equipment grounding conductor) is provided in the feeder cables to individual loads. This conductor is connected to all-exposed conducting material on the load that is not intended to normally conduct current. At the power panel directly connected to the distribution transformer, the protective earth conductor is connected to the ship's hull. Loads may have additional electrical connections to the ship's hull via bonding straps or mounting bolts.

Design of Shipboard Power System Grounding/Earthing, First Edition. Norbert Doerry, Mohammed M. Islam, and John Prousalidis.
© 2025 The Institute of Electrical and Electronics Engineers, Inc.
Published 2025 by John Wiley & Sons, Inc.

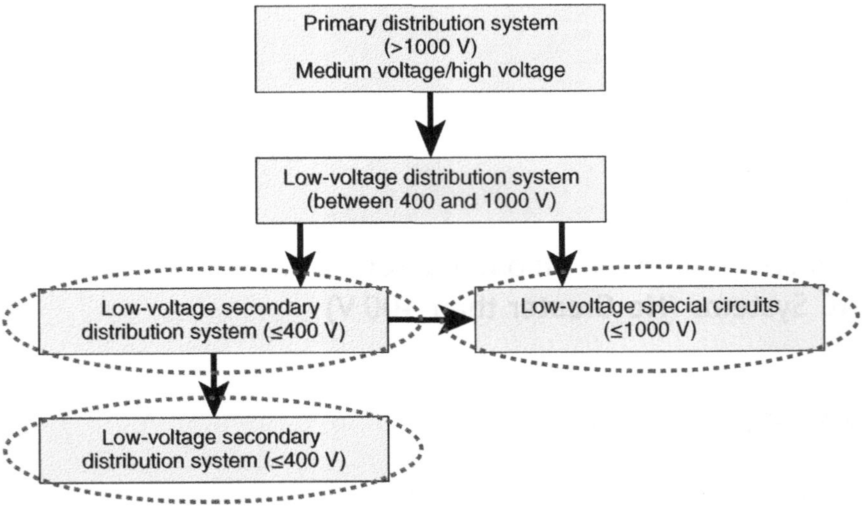

Figure 4.1 Applicability of an AC solidly grounded low-voltage distribution systems.

Within the United States, single-phase loads less than 1.4 kW are typically provided power at 120 V_{rms}. These loads usually will work in systems with nominal system voltages that range worldwide from 100 to 120 V. Loads greater than 1.9 kW (such as electric stoves and electric clothes dryers) are generally provided power at 240 V_{rms} with a center tap (neutral) that is grounded. In many shipboard applications, these 240 V_{rms} loads are provided 208 V_{rms} power using the neutral and two of the phases from the wye-connected secondary, as depicted in Figure 4.2. Loads between 1.4 and 1.9 kW may require either 120 or 240 V supply.

In Europe, single-phase loads are typically provided power at 240 V. These loads usually will work in systems with nominal system voltages that range worldwide from 220 to 240 V.

The circuit breakers for loads in Figures 4.2 and 4.3 do not include a circuit breaker pole for the neutral conductor since the neutral conductor is solidly grounded and should never have a voltage potential with respect to the ship's hull. While not required, in some installations, a circuit breaker pole is provided for the neutral to completely isolate the load equipment. The protective earth conductor should not have a circuit breaker pole.

The neutral conductor is normally grounded (power system ground) at the power panel connected to the distribution transformer to facilitate the installation of the current sensor. The current sensor detects if any current is leaking to the ship's hull and returning to the power system via the power system ground connection. This could occur, for example, if the neutral conductor to a load experiences a ground fault, or if an inappropriate connection between the neutral conductor and the ship's hull is installed in a subpanel or load. If there is sufficient

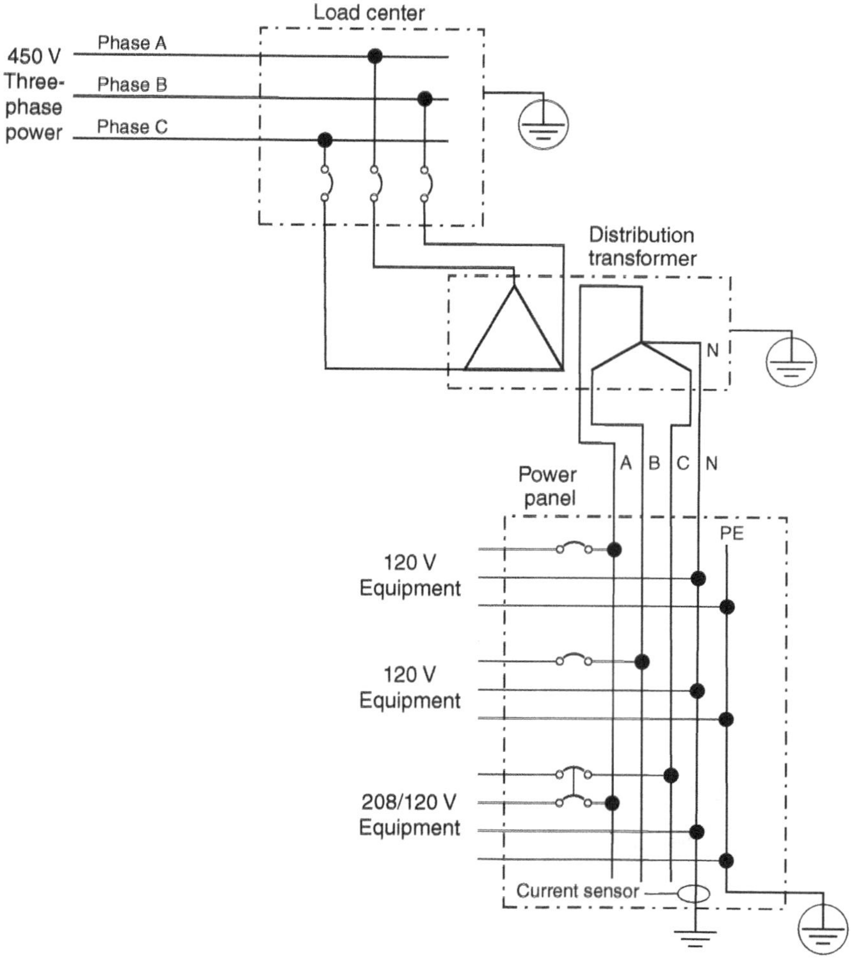

Figure 4.2 Solidly grounded single-phase 120/208 V from three-phase four-wire wye transformer secondary winding.

room to install the current sensor within the distribution transformer enclosure, power system grounding at the distribution transformer instead of at the power panel is also possible. The protective earth ground and the power system ground (neutral conductor) are often connected to the ship's hull at the same location; the current sensor should be installed such that it only measures the current into the neutral conductor from ground. Figure 4.4 depicts a current meter showing about 5 A flowing through the ground current sensor; the high level of ground current is indicative of a ground fault; the ground fault is likely on the neutral conductor.

An advantage of a solidly grounded system for non-vital and nonemergency loads is that a ground fault of a phase conductor, the conductor that is not the

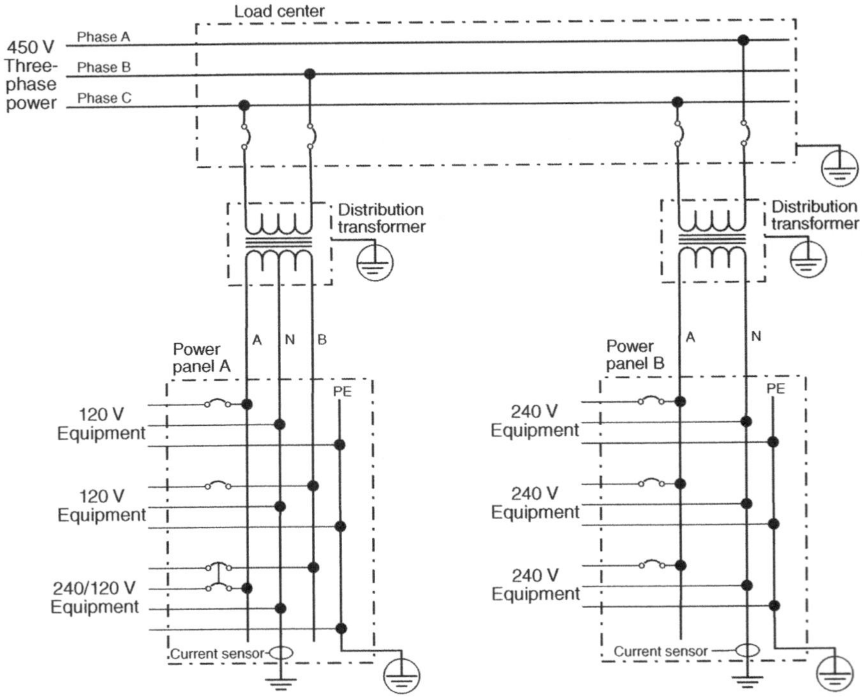

Figure 4.3 Solidly grounded single-phase 120/240 V and 240 V from single-phase transformer.

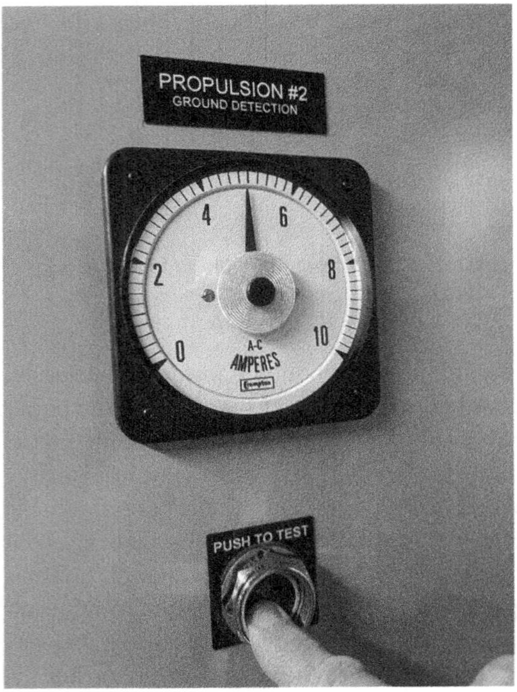

Figure 4.4 Ground detection current meter showing indication of ground fault. *Source:* Mohammed M. Islam.

"neutral" conductor, results in large fault currents that cause the appropriate circuit breaker to trip. By tripping, the ground fault is quickly isolated. The quick isolation of ground faults becomes a disadvantage when vital or emergency loads are served; it is often desirable to continue operation with a single ground fault.

A voltage may exist between the neutral conductor and the protective earth (equipment ground) conductor. While the two conductors are both connected to the ship's hull at the power panel, the protective earth conductor normally does not have any current flowing through it while the neutral conductor has the load current flowing through it. The load current flowing through the neutral conductor impedance results in a voltage between the neutral conductor and the protective earth conductor. Since the load current flows through both the phase conductor and the neutral conductor, the voltage drop across each should be nearly the same. Thus, the neutral-conductor-to-protective-earth-conductor voltage is about half the voltage drop measured at the load. Voltage drops are typically limited to 5% of the nominal system voltage; any neutral-conductor-to-protective-earth-conductor voltage in excess of 2.5% of nominal system voltage is an indicator the circuit is overloaded and has a larger than allowed voltage drop. For 120 V circuits, 2.5% corresponds to 3 V; for 240 V two-wire circuits, 2.5% corresponds to 6 V.

4.2 Design Considerations

4.2.1 General

Solidly grounded distribution systems at 120–240 V reduce the risk of incorporating commercial off-the-shelf equipment on the ship. Some commercial equipment will not operate if a neutral conductor is not at the same potential as the protective earth conductor. Solidly grounded distribution systems also enable the use of single-pole circuit breakers and switches; the neutral is directly connected to the protective earth conductor at the power panel and thus should not be a hazard. Solidly grounded distribution systems may be used to supply loads that are not mission critical and are not emergency loads. Typical examples of loads that may be served by solidly grounded systems include galley equipment, 120–240 V receptacles, and nonemergency lighting. Solidly grounded distribution systems are generally very localized in that they typically power loads in only a relatively few compartments on a single deck.

4.2.2 Transformers

If single-phase loads are powered from a three-phase four-wire system derived from a wye secondary winding of a transformer, the primary winding should be a delta winding. Figure 4.5 depicts a simple model of a delta-wye transformer with

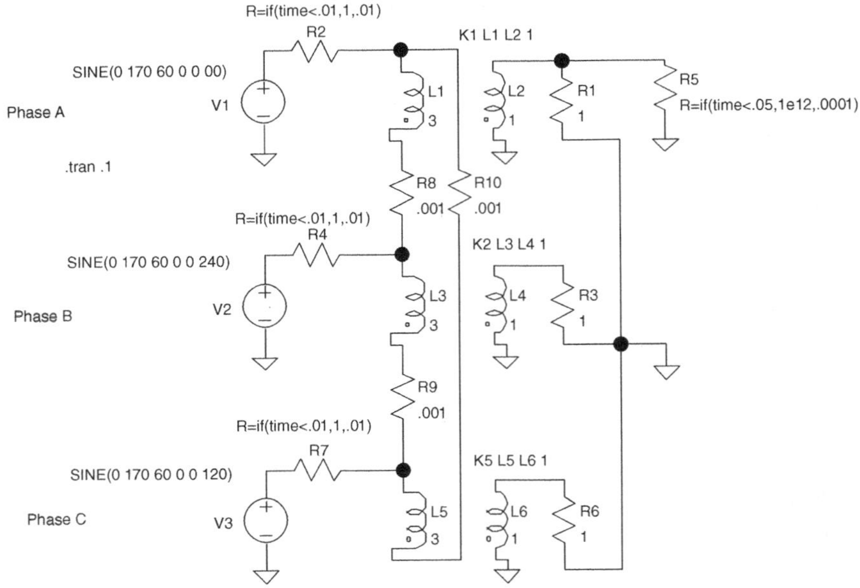

Figure 4.5 LTspice model of delta-wye transformer with secondary solidly grounded.

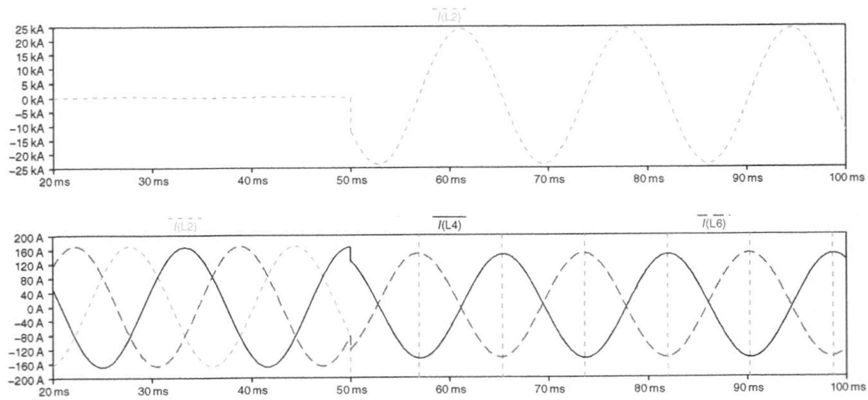

Figure 4.6 Secondary phase currents for delta-wye transformer with secondary solidly grounded.

a resistive load. As shown in Figure 4.6, at 50 ms, a ground fault is placed on phase *a*. As expected, the phase *a* secondary current becomes very large; its magnitude depends on the impedances between the generator and the ground fault. During the ground fault, the un-faulted secondary phase currents are also impacted;

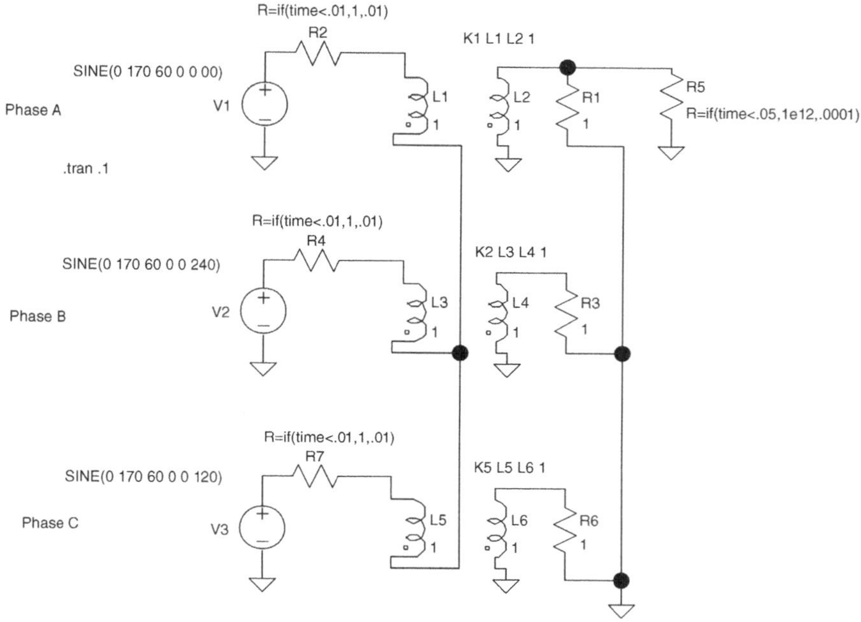

Figure 4.7 LTspice model of wye-wye transformer with secondary solidly grounded feeding three-phase four-wire distribution.

voltage drops in the primary circuit result in the un-faulted secondary phase current magnitudes being somewhat reduced, and their phase angles are no longer 120° apart. The large fault current enables circuit breakers to quickly trip and thereby isolate the ground fault.

Figure 4.7 depicts a schematic of a wye-wye transformer with the secondary solidly grounded with a three-phase four-wire resistive load. As shown in Figure 4.8, a ground fault on phase a results in a somewhat higher than normal current; in this

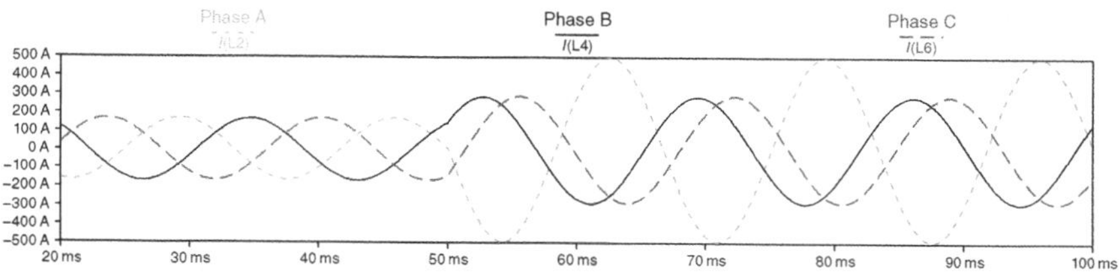

Figure 4.8 Secondary phase currents for wye-wye transformer with secondary solidly grounded.

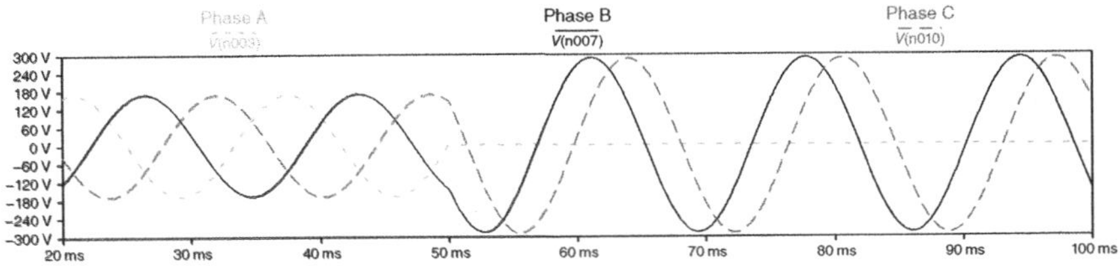

Figure 4.9 Secondary phase voltages for wye-wye transformer with secondary solidly grounded.

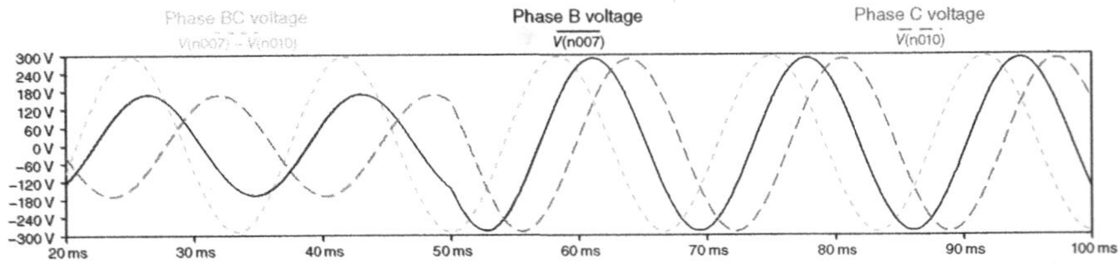

Figure 4.10 120/208 V load voltages for wye-wye transformer with secondary solidly grounded.

case, the fault current is roughly three times normal which may take the circuit breaker over a minute to trip. During the ground fault, the un-faulted phase voltages will be roughly 70% higher than normal, as depicted in Figure 4.9. This overvoltage may result in insulation failures and possibly degradation of MOV-based surge protective devices. Figure 4.10 depicts the voltage waveforms that a load using two of the three phases to provide 120 V rms line-to-ground and 208 V_{rms} line-to-line. The line-to-line voltage remains at roughly 208 V, but the 120 V line-to-ground voltages increase to 208 V. Having the line-to-ground voltage equal to the line-to-line voltage may result in the line-to-ground loads malfunctioning. For these reasons, a solidly grounded wye-wye transformer should not be used to supply a three-phase four-wire distribution system.

Figure 4.11 depicts a schematic of a wye-wye transformer with the secondary solidly grounded and with a 3-wire resistive load. As depicted in Figure 4.12 for the phase current and Figure 4.13 for the line-to-line voltages, the phase currents and line-to-line voltages are essentially unaffected by the ground fault; overcurrent protective devices will not trip to isolate the ground fault. As depicted in Figure 4.14, the line-to-ground voltages are nearly the same as for an ungrounded system. The ground fault current, as depicted in Figure 4.15, is small compared to the phase currents. Because overcurrent protective devices will not clear ground faults, a wye-wye transformer with the secondary solidly grounded should not be used with a three-wire distribution system.

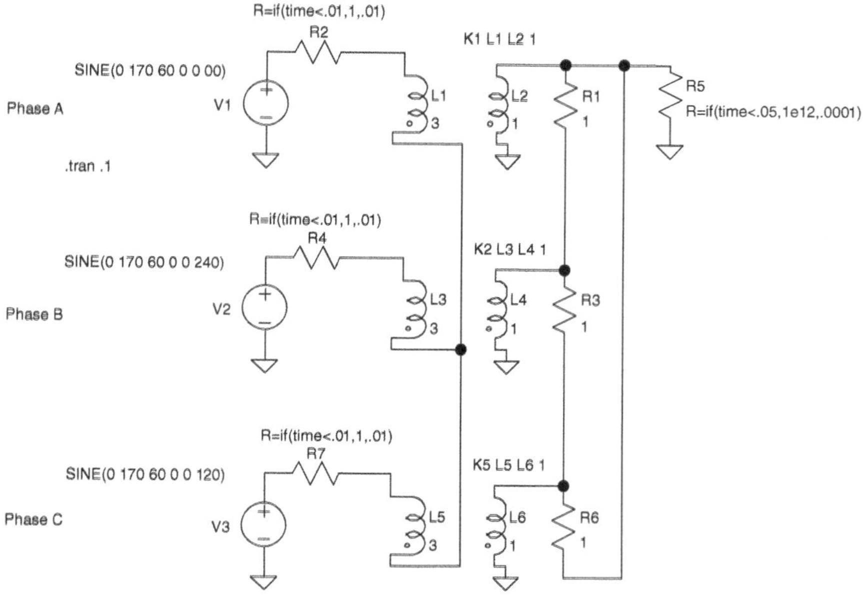

Figure 4.11 LTspice model of wye-wye transformer with secondary solidly grounded feeding three-wire distribution.

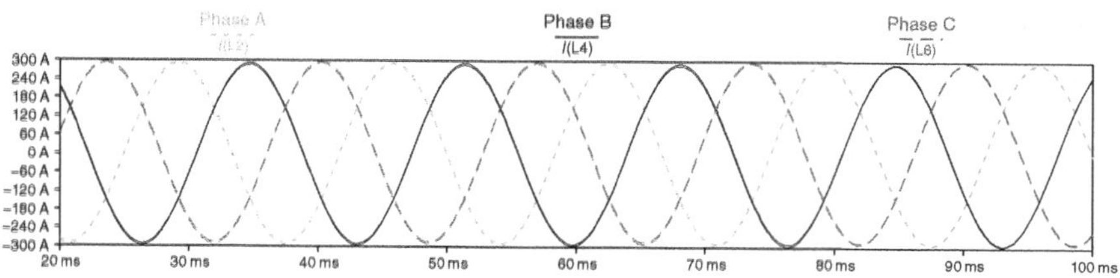

Figure 4.12 Secondary phase currents for wye-wye transformer with secondary solidly grounded feeding three-wire distribution.

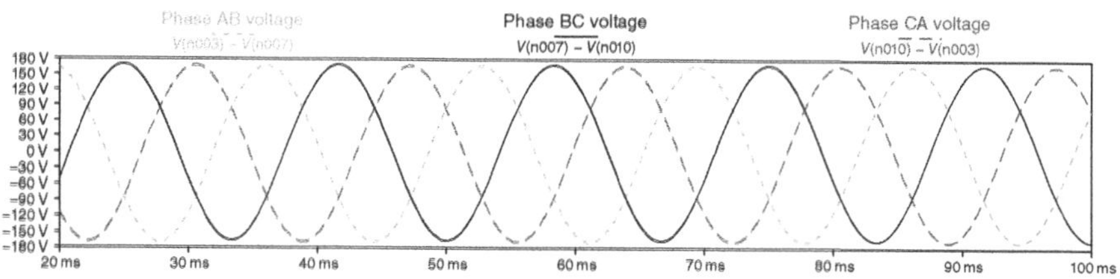

Figure 4.13 Secondary line-to-line voltages for wye-wye transformer with secondary solidly grounded feeding three-wire distribution.

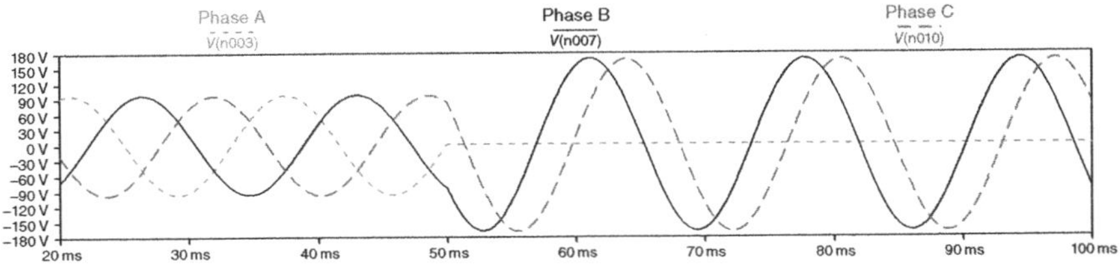

Figure 4.14 Secondary phase voltages for wye-wye transformer with secondary solidly grounded feeding three-wire distribution.

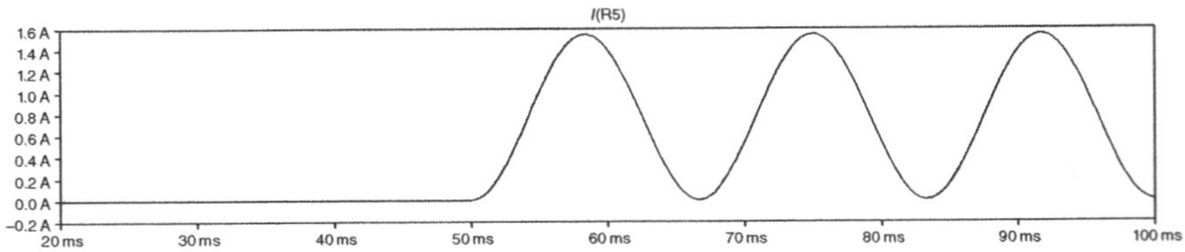

Figure 4.15 Fault current for wye-wye transformer with secondary solidly grounded feeding three-wire distribution.

4.2.3 Sub-power Panels

In the design of the low-voltage distribution system, it may at times be desirable to have one power panel feed power to a sub-power panel. Figure 4.16 depicts how a three-phase four-wire power panel that is connected to the distribution transformer should then connect to the sub-power panel. The three power phases are connected via a three-phase circuit breaker. The neutral and protective earth conductors are directly connected. Note that the subpanel does not have a power system ground connection for the neutral conductor but does have a protective earth ground for the protective earth conductor. Having only one power system ground ensures that its current sensor can accurately measure the ground current and prevents neutral current from flowing through the ship's hull.

4.2.4 Ground Fault Protection

Consideration should be given to providing ground fault protection to all 120–240 V receptacles for enhanced personnel safety. Ground fault protection for receptacles is typically provided by a GFCI, a Residual Current Detector (RCD), or a combination of RCD and miniature circuit breaker (RCBO). A GFCI will trip on either an overcurrent or if the current in the phase conductor and neutral conductor are mismatched by more than 4–6 mA. GFCIs are typically used in the United States. RCDs

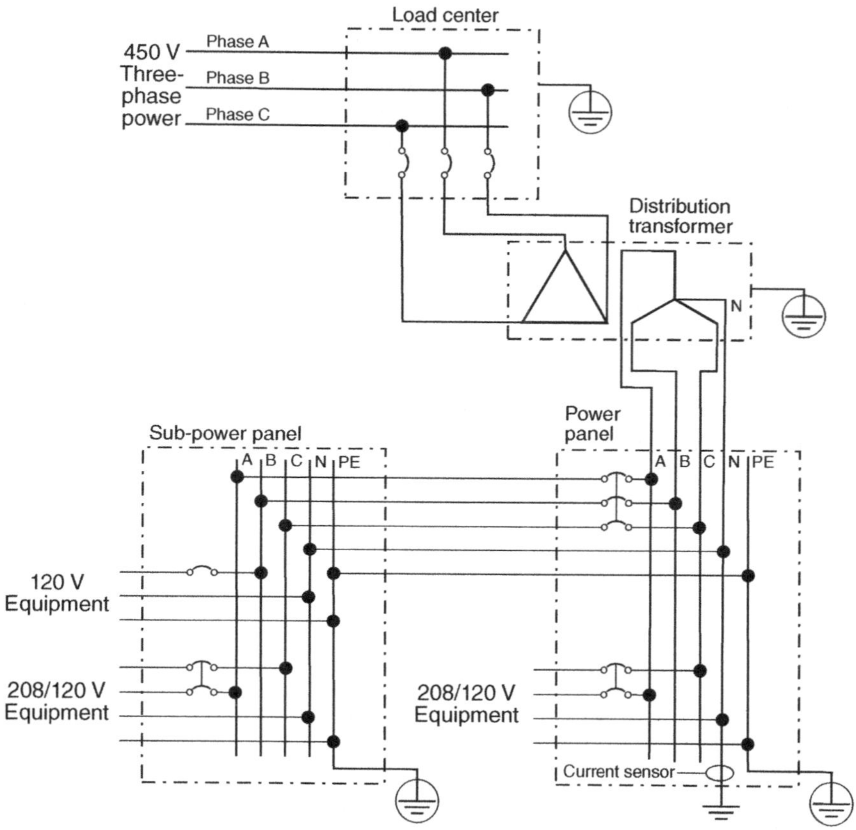

Figure 4.16 Solidly grounded systems with power panel and sub-power panel.

and RCBOs will trip if the current in the phase conductor and the neutral conductor are mismatched by more than 10 mA. RCDs do not include an overcurrent trip, while RCBOs do. RCDs and RCBOs are typically used in the European Union.

Equipment (not emergency loads and not mission critical) can be protected using a GFCI, RCD, RCBO, or an appliance leakage current interrupter (ALCI). An ALCI is functionally the same as a GFCI/RCBO but trips with a current mismatch of 30 mA. This higher level is intended to protect the equipment rather than the operator and will likely result in fewer nuisance trips.

Regulation or class rules may require a current sensor on the neutral ground connection. This current sensor, while less sensitive than a GFCI/RCD/ALCI, can provide an indication of a ground fault current with a magnitude less than that which would trip a traditional circuit breaker on overcurrent.

An inexperienced technician may reason that, since the neutral wire is connected to the ship's hull, a load can be operated by connecting it to the phase conductor and the ship's hull. With a few notable exceptions (such as the cathodic protection system), using the ship's hull as a power system conductor is to be avoided to prevent hull corrosion, localized heating, safety hazards, and EMI. Class rules and regulations often prohibit this practice. The neutral current sensor can detect this practice, and a GFCI/RCD/ALCI can prevent such loads from operating.

Care should be taken to limit the line-to-ground capacitance of circuits protected by a GFCI, RCD, RCBO, or ALCI. As shown in Figure 4.17 for a GFCI, if the aggregate line-to-ground capacitance is large enough (typically from EMI filters), then the difference in conductor currents through the GFCI could be larger than the GFCI trip setting under normal un-faulted operation. These nuisance trips are a potential source of many wasted hours looking for a ground fault that does not exist.

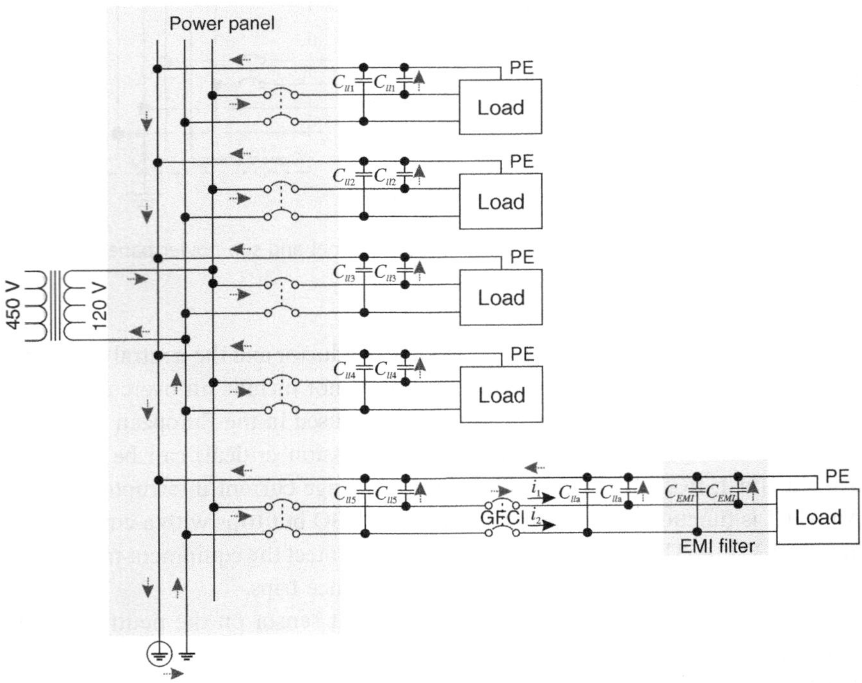

Figure 4.17 Solidly grounded systems with GFCI.

4.3 Cable Insulation Colors

Many color schemes for conductor insulation exist for power cables serving systems with nominal system voltages between 100 and 240 V. Table 4.1 lists the commonly used colors for power system conductors. In shipboard power systems, work on a ship may have occurred in many different locations in the world, resulting in nonstandard and inconsistent use of insulation colors. One should never assume the function of a conductor based solely on its insulation color.

Table 4.1 Conductor insulation colors for AC voltages no greater than 240 V.

Conductor	The United States	European Union	Other possibilities
Protective earth	Green Green-yellow stripes (bare)	Green-yellow stripes	
Neutral	White (silver) or gray	Blue	Black
Phases	Black (brass), red, orange, blue, brown, and yellow	Brown, black, and gray	White, purple

5

System Grounding: Shipboard HRG AC Primary Distribution Systems (Greater than 1 kV)

5.1 Characteristics

This chapter applies to AC primary distribution systems, as depicted in Figure 5.1. High-resistance grounded (HRG) AC primary distribution systems with a nominal system voltage greater than 1 kV are similar in characteristics to the HRG AC low-voltage distribution systems with a nominal system voltage no greater than 1 kV, detailed in Chapter 3. The differences include

- The line-to-ground parasitic capacitance of the higher voltage cable is usually much higher than that of the lower voltage systems because of the need for grounded conductor shields to control electric field stresses. This is particularly true for nominal system voltages above 4.16 kV.
- Special cable terminations are used to control the electrical fields around the conductors at the cable ends for primary distribution systems.
- The higher line-to-ground parasitic capacitance may result in higher-than-desired ground fault currents that require special design solutions of the primary distribution system.
- Transformers are typically used with grounding resistors and instrumentation to reduce the voltage to a lower level to enable the use of more commonly procured components and reduce the risk of damage to instrumentation from the higher voltage levels.
- Compared to low-voltage distribution systems, there are typically fewer sources and loads on an AC primary distribution system. This can simplify localizing a ground fault once it has been detected and potentially reduces the number of EMI filters that must be considered.

Design of Shipboard Power System Grounding/Earthing, First Edition. Norbert Doerry, Mohammed M. Islam, and John Prousalidis.
© 2025 The Institute of Electrical and Electronics Engineers, Inc.
Published 2025 by John Wiley & Sons, Inc.

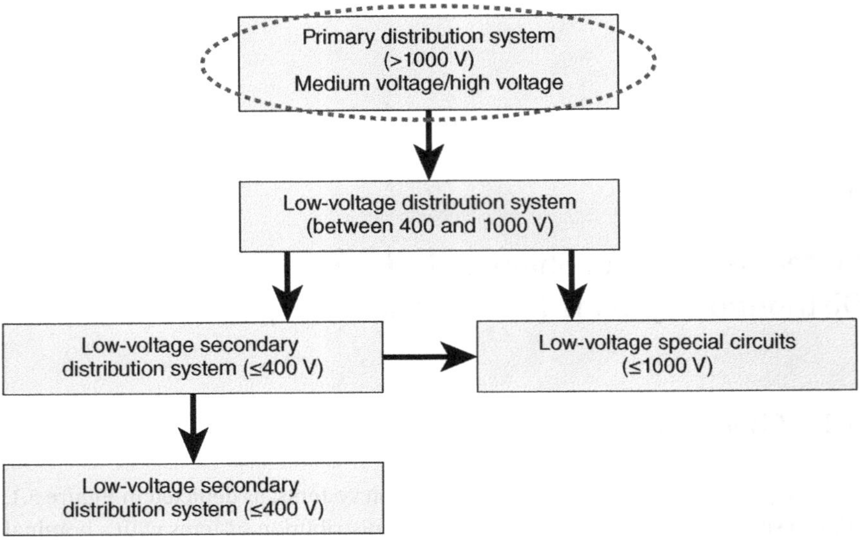

Figure 5.1 Applicability of HRG AC primary distribution systems.

5.2 Grounding Circuit

5.2.1 Grounding Resistor Transformer

The neutral ground, zigzag transformer, and wye-delta transformer grounding methods from Chapter 3 also apply to AC primary distribution systems. Usually, a transformer with a $n:1$ turns ratio is used as part of the grounding circuit to reduce the voltage on the grounding resistor to a low voltage so that specialized higher voltage resistors and instrumentation are not necessary; normally stocked grounding resistors and instrumentation can be employed. As depicted in Figure 5.2, the resistor on the secondary of the transformer should have a resistance equal to the desired grounding resistance divided by the turns ratio squared. The use of a transformer with a grounding resistor connected to a neutral created by a zigzag transformer is depicted in Figure 5.3. A transformer with a grounding resistor can similarly be used for the neutral ground and wye-delta transformer grounding methods.

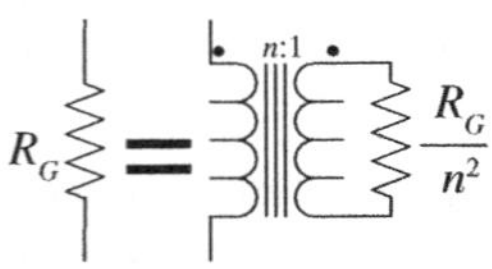

Figure 5.2 Use of transformer for grounding resistor.

5.2.2 Wye-Broken Delta Transformer

Figure 5.4 depicts a wye-broken delta transformer configuration. To keep the voltage on the grounding resistor low, each of the transformers typically has a turns ratio of $n:1$. The wye common point is hard grounded, and the grounding resistor is inserted into the delta secondary.

The equation for the delta connection is

$$\frac{v_a + v_b + v_c}{n} = v_{RG}$$

Rewriting this equation results in

$$\left(v_a - \frac{n}{3}v_{RG}\right) + \left(v_b - \frac{n}{3}v_{RG}\right) + \left(v_c - \frac{n}{3}v_{RG}\right) = 0$$

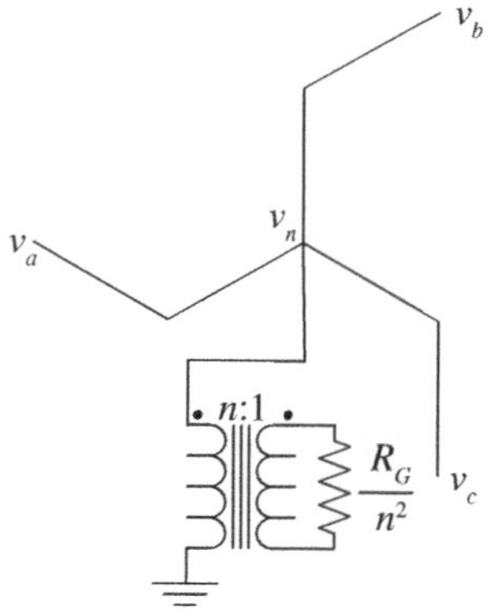

Figure 5.3 Use of transformer for grounding resistor with zigzag transformer.

Hence, $\frac{n}{3}v_{RG}$ is the neutral voltage. The currents through all of the secondary windings are identical and equal to $\frac{v_{RG}}{R_G}$. Since they are all equal, the current into the ground connection on the primary is $\frac{3}{n}\frac{v_{RG}}{R_G}$.

A grounding circuit employing a wye-broken delta transformer is typically depicted, as shown in Figure 5.5.

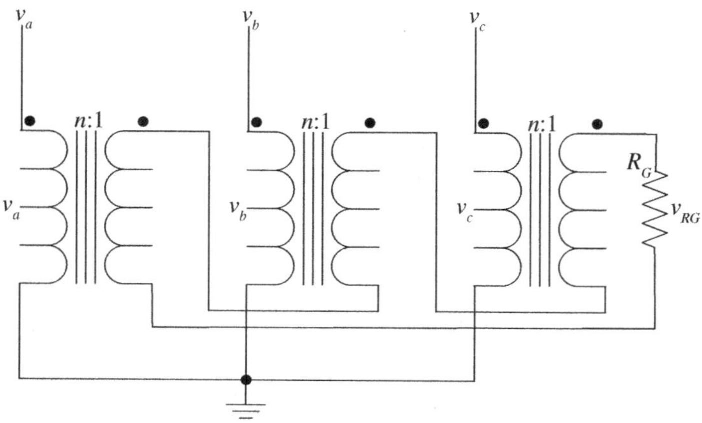

Figure 5.4 Wye-broken delta transformer configuration.

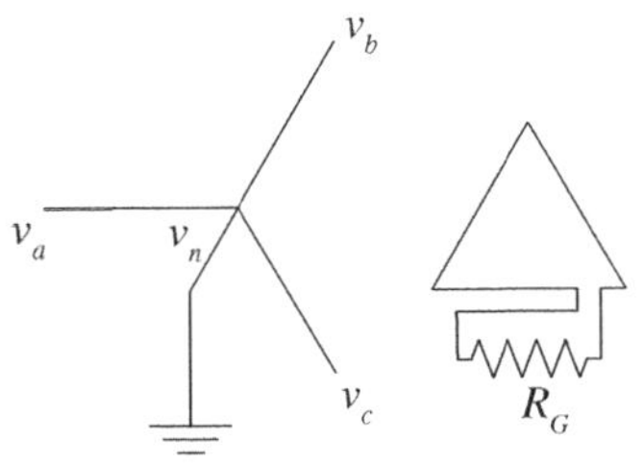

Figure 5.5 Wye-broken delta transformer grounding circuit.

One advantage of the wye-broken delta transformer in applications above 1 kV is that the voltages across the grounding resistor and instrumentation can be adjusted through the turns ratio n to be a low-voltage compatible with normally stocked grounding resistors and instrumentation; a second transformer is not required. Because this second transformer is not required, the wye-broken delta transformer is often the preferred choice for HRG systems for primary distribution systems.

5.2.3 Grounding Resistor Design

During a ground fault, the heat dissipated from a grounding resistor for an AC primary distribution system can range from below 10 to 80 kW or more. If an air-cooled resistor is used, the ground fault should be detected, localized, and cleared relatively quickly (within minutes) to preclude placing a significant heat burden on the machinery room ventilation system. An alternative is water cooling. Since the water cooling must be very reliable, its inclusion can add considerable cost and complexity.

Because of the thermal aspects of the grounding resistor design, the grounding resistors are typically installed in their own cabinets.

5.3 Modeling Shipboard HRG AC Primary Distribution Systems

This section provides guidance for modeling AC primary distribution systems with a nominal system voltage greater than 1 kV for the purpose of understanding ground fault transients.

5.3.1 Three-Conductor Shielded Cable Models

Cables for applications above 1 kV differ from cables for applications below 1 kV in that the higher voltage cables typically include shields around each conductor, as well as an overall cable shield. These shields are additional conductors and, through their capacitance and mutual inductance with the primary current-carrying conductors, impact the overall transient and ground fault response of the cable. Figure 5.6 depicts the geometry of a three-phase cable with conductor and cable shields.

where

D, overall diameter of the cable

ρ, radius of primary current-carrying conductor

d, distance between centers of the primary current-carrying conductors

r, outer radius of conductor insulation

s, thickness of cable shield

t, thickness of cable sheath

u, thickness of conductor shield.

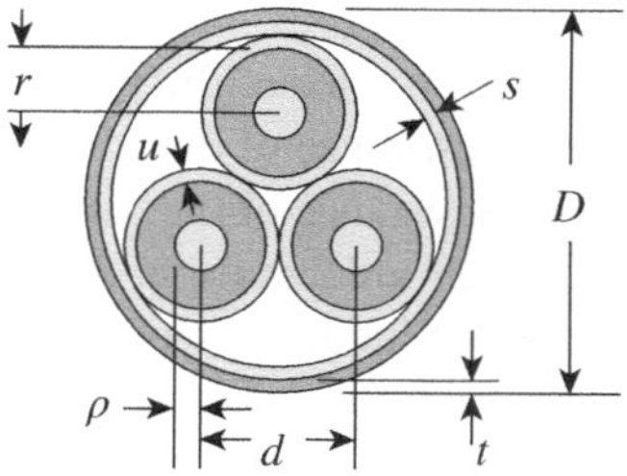

Figure 5.6 Three-phase shielded cable dimensions.

The modeling is simplified with the following assumptions:

- The four shields are not in contact with each other over the length of the cable.
- The conductor shields are connected to ground only at one end. Their primary purpose is to control the electric field in the conductor insulation to avoid localized electric field stresses and insulation failures.
- The cable shield is connected to ground at both ends. Its primary purpose is to conduct CM current to minimize the flow of CM current through the ship's hull.
- The only capacitances that need to be modeled are between the conductors and their conductor shield.
- The only mutual inductances that need to be calculated are between each primary current-carrying conductor and the other primary current-carrying conductors, its own conductor shield, and the cable shield.

These assumptions result in the cable model depicted in Figure 5.7.

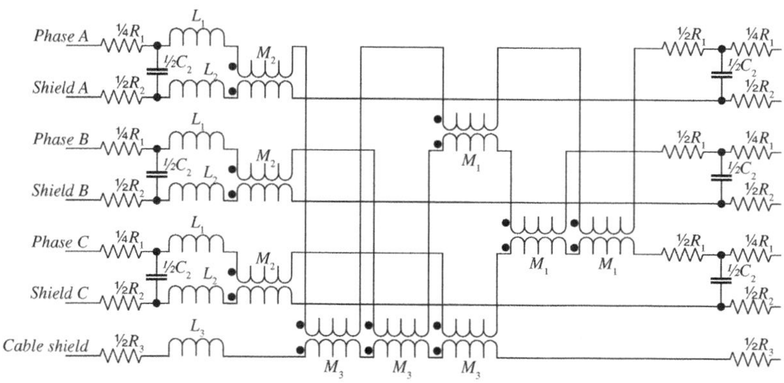

Figure 5.7 Schematic for cable model for shielded cable depicted in Figure 5.6.

where

R_1, resistance of the primary current-carrying conductors
R_2, resistance of the conductor shields
R_3, resistance of the cable shield
L_1, self-inductance of the primary current-carrying conductors
L_2, self-inductance of the conductor shields
L_3, self-inductance of the cable shield
M_1, mutual inductance between two of the primary current-carrying conductors
M_2, mutual inductance between a primary current-carrying conductor and its conductor shield
M_3, mutual inductance between a primary current-carrying conductor and the cable shield
C_2, capacitance between a primary current-carrying conductor and its conductor shield.

The self-inductance (μH) of a round wire of radius ρ m and length l m is given by equation (7) of Grover (2009):

$$L_1 = 0.2l\left[\ln\left(\frac{2l}{\rho}\right) - \frac{3}{4}\right]$$

The mutual inductance (μH) with respect to a conductor at a distance d m away is given by equation (1) of Grover (2009):

$$M_1 = 0.2l\left[\ln\left(\frac{l}{d\sqrt{2}} + \sqrt{1 + \frac{l^2}{2d^2}}\right) - \sqrt{1 + \frac{2d^2}{l^2}} + \frac{d\sqrt{2}}{l}\right]$$

If the length l is much greater than the conductor separation d, then Eqs. (2.3) and (2.4) apply.

$$L_1 = 0.2l\left[\ln(2) + \ln(l) - \ln(\rho) - \frac{3}{4}\right]$$

$$M_1 \approx 0.2l\left[\ln\left(\sqrt{2}\right) + \ln(l) - \ln(d) - 1\right]$$

The self-inductances (μH) of the shields are

$$L_2 = 0.2l\left[\ln\left(\frac{2l}{r+u}\right) + \ln(\xi) - 1\right]$$

$$L_3 = 0.2l\left[\ln\left(\frac{2l}{\frac{D}{2} - t}\right) + \ln(\xi) - 1\right]$$

If the thickness of the shield is much smaller than its radius, then $\ln(\xi)$ can be approximated by

$$\text{For } L_2\text{: } \ln(\xi) \approx 0.0167\left(20\frac{u}{r+u}\right) \tag{5.1}$$

For L_3: $$\ln(\xi) \approx 0.0167\left(20\frac{s}{\frac{D}{2}-t}\right) \tag{5.2}$$

The approximations of Eqs. (5.1) and (5.2) are based on table 4 of Grover (2009).

The mutual inductance (μH) calculations for M_2 and M_3 are based on equation (24) of Grover (2009):

$$L_1+L_2-2M_2=0.2l\left[\ln\left(\frac{r+u}{\rho}\right)+\frac{2\left(\frac{r}{r+u}\right)^2}{1-\left(\frac{r}{r+u}\right)^2}\ln\left(\frac{r+u}{r}\right)-0.75+\ln(\xi)\right]$$

$$L_1+L_3-2M_3=0.2l\left[\ln\left(\frac{\frac{D}{2}-t}{\rho}\right)+\frac{2\left(\frac{\frac{D}{2}-t-s}{\frac{D}{2}-t}\right)^2}{1-\left(\frac{\frac{D}{2}-t-s}{\frac{D}{2}-t}\right)^2}\ln\left(\frac{\frac{D}{2}-t}{\frac{D}{2}-t-s}\right)-0.75+\ln(\xi)\right]$$

where $\ln(\xi)$ is the same for M_2 as it is for L_2, and it is the same for M_3 as it is for L_3.

Using approximations and simplifications results in

$$L_1+L_2-2M_2 \approx 0.2l\left[\ln\left(\frac{r+u}{\rho}\right)+0.25+\ln(\xi)\right] \tag{5.3}$$

$$M_2 \approx \frac{1}{2}\left(L_1+L_2-0.2l\left[\ln\left(\frac{r+u}{\rho}\right)+0.25+\ln(\xi)\right]\right) \tag{5.4}$$

$$M_3 \approx \frac{1}{2}\left(L_1+L_3-0.2l\left[\ln\left(\frac{\frac{D}{2}-t}{\rho}\right)+0.25+\ln(\xi)\right]\right) \tag{5.5}$$

The resistance R_1 is usually provided in cable data sheets. IEEE Std 1580-2021 (2021), MIL-HDBK-299 (1991), and ASTM B8-11 (2017) may also be consulted for values.

The shield resistances, R_2 and R_3, can be approximated by the resistivity of the shield material. If we assume the shields are cylinders, then the resistances of the shields are given by

$$R_2 = R_{vol}\frac{l}{\pi(2r+u)u} \tag{5.6}$$

$$R_3 = R_{vol}\frac{l}{\pi(D-2t-s)s} \tag{5.7}$$

where R_{vol} is the resistivity of the shield material in ohm meters (Ωm).

The impact of temperature on resistivity is given by

$$R_{vol}(T) = R_{vol}(T_0) \times (1 + \alpha_0(T - T_0)) \tag{5.8}$$

where

$R_{vol}(T)$, resistivity (Ωm) at a given temperature T
$R_{vol}(T_0)$, resistivity (Ωm) at a reference temperature T_0
α_0, temperature coefficient (1/°C).

R_{vol} for solid annealed copper is about 1.72×10^{-8} Ωm at 20 °C. The temperature coefficient (α_0) of annealed copper is 0.00393/°C. Thus, at 40 °C, R_{vol} for annealed copper is about 1.86×10^{-8} Ωm. The anticipated operating temperature of the cable should be used.

Because a braid is not a solid material, the resistance of braided shields will be greater; lacking additional information, doubling, or even tripling the calculated resistance value for a braided shield is not unreasonable.

These calculations ignore the impact of the skin effect at high frequencies. The skin effect will increase the resistance values at higher frequencies.

The conductor-to-conductor shield capacitance is given by

$$C_2 = l\frac{2\pi\varepsilon_0\kappa}{\ln\left(\frac{r}{\rho}\right)} \tag{5.9}$$

where

ε_0, electric permittivity of free space

$$\varepsilon_0 = 8.85\ \text{pF/m}$$

κ, dielectric constant.

Chapter 2 provides nominal dielectric constants for some insulators.

An alternate method to approximate the line-to-ground capacitance for 13.8 kV cable is to use the guidance from Table 5.1. The line-to-ground capacitance per

Table 5.1 Estimates for line-to-ground capacitance (per phase) of 13.8 kV cable.

13.8 kV cable shielded three-phase: conductor size	I_C (A/1000 ft)	C_p (μF/1000 ft) conductor-to-conductor shield
1000 MCM	1.15	0.128
750 MCM	0.93	0.103
350 MCM	0.71	0.079
4/0 AWG	0.65	0.072
2/0 AWG	0.55	0.061

Source: Adapted from Bridger (1983).

phase is calculated from the frequency (ω), charging current per unit length (I_C), and the line-to-line voltage ($V_{line-line}$) using

$$C_P = \frac{C_2}{l} = \frac{I_C}{\sqrt{3}\omega V_{line-line}} \tag{5.10}$$

5.3.2 Single-Conductor Cable and Insulated Bus Pipe Models

In AC primary distribution systems with required bus ratings above 300 A, it may be advantageous to use sets of three single-conductor cables. The ampacity of each single-conductor cable can be considerably higher than for the three-conductor cable for the same bend radius. Because the total number of conductors and shields is considerably less for single-conductor cables compared to three-conductor cables, there are far fewer conductor and shield terminations; installation costs of the cables are greatly reduced.

For current levels in the thousands of amps, insulated bus pipe (IBP), as depicted in Figure 5.8, may prove cost-effective. IBP consists of either a solid rod or solid pipe that is appropriately insulated; a conductor shield provides electric field control for the insulation. Fittings, much like pipe elbows, can be used to implement a

Figure 5.8 Insulated bus pipe with shield cross-section view. *Source:* Norbert Doerry.

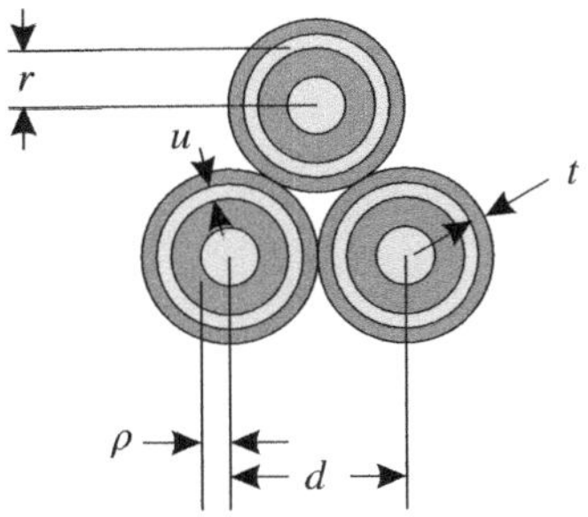

Figure 5.9 Triad arrangement for a set of single-conductor shielded cables or shielded insulated bus pipes.

very tight turning radius; routing of the distribution system in crowded machinery rooms is greatly facilitated by these tight turning radii.

Both single-conductor cable and bus pipe may be arranged as triads (Figure 5.9) or in a single bank (Figure 5.10). The triad has the advantage of balanced mutual inductances among the conductors. For the single bank, the mutual inductance between the inner conductor (B) and the two outer conductors (A and C) is different from the mutual inductance between the two outer conductors. The ampacity of the cables may be somewhat higher for the single bank because of improved airflow.

Figure 5.11 depicts a model of a set of three single-conductor cables. The parameters, R_1, R_2, L_1, L_2, M_2, and C_2 are calculated in the same manner as in Section 5.3.1. The mutual inductances between the conductors are given by

$$M_{AB} = M_{BC} \approx 0.2l\left[\ln\left(\sqrt{2}\right) + \ln(l) - \ln(d) - 1\right] \tag{5.11}$$

For the triad arrangement (Figure 5.9), M_{CA} is given by

$$M_{CA} = M_{AB} \tag{5.12}$$

For the single-bank arrangement (Figure 5.10), M_{CA} is given by

$$M_{CA} \approx 0.2l\left[\ln\left(\sqrt{2}\right) + \ln(l) - \ln(2d) - 1\right] \tag{5.13}$$

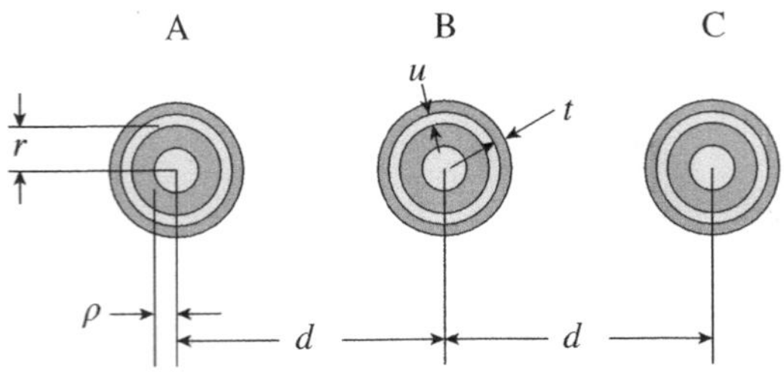

Figure 5.10 Single-bank arrangement for a set of shielded single-conductor shielded cables or shielded insulated bus pipes.

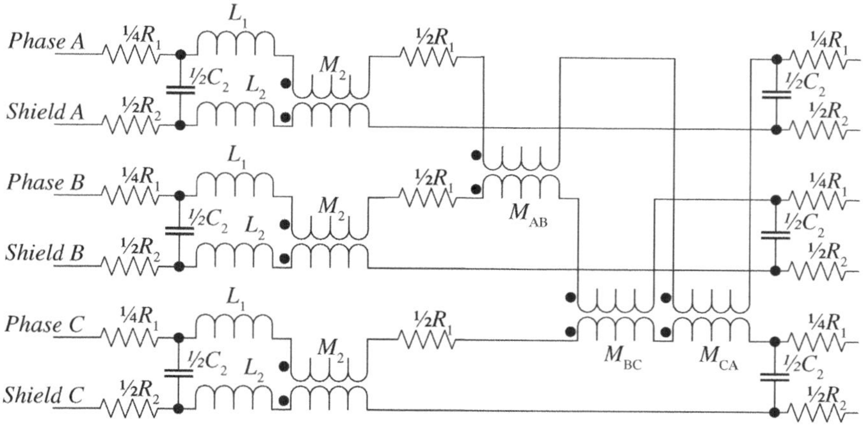

Figure 5.11 Schematic for cable model.

One implication of the single-bank arrangement is that the cable impedances are not entirely symmetrical; the asymmetry can result in CM voltages under normal operation, as detailed in Chapter 13.

In both arrangements, each of the conductor shields is normally grounded only on one end. Since both arrangements lack a shield that is grounded on both ends, all CM current associated with the three conductors will likely have a return path through the ship's hull. If this CM current is not large, this may not be an issue.

5.3.3 EMI and Harmonic Filters

Manufacturers should be consulted to determine the line-to-ground capacitances of EMI and harmonic filters.

5.3.4 Generators and Motors

Generator and motor data sheets typically do not provide the line-to-ground capacitance. Figures 5.12 through 5.15 may be used as guidance for line-to-ground capacitance estimates. Analyses should consider a range of capacitance values to account for estimation tolerances and design changes in the 60 or more years since the graphs were developed.

Figure 5.12 depicts capacitance per phase-to-ground curves for salient-pole generators and motors for voltages between 2.3 and 13.8 kV. These curves are generally useful for diesel generator sets.

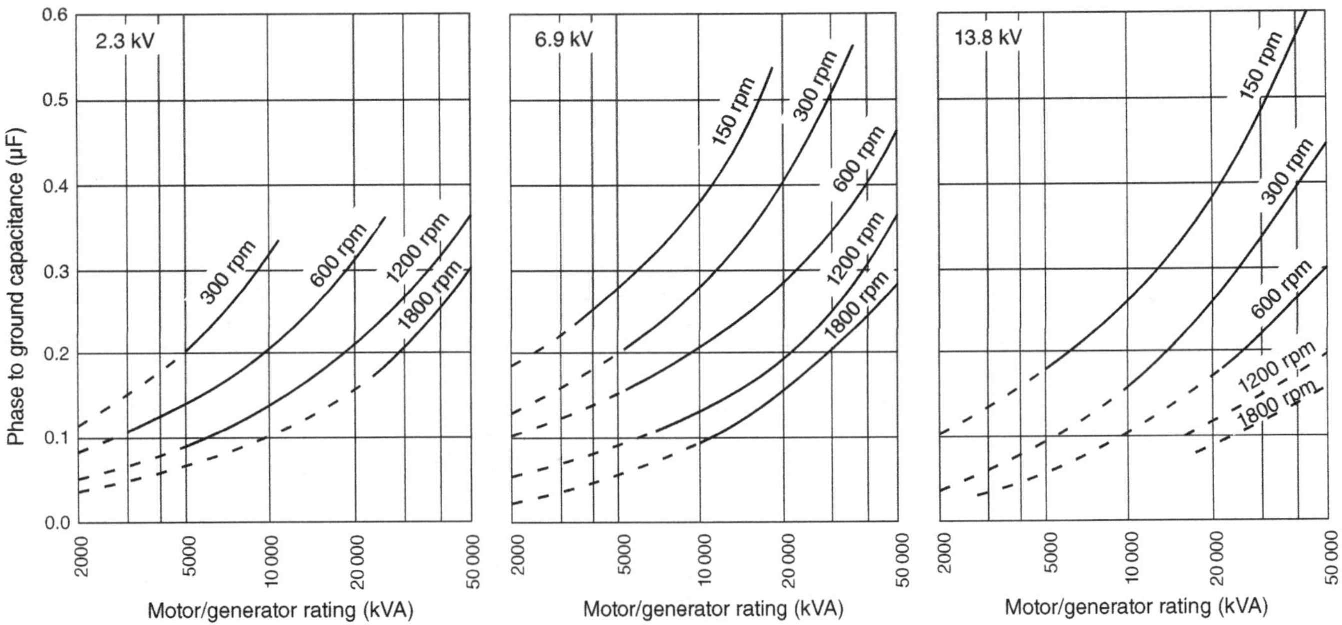

Figure 5.12 Phase-to-ground capacitance for salient-pole generators and motors. *Source:* Adapted from Westinghouse Electric Corporation Central Station Engineers (1964).

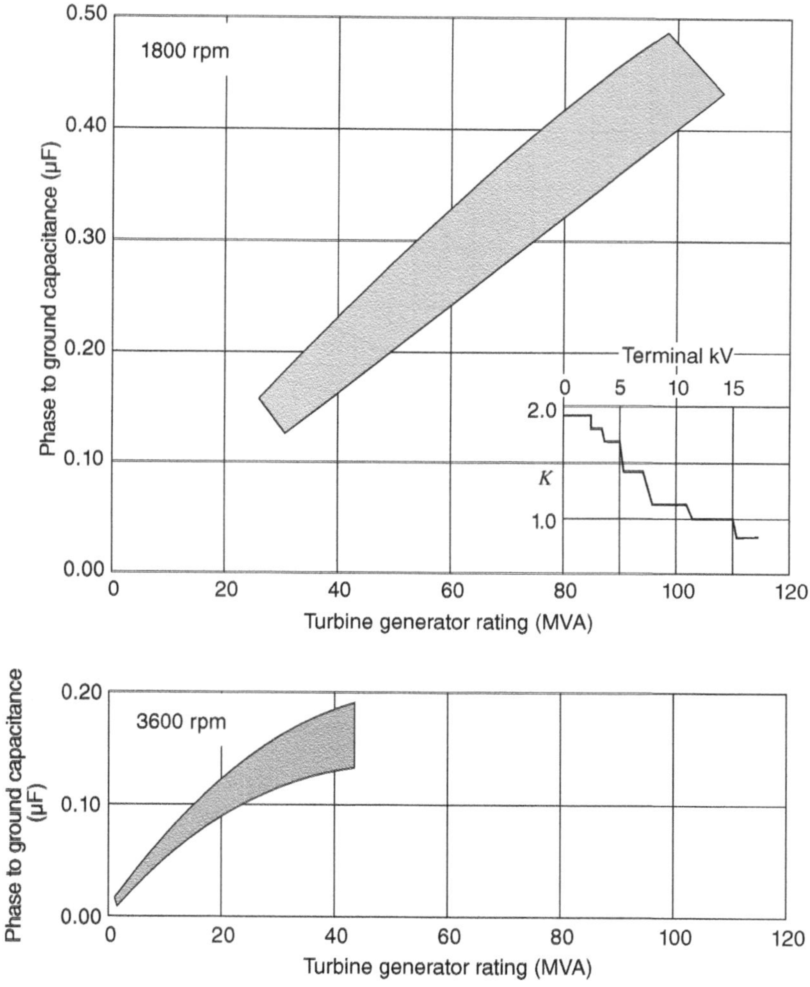

Figure 5.13 Estimated capacitance-to-ground per phase of turbine generator (13.2 kV). *Source:* Adapted from Westinghouse Electric Corporation Central Station Engineers (1964).

Figure 5.13 provides line-to-ground capacitance estimates for 13.2 kV turbine generators; a *K* factor is provided to adjust the curve values by multiplication for other voltages.

Figure 5.14 provides capacitance per phase-to-ground for 2.3 kV synchronous motors; the capacitance per phase should not vary more than about 15% for synchronous motors between 2.3 and 6.6 kV.

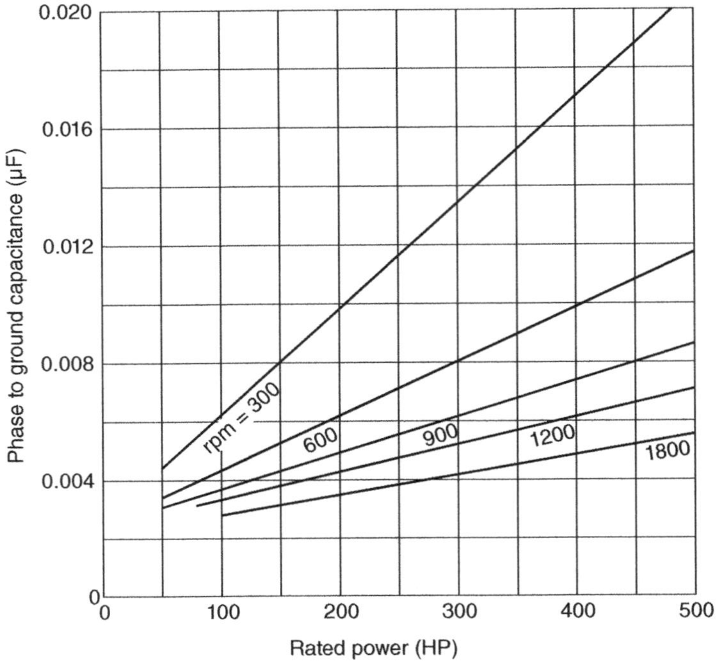

Figure 5.14 Estimated capacitance-to-ground per phase of 2.3 kV synchronous motors. *Source:* Adapted from Westinghouse Electric Corporation Central Station Engineers (1964).

Figure 5.15 provides capacitance per phase-to-ground for 2.3 kV induction motors; the capacitance per phase should not vary more than about 15% for induction motors between 2.3 and 6.6 kV.

Table 5.2 provides alternate data with the line-to-ground capacitance calculated from I_C.

Manufacturers should be consulted if more accurate data is required.

5.3.5 Transformers

Barker (1979) and Bridger (1983) suggest that for industrial applications, the capacitance to ground of primary distribution system transformers is negligible. The use of a grounded transformer shield may result in a non-negligible capacitance. The transformer manufacturer should be consulted for guidance as to the value of capacitance to use.

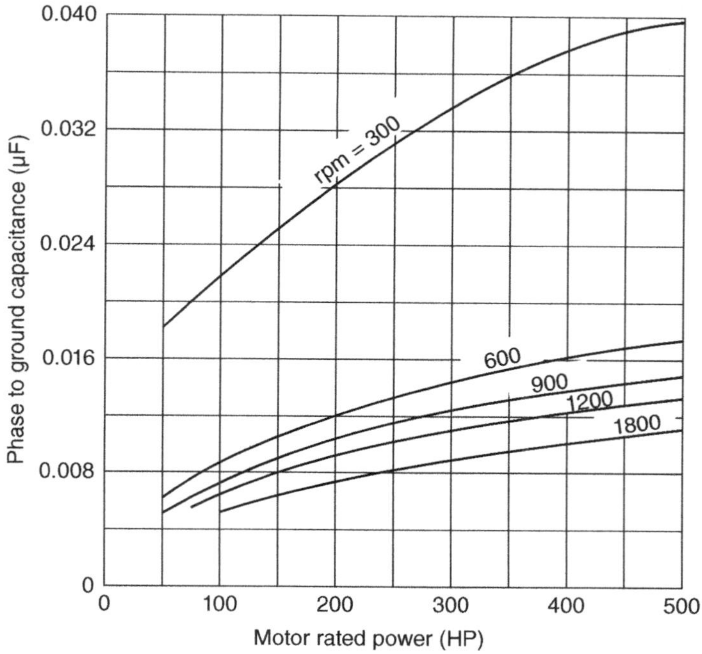

Figure 5.15 Estimated capacitance-to-ground per phase of 2.3 kV induction motors. *Source:* Adapted from Westinghouse Electric Corporation Central Station Engineers (1964).

Table 5.2 Estimates for line-to-ground capacitance (per phase) of medium voltage motors.

	I_C (A/1000 HP)	C_p (μF/1000 HP) line-to-ground per phase
13.8 kV motor – estimated	0.15	0.0166
4.16 kV motor – estimated	0.05	0.0055
2.4 kV motor – estimated	0.03	0.0033
2.4 kV motor with cables – tested	0.06	0.0067

Source: Adapted from Bridger (1983).

5.3.6 System Simulation

Figure 5.16 depicts an LTspice model of an AC primary distribution system HRG system. The assumptions for this model are

Cable:		
r	16 mm	Radius of conductor insulation
u	0.250 mm	Thickness of conductor shield
s	0.750 mm	Thickness of conductor shield
ρ	8 mm	Radius of conductor
d	32.5 mm	Distance between conductor centers
t	4 mm	Thickness of outer sheath
D	92 mm	Overall diameter of the cable

	Cable 1	**Cable 2**	
l	35 m	65 m	Length of cable

Calculated values:

L1	58.3 μH	116.3 μH	Conductor self-inductance
L2	51.6 μH	103.9 μH	Conductor shield self-inductance
L3	39.8 μH	82.0 μH	Cable shield self-inductance
M1	44.3 μH	90.3 μH	Conductor-to-conductor mutual inductance
M2	52.1 μH	104.7 μH	Conductor-to-conductor shield mutual inductance
M3	42.3 μH	86.7 μH	Conductor to cable shield mutual inductance
R1	1.64 mΩ	3.05 mΩ	Conductor resistance
R2	13.1 mΩ	24.2 mΩ	Conductor shield resistance
R3	1.69 mΩ	3.13 mΩ	Cable shield resistance
C2	0.00646 μF	0.011993 μF	Conductor-to-conductor shield capacitance

Note that for the model of the cable, the capacitances and resistances are split equally into two parts.

Generator:

V	13.8 kV line-to-line voltage (11.3 kV line-to-ground peak voltage)
C	0.15 μF
L	1.4 mH (calculated from 0.1 PU sub-transient reactance for 36 MVA base power)

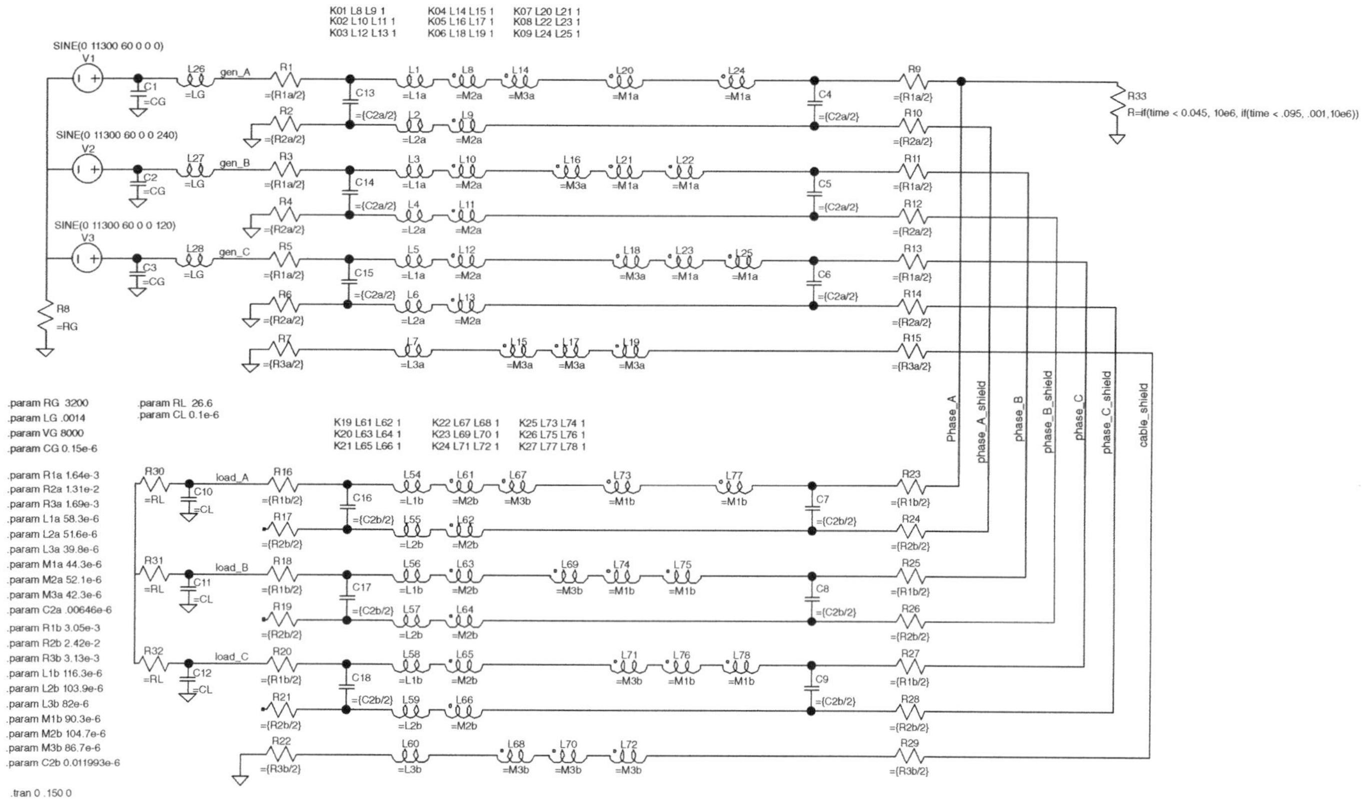

Figure 5.16 LTspice model of an AC primary distribution system with HRG system and phase *a* ground fault.

Load:

RL	26.6 Ω
CL	0.1 μF

The results of the simulation are presented in Figures 5.17 through 5.19. The voltage transients at the initiation of the ground fault and at the termination of

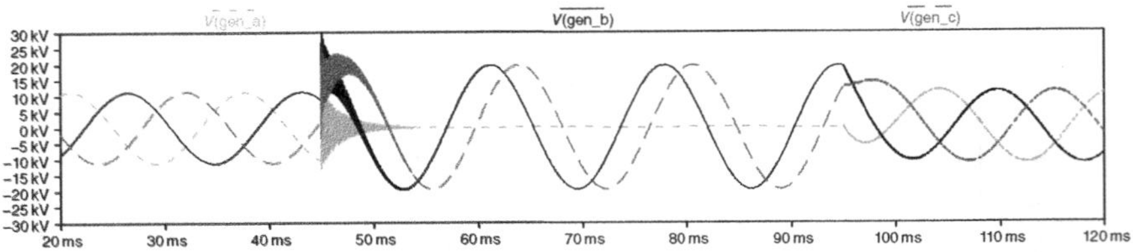

Figure 5.17 Phase voltages at the generator terminals with phase *a* ground fault.

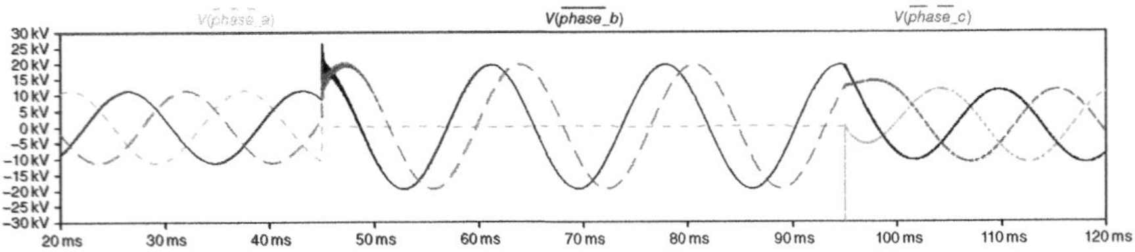

Figure 5.18 Phase voltages at the fault with phase *a* ground fault.

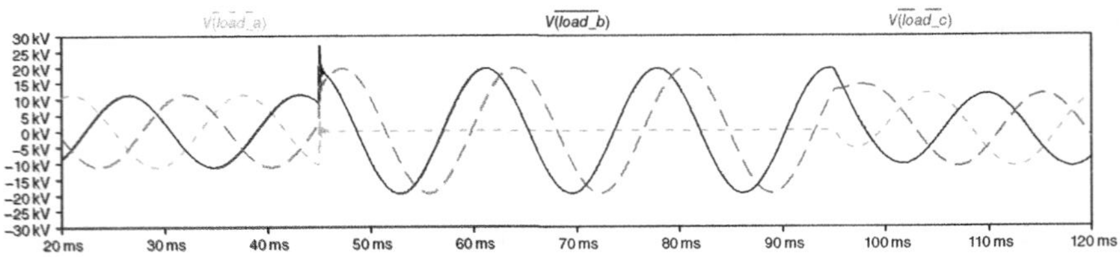

Figure 5.19 Phase voltages at the load with phase *a* ground fault.

the ground fault differ depending on the location. The transient at the initiation of the ground fault, in this case, is greatest at the generator end and the smallest at the load end. At the termination of the ground fault, the fault location experiences a spike that is not present at the other locations; this spike is due to the local responses of the cable inductances and capacitances.

5.4 Location of Grounding Circuit

As with the low-voltage distribution system AC HRG systems, consideration should be given to employing a grounding circuit at each switchboard based on the line-to-ground capacitance of the equipment and cables connected to it. This enables matching the grounding resistor value to the distributed capacitance of the distribution system connected to the switchboard. As switchboards are paralleled, the resultant grounding resistance will automatically match the impedance of the line-to-ground capacitance.

If regulation or class rules require only a single point ground, consideration should be given to implementing a grounding bus between all of the switchboards. This grounding bus will serve as the ground plane for all of the switchboards and is only connected in one location at a time to the ship's hull to implement the single point ground. The grounding resistors (or wye neutral of a wye-broken delta grounding transformer) connect to this grounding bus. Each switchboard would have a local grounding switch to connect the grounding bus to the ship's hull, but only one would normally be closed. If the grounding bus were damaged, other switchboard switches could be closed as needed. The grounding bus should be rated to continuously carry the maximum ground fault current and to carry the maximum line-to-line fault current for the maximum time needed to clear a line-to-line short circuit. National regulations and class rules may require a minimum conductor size. Each switchboard should measure the voltage or impedance between the grounding bus and the ship's hull to detect and report to the operator if a solid connection between the two is not present. The current through the grounding switch connecting the grounding bus to the ship's hull should also be monitored to detect and report to the operator the presence of a ground fault.

If the grounding circuit is connected at each switchboard, consideration should be given to installing transient surge suppression between the terminals of the generator and the grounding bus or ground. When the generator breaker is open, the generator would no longer be connected to the grounding resistor. Any line-to-ground transients caused by arcing across the circuit breaker contacts during opening or closing would be suppressed.

An alternate approach to protecting wye-connected generators would be to insert a second grounding resistor to the generator neutral conductor. This resistor would have a very high value corresponding to the impedance of the parasitic capacitance of the generator by itself. This method has the advantage of being able to measure the voltage across this grounding resistor to detect a stator line-to-ground fault when the generator is operating, but the generator breaker is open. A disadvantage is that any CM voltage produced by the generator windings will result in an increased CM current circulating between the switchboard grounding resistors and the generator grounding resistor.

5.5 Ground Fault Detection

In an AC primary distribution system with HRG, ground fault detection is usually accomplished by measuring the voltage across and the current through the grounding resistor. The voltage across the grounding resistor (other than when using a wye-broken delta grounding transformer) during a ground fault will be roughly equal to the line-to-neutral voltage of the source divided by the turns ratio of the grounding transformer. The faulted phase is typically identified by measuring the line-to-ground voltage for each phase; the faulted phase will have a very low or zero line-to-ground voltage.

Because of the lack of integration with the machinery monitoring and control system and the potential personnel hazard of the greater than 1 kV pushbutton switch, the traditional three-lamp ground fault detection circuit is generally not used in AC primary distribution system applications. If employed, a dedicated transformer for each lamp enables using commonly available low-voltage lamps. Additionally, a low-voltage circuit may be employed, with the push button activating a relay to connect the lamp transformers to ground. This relay circuit, however, requires a power source for the control power.

Insulation monitoring devices are generally incompatible with HRG systems; the insulation monitoring device would measure the resistance of the grounding resistor and not of the insulation.

5.6 Ground Fault Localization

All of the ground fault localization methods for AC low-voltage distribution systems with HRG systems described in Chapter 3 also apply to AC primary distribution systems with HRG systems.

5.7 Electrical Insulation Impact

5.7.1 Insulation Voltage Rating

Because HRG systems can operate for extended periods of time with a ground fault, the insulation between conductors and ground should be rated for the full line-to-line voltage (instead of line-to-neutral voltage). The rated voltage for the insulation should be at least $\sqrt{3}$ times the nominal system voltage to account for the steady-state line-to-ground voltage during the fault. The design of the insulation system should also account for the line-to-ground voltage spikes and transients that occur when a ground fault is initiated and cleared.

5.7.2 Partial Discharge

Cables for AC primary distribution systems are susceptible to partial discharge (PD). PD occurs when the voltage on opposite sides of a void inside the conductor insulation results in the electric field strength within the void exceeding the breakdown electric field strength of the air (or other gas within the void). The voltage difference results from the insulation being between the conductor at the line voltage and the conductor shield, which is grounded. The voids are due to defects in the insulation and can grow due to the heat and pressure from the arcing caused by the excessive electric field strength. This breakdown is called "partial" because it occurs for only part of the path between the conductor and the shield. Initially, PD occurs in a void when a voltage surge between the conductor and ground results in the electric field strength within the void exceeding the breakdown electric field strength. As a result of each voltage surge, the void becomes a little larger, and the line-to-ground voltage needed to exceed the breakdown voltage becomes smaller. Eventually, the void grows large enough that breakdown occurs under normal voltage conditions when the voltage peak of the line-to-ground voltage is greater than that needed to exceed the breakdown electric field strength. Once this happens, the void grows rapidly until the insulation fails completely, and a line-to-ground fault occurs.

PD is characterized by a series of current spikes between the conductor and ground. Each spike typically lasts on the order of nanoseconds to a microsecond. The series of spikes will occur over the time interval when the line-to-ground voltage results in the electric field strength within the void exceeding the breakdown electric field strength. No PD will occur during the normal sine wave when the line-to-ground voltage is insufficient for breakdown. For power systems with a nominal system frequency of 50–60 Hz, if the time between bursts of PD is less than about 2 ms, the cable is likely to fail quickly.

PD can be detected in a number of different ways. Commercial online partial discharge monitoring systems are now available and should be considered for primary distribution systems. See IEEE Std 400.3-2022 (2022) for a more complete discussion of the many ways to detect PD.

5.8 Limiting Ground Fault Current

The same techniques for limiting ground fault current for AC low-voltage distribution systems also apply to AC primary distribution systems. Because the line-to-ground capacitance of the cables will be higher, particular attention should be paid to limiting the line-to-ground capacitance of EMI filters.

5.9 Cable Terminations

Cable terminations for AC primary distribution systems must be made more carefully than for AC low-voltage distribution systems; improper termination can result in excessive electric field stress on the insulation at the terminations and lead to insulation failure.

The purpose of the conductor shield (along with semiconducting insulation layers) is to ensure a uniform electric field surrounds the conductor insulation. This shield, however, has a parasitic capacitance between it and the conductor; hence a current will flow through this shield from the conductor to ground. Under balanced operation, the three-conductor shield currents will cancel; no current will actually flow in the ship's hull. If a ground fault occurs, then the conductor shield currents will not cancel and a hull current will flow between the ground fault connection to the ship's hull and the conductor shield's connection to the ship's hull.

If a ground fault occurs between a conductor shield and its conductor on one cable, then the ground fault current will flow from where the grounding resistor is grounded and from where conductor shields of other cables are grounded, through the hull, to the location where the faulted cable conductor shields are grounded. To minimize the distance that the ground fault current flows through the ship's hull for this case, the grounding points for all the shields should be close together, and close to where the grounding resistor is located. For most architectures, this translates into grounding the conductor shields at the source (generator) end of the cable. For bus ties between generators, the choice of which end to connect the conductor shield to ground is arbitrary.

The purpose of the cable shield is to divert some of the CM current from the hull; hence it should be grounded at both ends. While the resistance of the cable shield is likely much greater than the ship's hull, its mutual inductance with the conductors is also much greater than between the conductors and the ship's hull. The net result is that the CM impedance of the current path through the cable shield and drain wires is likely to be about the same magnitude as the impedance of the ship's hull.

Figure 5.20 depicts a typical set of termination kits installed on the load end of a three-phase cable; the appearance of the termination kits varies among manufacturers and voltage applications. The cable shield is brought out of the cable and connected to ground. The conductor shields are not brought out. The termination kit includes layers of semiconducting materials to even out the electric field. Each termination kit comes with manufacturer's installation procedures that should be followed; deviating from the installation procedures may result in insulation failure.

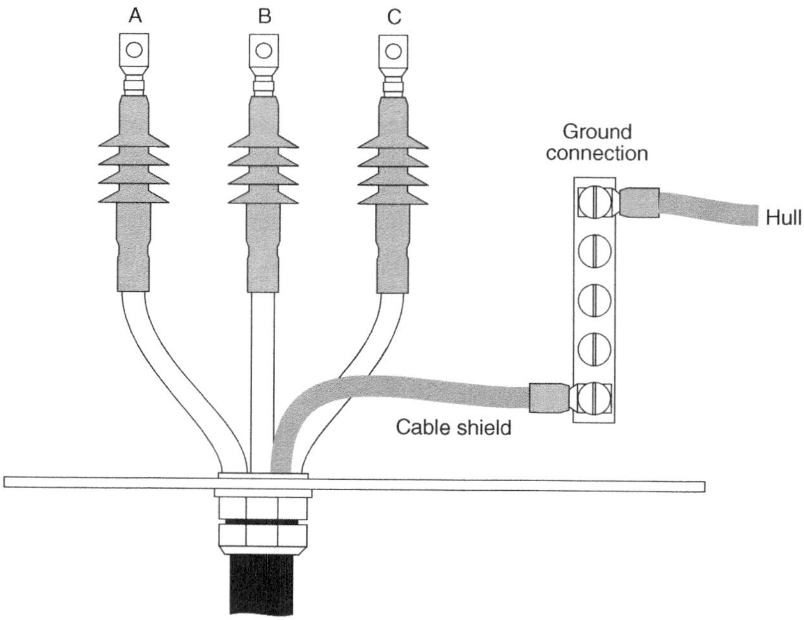

Figure 5.20 Cable termination at the load end of an AC primary distribution system cable.

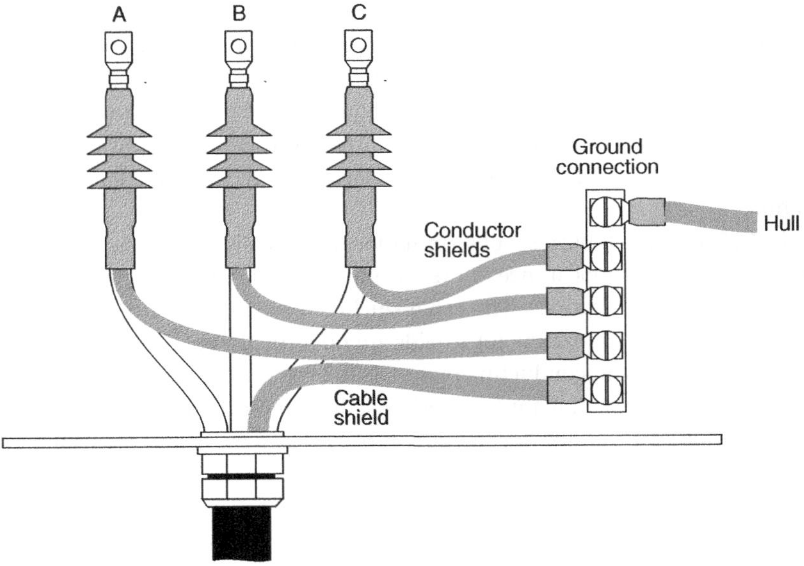

Figure 5.21 Cable terminations at the generator end of an AC primary distribution system cable.

Figure 5.21 depicts the generator end of the cable; the termination is the same as for Figure 5.20 except connections to each of the conductor shields are brought out of the cable and connected to ground.

The connection boxes or compartments should have sufficient volume for routing and terminating the main power conductors, as well as the conductor shields and cable shields. Care should be taken to ensure creepage and clearance distance requirements are met. With covers off or doors open, workers should have clear access to all screw terminals and cable terminations.

References

ASTM B8-11 (2017). *Standard Specification for Concentric-Lay-Stranded Copper Conductors, Hard, Medium-Hard, or Soft*. Philadelphia, PA: ASTM International. Available from http://webstore.ansi.org.

Barker, David S. (1979). Charging Current Data for Guesswork-Free Design of High Resistance Grounded Systems. *IEEE Transactions on Industry Applications* 1A-35 (2): 136–140. https://doi.org/10.1109/TIA.1979.4503628.

Bridger, Baldwin (1983). High-Resistance Grounding. *IEEE Transactions on Industry Applications*, 1A-19 (1): 15–21. https://doi.org/10.1109/TIA.1983.4504149.

Grover, Frederick W. (2009). *Inductance Calculations*, Mineola, NY: Dover Publications, Inc.

IEEE Std.400.3-2022 (2022). *IEEE Guide for Partial Discharge Field Diagnostic Testing of Shielded Power Cable Systems*. New York: Institute of Electrical and Electronics Engineers. Available from http://ieeexplore.ieee.org.

IEEE Std 1580-2021 (2021). *IEEE Recommended Practice for Marine Cable for Use on Shipboard and Fixed or Floating Facilities*. New York: Institute of Electrical and Electronics Engineers. Available from http://ieeexplore.ieee.org.

MIL-HDBK-299 (1991). *Cable Comparison Handbook, Data Pertaining to Electric Shipboard Cable*. Washington, DC: Department of Defense. Available from http://quicksearch.dla.mil.

Westinghouse Electric Corporation Central Station Engineers (1964). *Electrical Transmission and Distribution Reference Book*. East Pittsburgh, PA: Westinghouse Electric Corporation.

6

System Grounding: Shipboard Ungrounded DC Systems (No Greater than 1 kV)

6.1 Characteristics

As shown in Figure 6.1, DC power systems with nominal system voltages below 1 kV may be ungrounded. An ungrounded power system is not recommended for a DC primary distribution system.

Because the parasitic capacitance of a DC system has infinite impedance with DC, the resistance to ground is of importance. Typically, insulation systems are designed to have a resistance of at least 1 MΩ/1 kV of nominal system voltage (with 1 MΩ being the minimum for insulation of systems below 1 kV). Normally, the line-to-neutral voltage is half the line-to-line voltage. The line-to-ground capacitances of both conductors are charged to half the line-to-line voltage. During a ground fault, the line-to-ground capacitance of the faulted conductor discharges quickly to zero volts, while the line-to-ground capacitance of un-faulted conductor charges rapidly to the full line-to-line voltage. When the fault clears, the line-to-ground capacitance of the un-faulted conductor discharges slowly through the line-to-ground insulation resistance, while the line-to-ground capacitance of the previously faulted conductor slowly charges. If the line-to-ground capacitances are on the order of microfarads and the line-to-ground insulation resistances are on the order of tens of megaohms, then the RC time constant is on the order of seconds to tens of seconds. The cable also has self and mutual inductance, which may cause a resonance with the capacitance in the form of ringing in response to the beginning of a ground fault. Higher capacitance will generally lead to higher magnitude ringing. Figure 6.2 depicts the voltage waveforms at a load for a ground fault in its feeder cable. Figure 6.3 depicts the ground fault current.

DC power systems are generally implemented by using either a passive rectifier or an active rectifier to convert AC power to DC power. Because of maintenance concerns and voltage limitations (limited to below 1 kV), commutated DC

Design of Shipboard Power System Grounding/Earthing, First Edition. Norbert Doerry, Mohammed M. Islam, and John Prousalidis.
© 2025 The Institute of Electrical and Electronics Engineers, Inc.
Published 2025 by John Wiley & Sons, Inc.

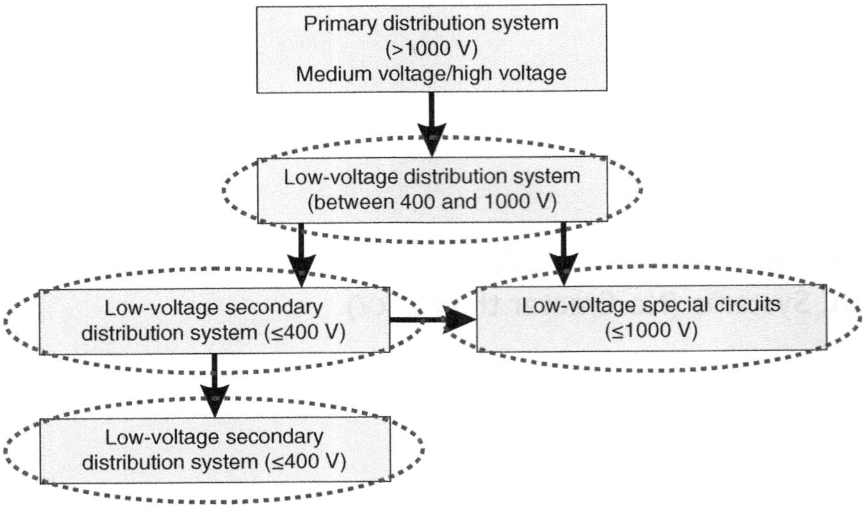

Figure 6.1 Applicability of ungrounded DC power systems.

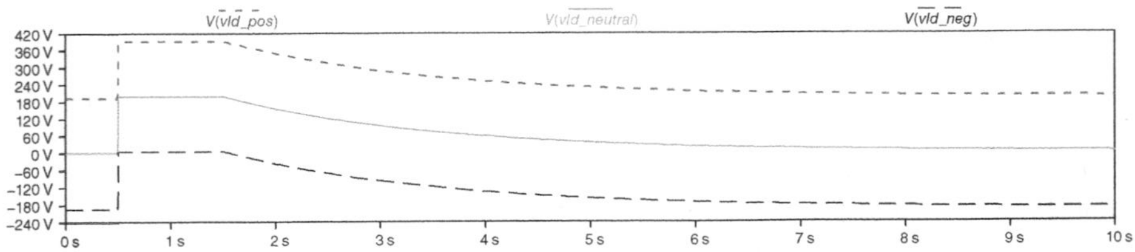

Figure 6.2 Line-to-ground voltages and neutral-to-ground voltages of a ground fault from 0.5 to 1.5 seconds for ungrounded DC power system.

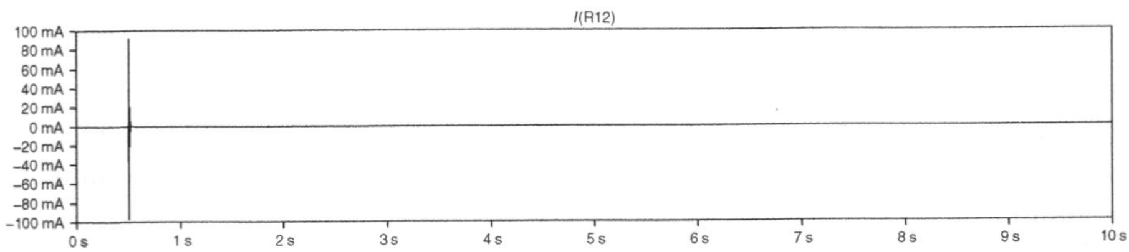

Figure 6.3 Ground fault current with ground fault from 0.5 to 1.0 seconds for ungrounded DC power system.

generators are seldom used in modern systems. Controlled rectifiers that employ silicon-controlled rectifiers (SCRs) are becoming less common due to the high level of voltage ripple and current. Some systems are still battery-based, where power converters are used to supply current to keep the state of charge on the batteries relatively constant.

Most rectifier and DC-to-DC converter outputs have voltage ripple (line-to-line). The voltage ripple also contributes to ground fault current. The frequency components of the ripple can range from the tens of Hz to MHz. Typically, the magnitude is much smaller than the DC line-to-line voltage. If one of the significant components of the voltage ripple happens to be at the same frequency as the resonant frequency of the system, a relatively high fault current at that frequency is theoretically possible.

Passive rectifiers use diodes for rectification. The DC voltage is proportional to the AC voltage. To reduce voltage ripple, the three-phase AC power is often converted to six-phase power by employing phase-shifting transformers. The transformers also decouple the neutral-to-ground voltage on the AC side from the neutral-to-ground voltage on the DC side. Another benefit is that the total waveform distortion for the input AC current is much reduced compared to rectifying three-phase AC power directly. If the rectifier receives power directly from a generator, a six-phase high-frequency generator may be employed. The higher AC frequency reduces the size of passive filters needed to reduce the voltage ripple.

Figures 6.4 and 6.5 depict the line-to-ground voltages at the load and the ground fault current of a two-conductor cable with a DC source having a relatively small 4000 Hz ripple and a resistive load. As shown in Figure 6.6, both the source and the load have a 0.1 μF line-to-ground capacitance. The ground fault occurs 35% of the length of the cable between the source and load. On the inception of the ground fault, there are noticeable transient responses of both the line-to-ground voltages and the fault current. After about 30 ms, the transients die out, and the steady-state fault conditions remain. After the transients have decayed, the fault current is very

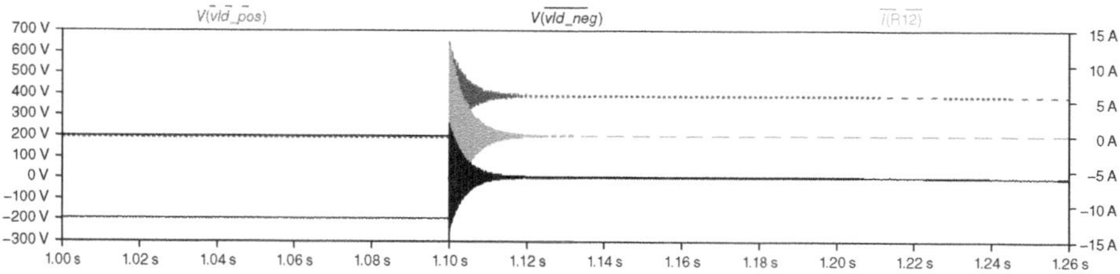

Figure 6.4 Two-conductor ungrounded cable with ground fault from 1.1 to 1.2 seconds in ungrounded DC power system.

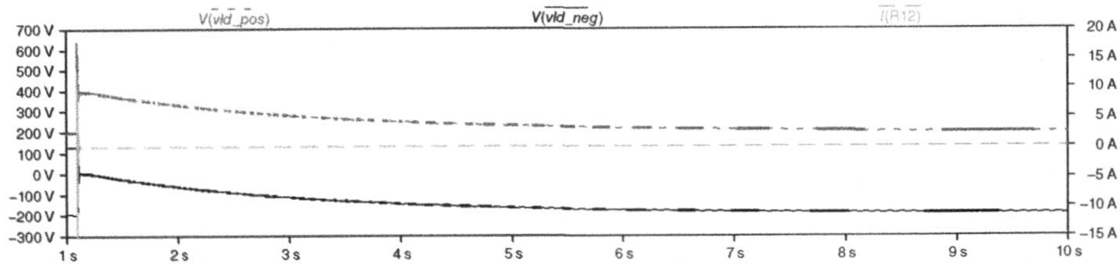

Figure 6.5 Two-conductor ungrounded cable recovery from ground fault from 1.1 to 1.2 seconds in ungrounded DC power system.

small. Unlike AC systems, when the fault clears, there is essentially no transient in the DC system. The line-to-ground voltages decay to their steady-state un-faulted condition over a time duration of seconds. For this reason, it is unlikely that the transients from a ground fault will cause an intermittent ground fault and a progressively growing ringing voltage magnitude.

6.2 Modeling

6.2.1 Two-Conductor Unshielded Cable

Two-conductor unshielded cable is appropriate for DC systems with voltages no greater than 1 kV and for applications requiring less than about 100 A. Cables for higher current levels should employ four-conductor to limit magnetic fields, and higher voltage cables should use shielded cables to reduce the electric field stress on the insulation.

Figure 6.7 depicts the cross section of a notional two-conductor cable. Figure 6.8 depicts a model of a segment of two-conductor cable of length l suitable for transient analysis.

As shown in Chapter 2, the self-inductance (μH) of a round wire of radius ρ m and length l m is given by Eq. (2.3).

$$L_S = 0.2l\left[\ln(2) + \ln(l) - \ln(\rho) - \frac{3}{4}\right]$$

The mutual inductance (μH) with respect to a conductor at a distance d m away is approximated by Eq. (2.4).

$$M \approx 0.2l\left[\ln\left(\sqrt{2}\right) + \ln(l) - \ln(d) - 1\right]$$

With an estimate for the length of the cable (l), L_S and M (μH) can be estimated using Eqs. (2.3) and (2.4).

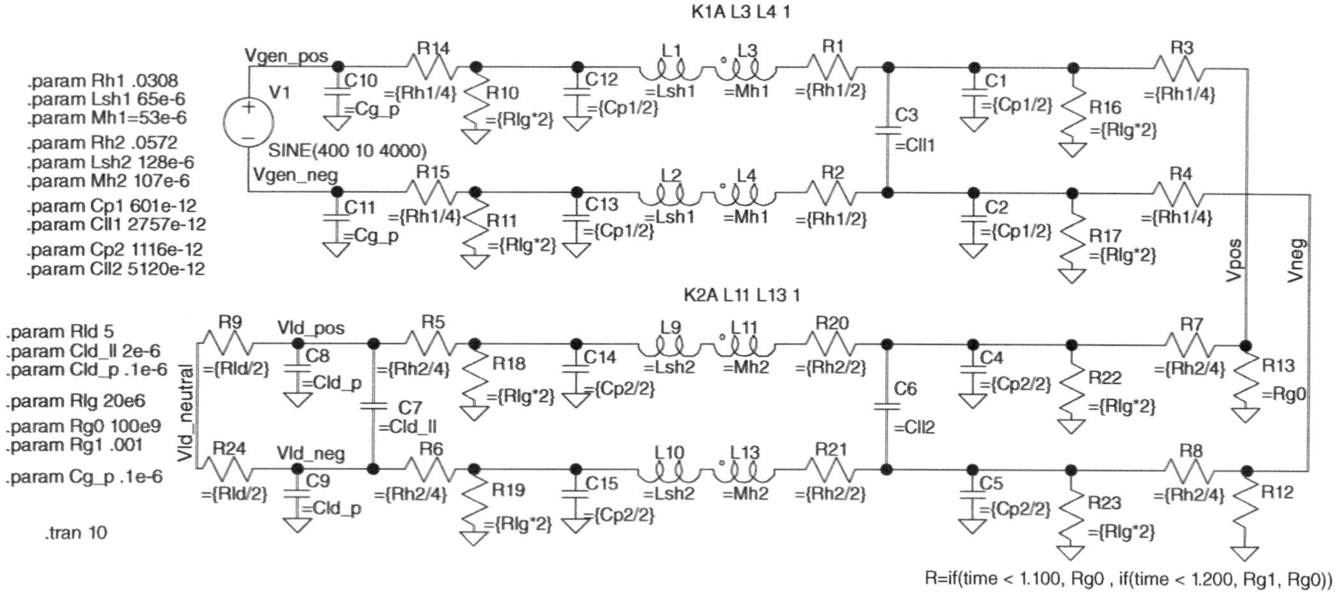

Figure 6.6 LTspice model of two-conductor ungrounded cable with ground fault from 1.1 to 1.2 seconds.

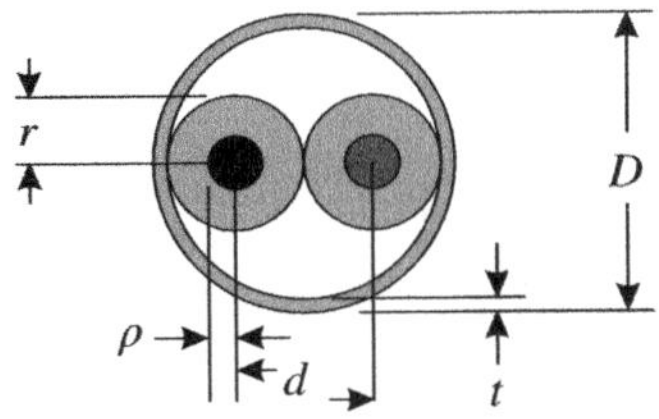

Figure 6.7 Two-conductor unshielded DC cable.

The resistance R is usually provided in the cable data sheets. IEEE Std 1580-2021 (2021), MIL-HDBK-299 (1991), and ASTM B8-11 (2017) may also be consulted for values.

The parasitic capacitance C_p depends on the geometries of the cable conductors and the grounded conductors; therefore, calculating the line-to-ground capacitance is very complex. The ground conductors are typically a combination of the ship structure and the elements of the cable trays or cable hangers. Ideally, a finite element analysis program would be used to develop estimates of the line-to-ground capacitances. Of course, early in the design process, the details necessary to produce a finite element model likely do not exist.

Using Eq. (2.9), one can approximate the line-to-ground capacitance of an unshielded cable by treating the capacitance as two capacitances in series: the conductor to the cable outer sheath capacitance (conductor-to-sheath capacitance), and the capacitance from the outer sheath to the structure (sheath-to-ground capacitance).

$$\frac{l}{C_p} = \frac{1}{C_1} + \frac{1}{C_2}$$

where

C_1, conductor-to-sheath capacitance (F)
C_2, sheath-to-ground capacitance (F).
C_1 and C_2 may be estimated using Eqs. (2.12) and (2.13):

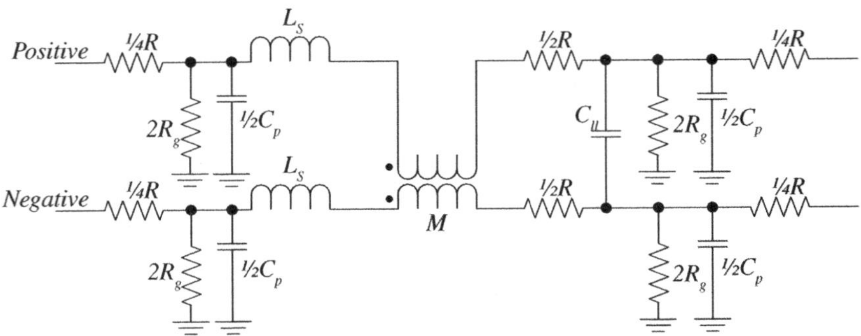

Figure 6.8 Unshielded DC cable transient model.

$$C_1 \approx l \frac{2\pi\varepsilon_0\kappa}{\ln\left(\frac{\sqrt{\frac{rD}{2}}}{\rho}\right)}$$

$$C_2 \approx l \frac{2\pi\varepsilon_0}{\operatorname{acosh}\left(\frac{2h}{D}\right)}$$

where

h, representative distance (m) of the center of the cable to the grounded conducting plane.

The value of h should be based on the type of cable hangers employed. If details on the cable hangers are not known, a value of 0.20 m may be used until better information is obtained. One may vary this value to determine the sensitivity of the transient response to this value.

The value of C_p obtained by using Eqs. (2.9) through (2.13) should be considered very crude. The actual value can easily vary from 80% to 150% of the calculated value when the length of the cable is known with certainty.

One can approximate the capacitance between the conductors using Eq. (2.14):

$$C_{ll} \approx l \frac{\pi\varepsilon_0\kappa}{\operatorname{acosh}\left(\frac{d}{2\rho}\right)}$$

The resistance R_g can be estimated as 1 MΩ/1 kV of nominal system voltage plus 1 MΩ.

6.2.2 Four-Conductor Unshielded Cable

To reduce the magnetic field from DC cables, four-conductor unshielded cables, as depicted in Figure 6.9, may be employed. Conductors on opposite sides of the cable are connected in parallel and are assumed to carry equal currents. With these conditions, the model depicted in Figure 6.8 may also be employed for four-conductor cable. The values for this model can be calculated based on determining their values for one-conductor, then adjusting it for being in parallel with another conductor (resistances and inductances are halved, capacitances are doubled).

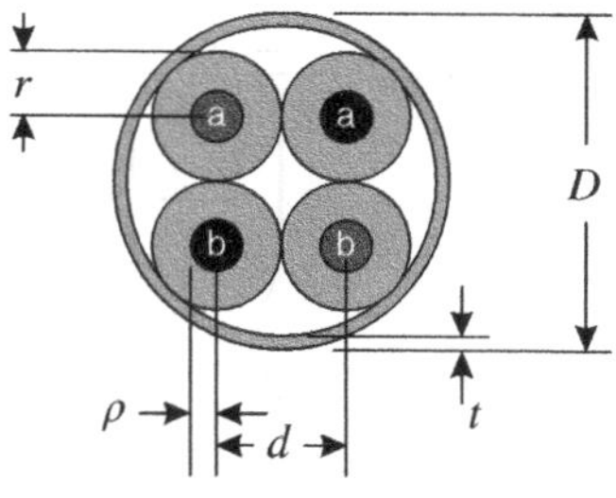

Figure 6.9 Four-conductor unshielded DC cable.

As with two-conductor cables, the resistance of a single conductor can be determined from data sheets or other references listed. The value R for the two conductors in parallel is half that of the single conductor.

The self-inductance L_{Sa} (μH/m) for a single conductor should also include the impact of the mutual inductance with the other conductor of the same polarity. The distance between the two conductors is $\sqrt{2}d$ and $\ln(\sqrt{2}d) = \ln(\sqrt{2}) + \ln(d)$.

$$L_{Sa} = 0.2l\left[\ln(2) + \ln(l) - \ln(\rho) - \frac{3}{4}\right] + 0.2l[\ln(l) - \ln(d) - 1]$$

The self-inductance L_S (μH/m) of the two conductors in parallel is half that for a single conductor

$$L_S = 0.2l\left[\frac{1}{2}\ln(2) + \frac{1}{2}\ln(l) - \frac{1}{2}\ln(\rho) - \frac{3}{8}\right] + 0.2l\left[\frac{1}{2}\ln(l) - \frac{1}{2}\ln(d) - \frac{1}{2}\right]$$

$$L_S = 0.2l\left[\ln\left(\sqrt{2}\right) + \ln(l) - \frac{1}{2}\ln(\rho) - \frac{1}{2}\ln(d) - \frac{7}{8}\right] \tag{6.1}$$

The mutual inductance M_{ab} of a conductor of one polarity with a conductor of the opposite polarity is

$$M_{ab} \approx 0.2l\left[\ln\left(\sqrt{2}\right) + \ln(l) - \ln(d) - 1\right]$$

Since there are two conductors of the opposite polarity, the effective mutual inductance of a conductor with the conductors of the opposite polarity is twice M_{ab}. Since there are two conductors in parallel, the inductance of both conductors of one polarity with both conductors of the opposite polarity is thus M_{ab}.

$$M = M_{ab} \approx 0.2l\left[\ln\left(\sqrt{2}\right) + \ln(l) - \ln(d) - 1\right] \tag{6.2}$$

The parasitic capacitance may again be calculated using Eq. (2.9). However, the values for C_1 and C_2 are doubled because the cable has two conductors of the same polarity.

$$C_1 \approx 2l\frac{2\pi\varepsilon_0\kappa}{\ln\left(\frac{\sqrt{\frac{rD}{2}}}{\rho}\right)} \tag{6.3}$$

$$C_2 \approx 2l\frac{2\pi\varepsilon_0}{\operatorname{acosh}\left(\frac{2h}{D}\right)} \tag{6.4}$$

The line-to-line capacitance is four times that calculated for two conductors because each of the two paralleled conductors has a capacitance with respect to each of the opposite polarity conductors.

$$C_{ll} \approx 4l \frac{\pi \varepsilon_0 \kappa}{\operatorname{acosh}\left(\frac{d}{2\rho}\right)} \tag{6.5}$$

6.2.3 Passive Rectifiers

Figure 6.10 depicts a simple system consisting of a six-phase generator, a 12-pulse rectifier, two segments of cable, and a resistive load. A ground fault is inserted at the junction of the two segments of cable. The system without a fault exhibits common mode (CM) currents and voltages; the passive rectifier is a CM voltage source (see Chapter 13). If one ignores the high-frequency component of the voltages with respect to ground, the response of the passive rectifier is very similar to Figure 6.2. The top waveform in Figure 6.11 is the line-to-line voltage at the load; it is not significantly impacted by the ground fault.

The ground fault current (Figure 6.12), however, is different because the high-frequency components of the CM voltage interact with the parasitic capacitances to produce a sustained fault current. The CM current at the output of the rectifier (Figure 6.13) is impacted by the ground fault but is dominated by the high-frequency components. The line-to-ground voltages have significant ripple, which is also a property of the CM voltage with respect to ground.

For an ungrounded system, many of the circuit parameters that govern the CM and ground fault voltages and currents are parasitic and can change significantly under different environmental conditions. Insulation resistance, for example, is strongly impacted by temperature and humidity. This means that measurements taken of the same system may differ significantly due to local differences in temperature and humidity.

6.2.4 Active Rectifiers

There are many topologies of active rectifiers. Each topology will have its own CM voltage characteristics and ground fault response. In many cases, the waveforms depicted for the passive rectifier will be similar to the waveforms for active rectifiers. See Chapter 13 for more details on CM modeling.

6.3 Ground Fault Detection

DC systems with a nominal system voltage no greater than 1 kV can employ ground lights, as depicted in Figure 6.14. As with the AC counterpart, the lights are normally both dim with no ground fault. If a ground fault exists, the light connected to the faulted conductor will extinguish and the other light will be bright. If the test switch is depressed, opening the circuit, both lights should be dim. If the lights go out, then one or both of them are burned out. The circuit in Figure 6.14 is

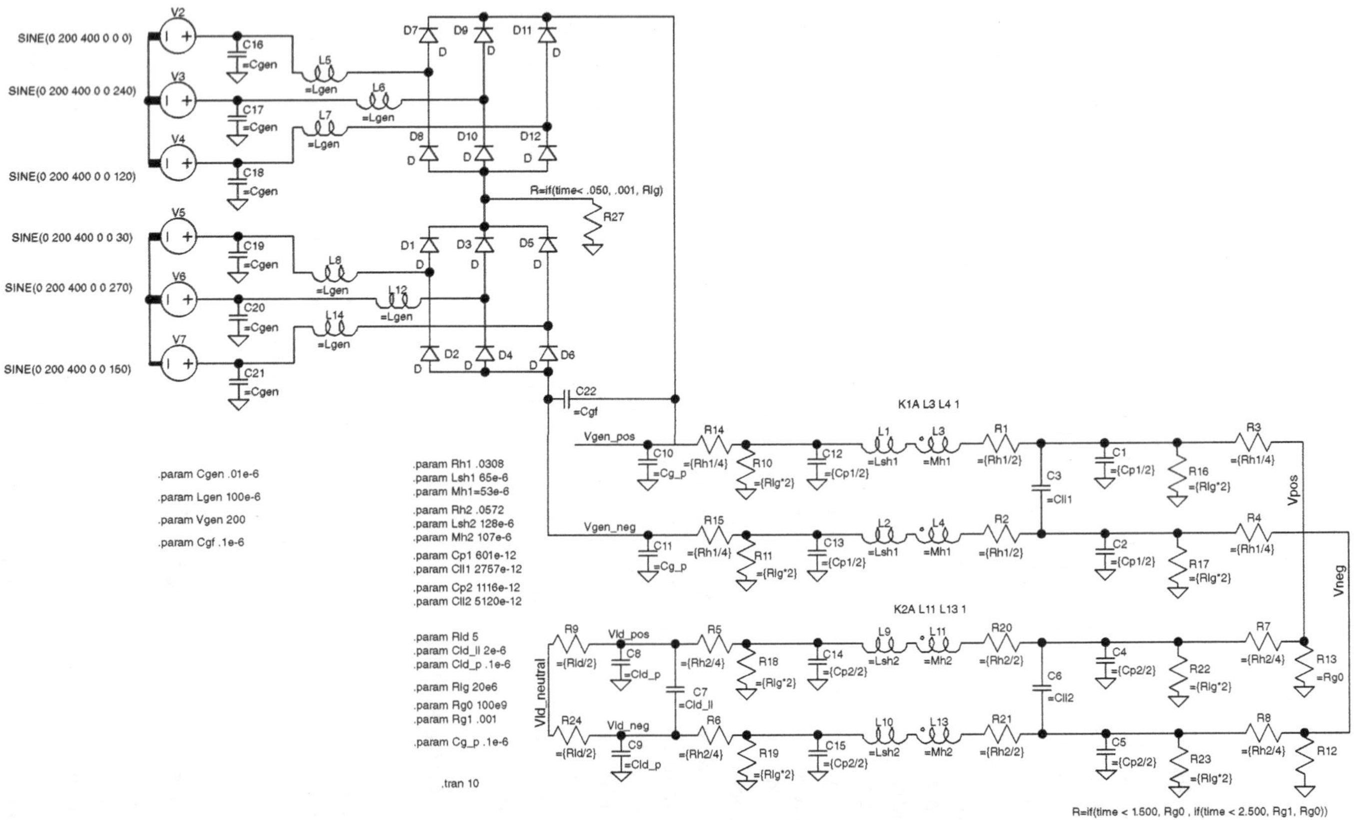

Figure 6.10 LTspice model of an ungrounded DC system powered from passive rectifier and six-phase generator.

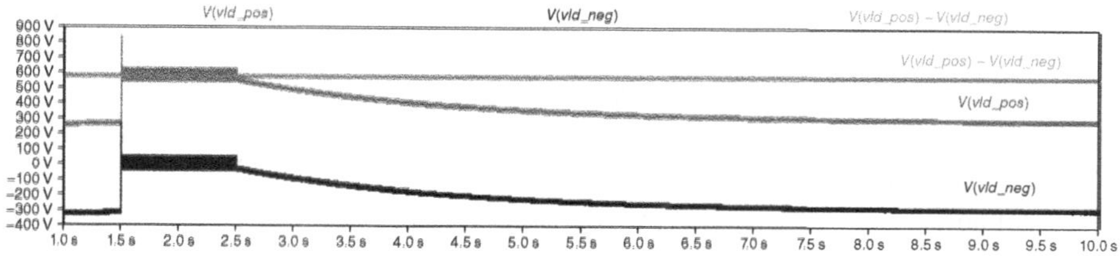

Figure 6.11 Ungrounded DC system powered from passive rectifier – voltages at the load.

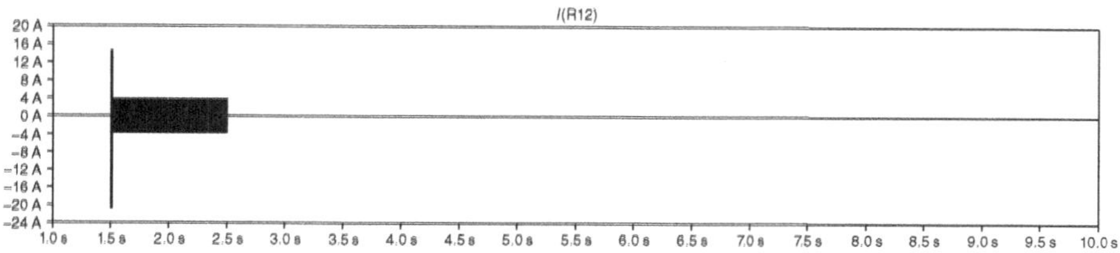

Figure 6.12 Ungrounded DC system powered from passive rectifier – current through the ground fault.

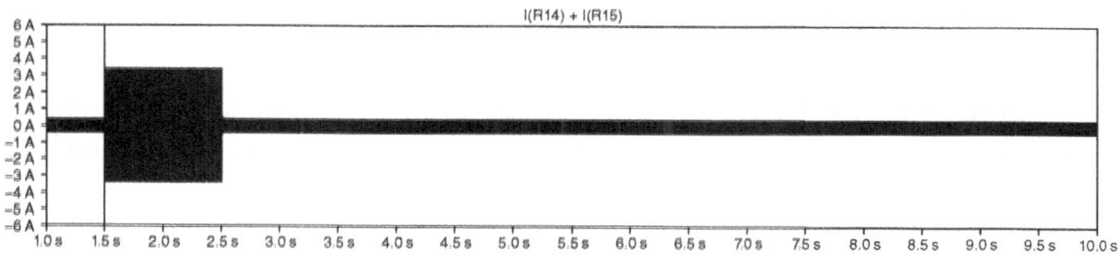

Figure 6.13 Ungrounded DC system powered from passive rectifier – CM current at output of rectifier.

Figure 6.14 DC ground lights in an ungrounded DC power system.

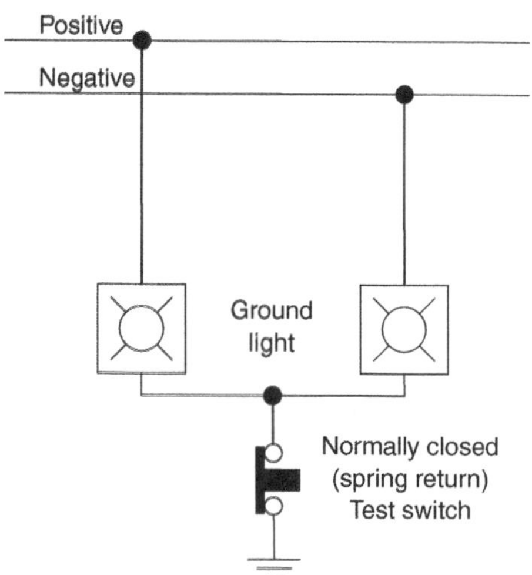

generally not practical for systems with a nominal system voltage greater than about 280 V; lamps at these higher voltages are not generally available. Fixed ratio DC-to-DC converters are available that could enable higher voltage operation (analogous to using a transformer for AC lamps), however, this solution is likely not cost-effective.

Voltage sensors from each of the conductors to ground can also be used to detect a ground fault.

Insulation monitoring devices can also be used when the system voltage is no greater than 1 kV. These devices inject either a voltage or current signal between the power system conductors and ground and evaluate the conductor insulation resistance by measuring the current or voltage impact of the injected signal. If the insulation resistance falls beneath a preset value, usually on the order of 50 kΩ, the operator is notified of the low insulation resistance.

For many insulation monitoring devices, only one should be installed per system to avoid interference among multiple units. This includes ground detection lamps since their resistance to ground may be interpreted as low insulation resistance.

6.4 Ground Fault Localization

Figure 6.15 depicts the schematic of a passive rectifier-based DC system with three feeder cables and loads. Figures 6.16 through 6.21 depict waveforms for a fault that is applied to one feeder (at the junction of Cable 1 and Cable 2) at time 0.5 second and cleared at time 1 second. As shown in Figure 6.16, one cannot use the magnitude of the CM current at the DC source bus alone to determine which feeder is faulted; the rectifier, acting as a CM voltage source, causes a CM current to flow under normal operation, which can mask the ground fault induced CM current. However, if one compares the CM current of a feeder cable at the source bus to the CM current entering the load, the faulted bus will usually show a significant difference, as depicted in Figure 6.17. For un-faulted cables, the CM current entering the feeder cable at the source bus will not differ significantly from the CM current entering the load, as depicted in Figure 6.18.

The CM current in a feeder cable is often dominated by the line-to-ground capacitance of the load. If the load capacitance of the faulted feeder is greatly reduced, the difference in CM current of the faulted feeder cable between its connections at the source bus and at the load will still be significant, as shown in Figure 6.19. Figure 6.19 shows that the CM current drops significantly in the feeder past the location of the ground fault. These features indicate that a clamp-on ammeter may be employed to locate the ground fault. For each feeder, the CM current at

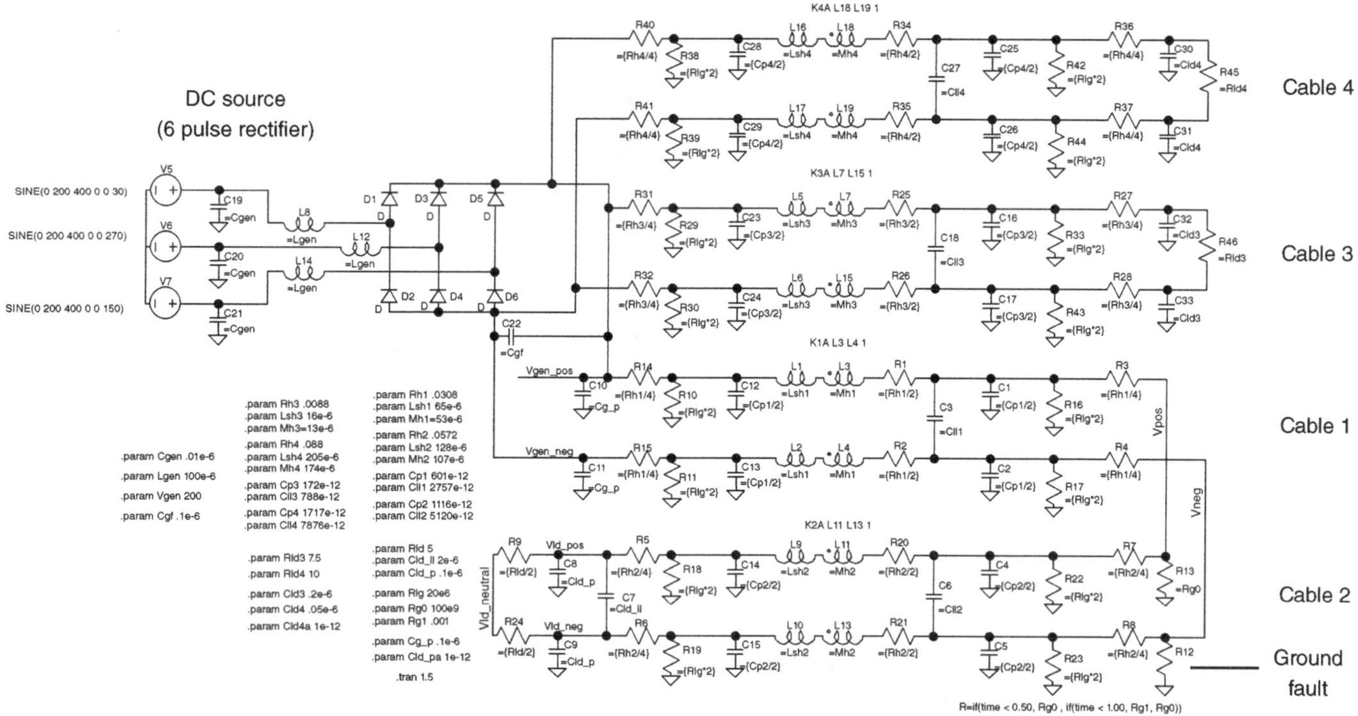

Figure 6.15 LTspice model of a passive rectifier-based ungrounded DC system with three feeders.

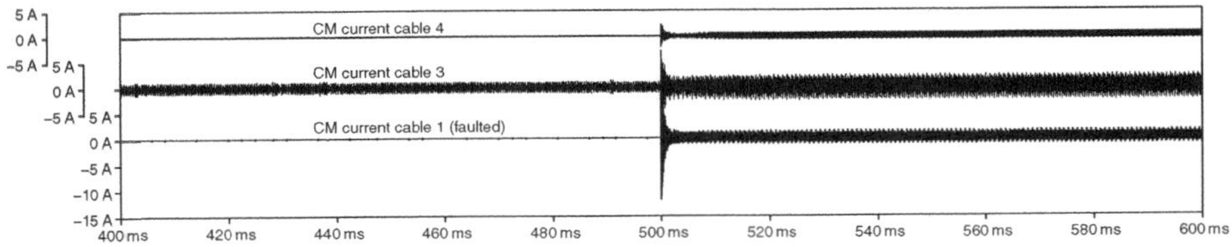

Figure 6.16 Passive rectifier-based DC system CM current of three feeders (Cable 1, Cable 3, and Cable 4).

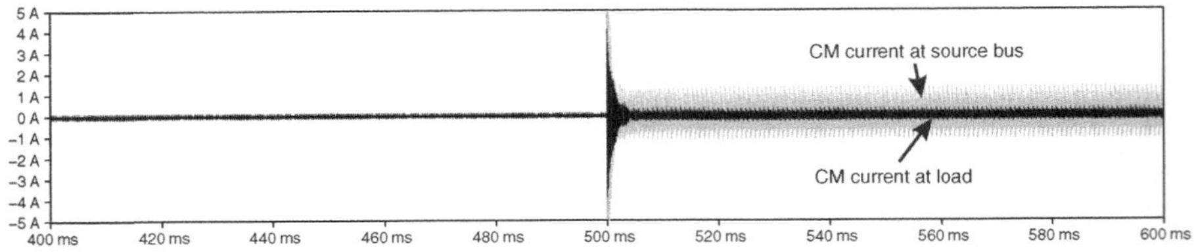

Figure 6.17 Passive rectifier-based DC system CM current on faulted feeder (Cables 1 and 2) at the source bus and at the load.

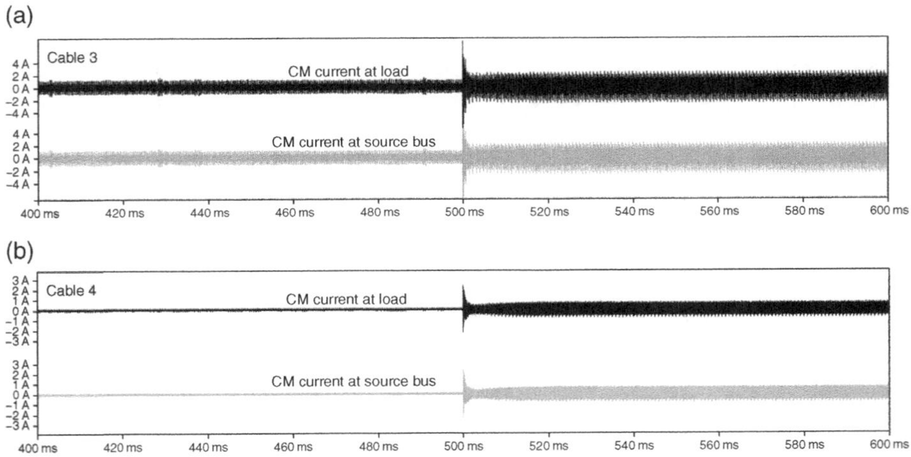

Figure 6.18 Passive rectifier-based DC system CM current on un-faulted feeder at source end and load end. (a) Cable 3; (b) Cable 4.

the source end and at the load end is compared. The magnitude of the CM current will drop significantly along the faulted feeder at the location of the ground fault.

Figure 6.20 demonstrates that the line-to-line voltage of the DC bus is not significantly impacted by the ground fault. The CM voltage of the DC bus is impacted; its average value rises to half the average line-to-line voltage.

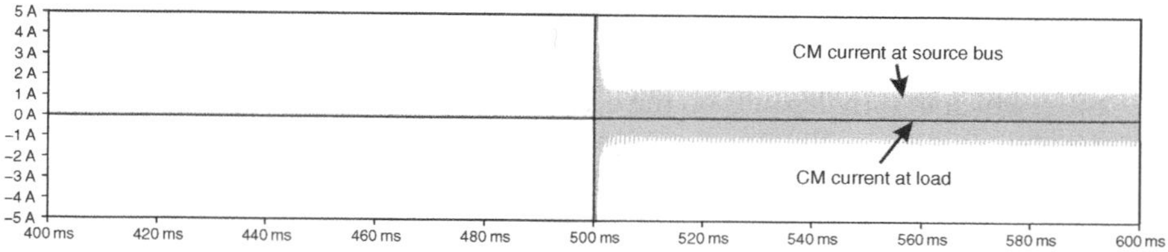

Figure 6.19 Passive rectifier-based DC system CM current on faulted feeder before and after fault with low load line-to-ground capacitance.

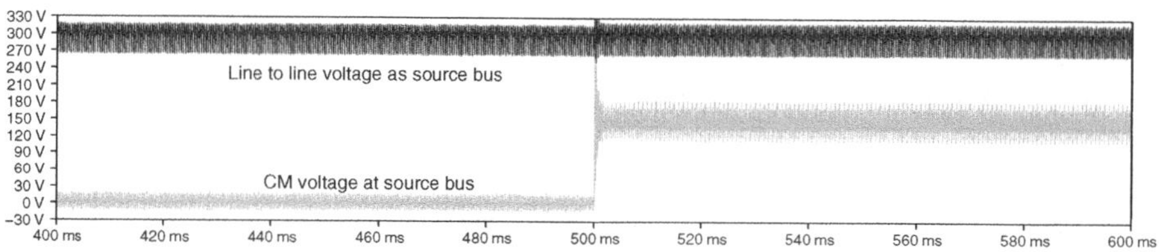

Figure 6.20 Passive rectifier-based DC system CM voltage and line-to-line voltage on bus.

Figure 6.21 shows that the CM voltage difference between the input AC generator connection and the output DC source bus is not significantly impacted by the ground fault. The ground fault does determine the relative magnitudes of the AC side CM voltage to ground and the DC side CM voltage to ground; the ground fault impacts the overall CM circuit such that the CM voltages to ground are different from their un-faulted values.

If an active filter is employed, or if the system is battery-based, then the CM voltages on the input and output of the rectifier will depend on the particular implementation and will likely vary from those depicted in Figures 6.16 through 6.21.

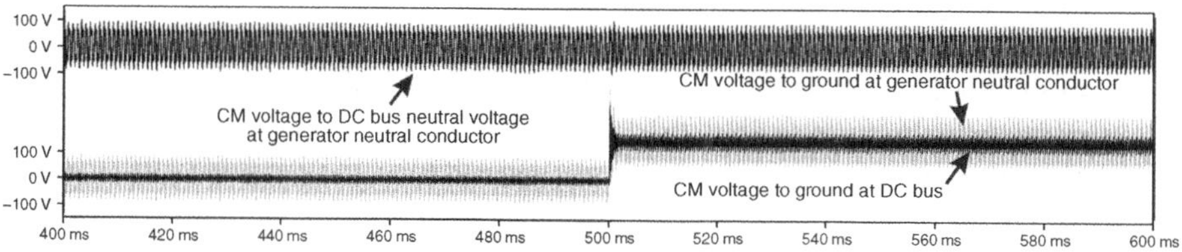

Figure 6.21 Passive rectifier-based DC system CM voltages at bus and at wye point of generator compared to CM voltage difference between input and output of rectifier.

The waveforms will depend on the parasitic capacitances and insulation resistances, in addition to the CM voltage difference between the input and output of the rectifier, which in turn depends on the design of the rectifier. See Chapter 10 for more details.

Commercial ground fault localization methods also exist. Some insert a square waveform current into the faulted conductor with a magnitude between about 5 and 30 mA with a period of about 10 Hz. This signal can be detected with clamp-on sensors to determine which feeder is faulted, and where on the feeder the fault is located.

6.5 Auctioneering Diodes

One of the advantages of DC distribution systems is the ability to use auctioneering diodes to seamlessly transfer power from one DC source to another. Auctioneering diodes are typically implemented using an asymmetrical configuration, as depicted in Figure 6.22. The voltage source (V_1 or V_2) with the higher voltage provides power to all of the loads. It is also possible for V_1 and V_2 to employ a droop characteristic to share power. Should either V_1 or V_2 stop functioning, then the other will immediately provide power to all the loads; both voltage sources must be rated to be able to supply all of the loads.

As described by Doerry and Ashton (2019), one disadvantage of the asymmetrical configuration is that there are multiple return current paths, as depicted in Figure 6.22. This results in CM currents in all of the feeder cables.

To avoid the CM currents, one could use the symmetrical configuration depicted in Figure 6.23. The additional diodes only allow a return current on the same circuit as the supply current. Unfortunately, the symmetrical configuration is susceptible to a fault condition where a particular double ground fault, as depicted in Figure 6.24, can result in the two voltage sources being in series; the loads are subjected to a line-to-line voltage nearly double the nominal system voltage. For this reason, the symmetrical configuration is not often used.

Doerry (2019) documents a modified symmetrical configuration proposed by Ashton and depicted in Figure 6.25. This configuration prevents multiple return current paths because all but the supply return path must pass through three diodes, while the supply return path only passes through one diode. When paralleled, the impedance of the three series diodes is orders of magnitude greater than the impedance of one diode. This configuration is not susceptible to the double ground fault depicted in Figure 6.24; the double ground fault would result in one of the sources being short-circuited and tripping on overcurrent.

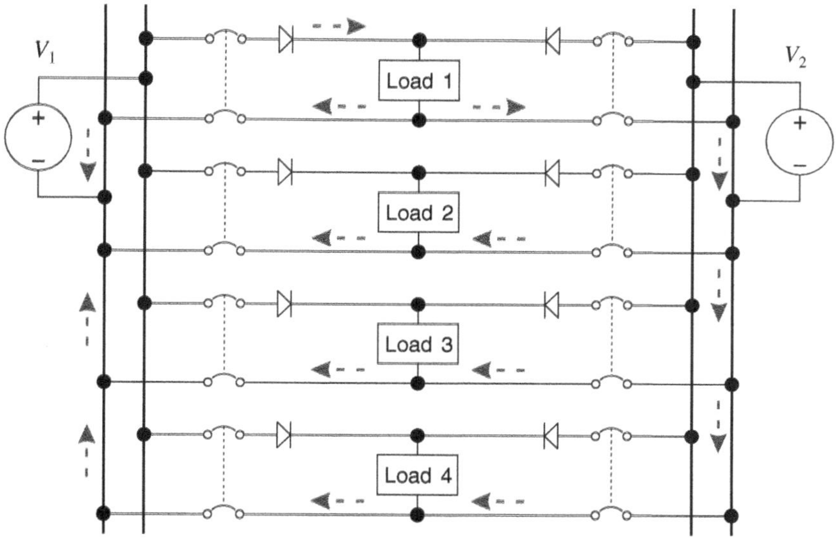

Figure 6.22 Asymmetric auctioneering diode configuration with multiple return current paths.

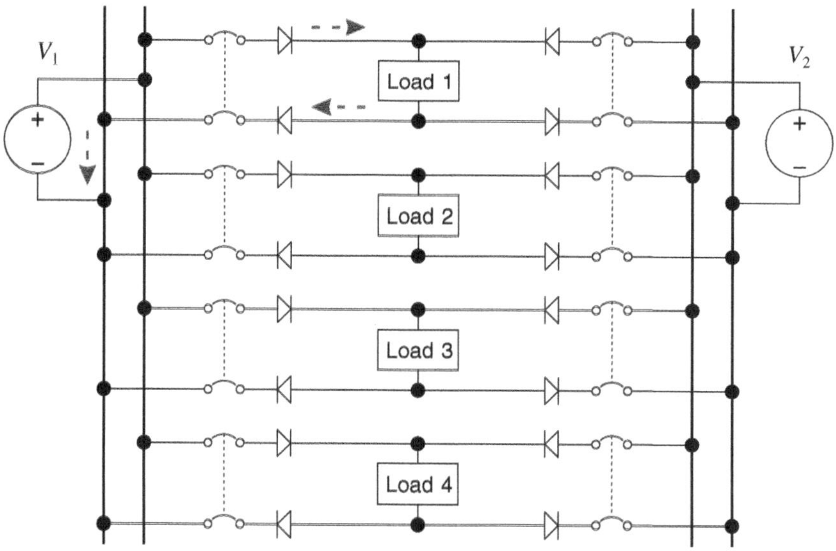

Figure 6.23 Symmetric auctioneering diode configuration with single return current path.

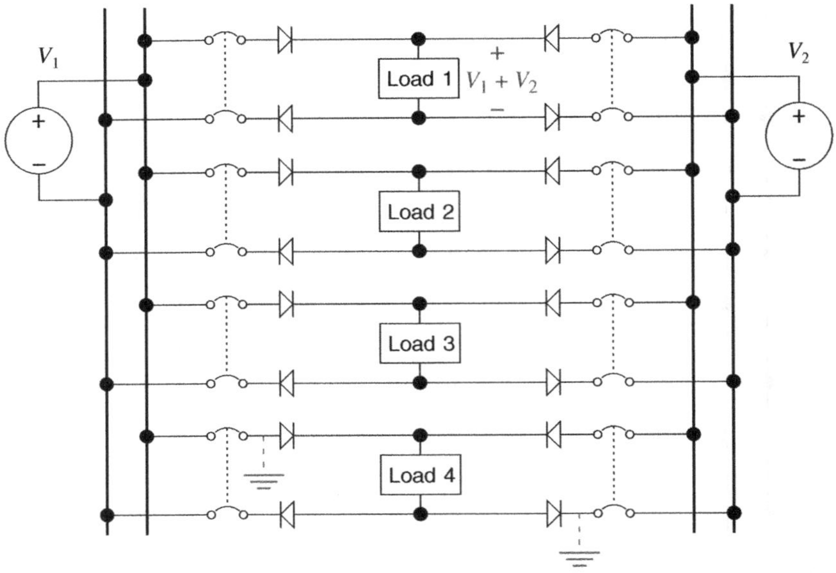

Figure 6.24 Symmetric auctioneering diodes configuration with double ground fault resulting in overvoltage.

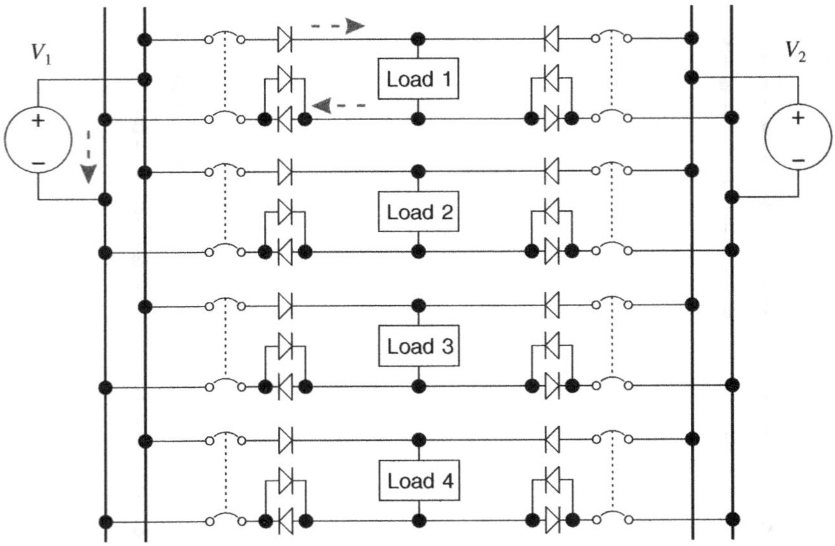

Figure 6.25 Modified symmetric auctioneering diodes configuration.

6.6 Electrical Insulation Impacts

The insulation for each conductor should be rated to take the nominal system voltage (line-to-line) as well as the superimposed CM voltage. This CM voltage will depend on the details of the rectifier design as well as the equipment and system feeding the rectifier. A rule of thumb is to multiply the nominal system voltage by 1.2, then use the next higher standard DC insulation rating. Cables designed for AC use generally have a DC line-to-line voltage rating of approximately 150% of the AC rating for line-to-neutral voltage.

6.7 Cable Insulation Colors

DC cable insulation for ungrounded systems will typically use the following insulation color convention:

- Positive conductor: red or brown.
- Negative conductor: black, gray, blue, or light blue.
- Protective ground: green, green with yellow stripe, or bare.

The insulation color should never be considered absolute; the voltage of a conductor should be verified through measurement.

References

ASTM B8-11 (2017). *Standard Specification for Concentric-Lay-Stranded Copper Conductors, Hard, Medium-Hard, or Soft*. Philadelphia, PA: ASTM International. Available from http://webstore.ansi.org.

Doerry, Norbert and Ashton, Robert (2019). Auctioneering Diodes: Pros and Cons. *Proceedings of the 2019 IEEE Electric Ship Technologies Symposium (ESTS 2019)*, Arlington, VA, (14–16 August 2019).

Doerry, Norbert (2019). Modified Symmetric Auctioneering Diode Configuration, Burke, VA. 24 September 2019. Available from http://doerry.org/norbert/papers/papers.htm.

IEEE Std 1580-2021 (2021). *IEEE Recommended Practice for Marine Cable for Use on Shipboard and Fixed or Floating Facilities*. New York: Institute of Electrical and Electronics Engineers. Available from http://ieeexplore.ieee.org.

MIL-HDBK-299 (1991). *Cable Comparison Handbook, Data Pertaining to Electric Shipboard Cable*. Washington, DC: Department of Defense. Available from http://quicksearch.dla.mil.

7

System Grounding: Shipboard HRG DC Systems

7.1 Characteristics

As shown in Figure 7.1, high-resistance grounded (HRG) DC power systems are applicable to primary distribution systems, low-voltage distribution systems, and some low-voltage special circuits (depending on the special needs of the low-voltage special circuit).

The primary purpose of an HRG system for DC systems is to provide a means for equalizing the voltage across the parasitic capacitances between ground and the conductors under normal operating conditions (without a ground fault). Equalizing the voltages is particularly important following the removal of a ground fault to reduce the voltage stress on insulation systems. Operating the power system with equal magnitude voltages between ground and each of the conductors also helps minimize common mode (CM) currents, as will be described in Chapter 13.

The normal resistance of cable and equipment insulation, along with line-to-ground capacitor leakage current, will equalize the voltage to some extent. Insulation is generally recommended to have a resistance of at least 1 MΩ/1 kV of nominal system voltage with a minimum of 1 MΩ. The insulation resistance is a function of both temperature and humidity. A rule of thumb is that for every 10 °C increase in temperature, the insulation resistance is halved. Water absorption in the insulation from either immersion or condensation can also reduce the resistance significantly. Because these factors can impact the insulation of the positive and negative conductors differently, the line-to-ground resistance of the insulation of the two conductors may not be equal, resulting in the neutral point voltage (voltage midway between the positive and negative conductors) not being equal to the ground potential and drifting over time.

Figure 7.2 depicts an HRG DC system. This model represents the voltage ripple in the system as a sine wave with frequency of 1 kHz. Figure 7.3 depicts the positive

Design of Shipboard Power System Grounding/Earthing, First Edition. Norbert Doerry, Mohammed M. Islam, and John Prousalidis.
© 2025 The Institute of Electrical and Electronics Engineers, Inc.
Published 2025 by John Wiley & Sons, Inc.

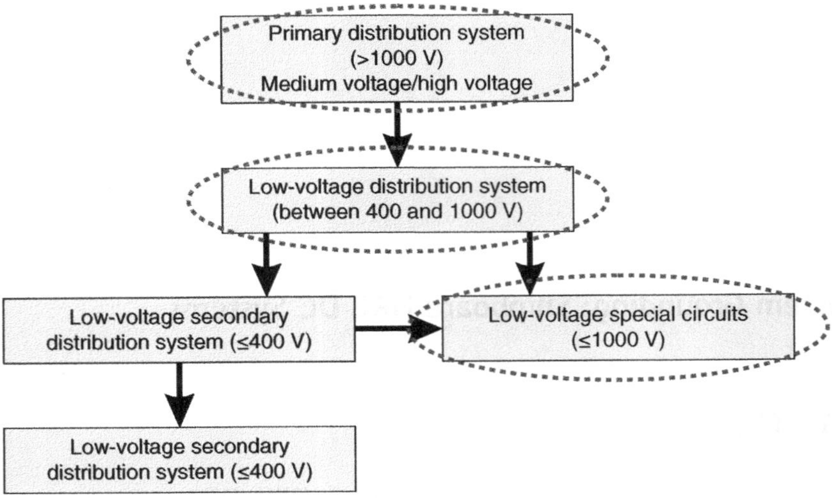

Figure 7.1 Applicability of HRG DC power systems.

and negative conductor voltages with respect to ground, as well as the neutral voltage. A ground fault is applied at 0.5 second and remove after 1 second. Figure 7.4 depicts the transient fault current. The fault current initially rings with a frequency in the kHz range but decays rapidly with a time constant on the order of milliseconds. Figure 7.5 depicts the current through the grounding resistor; unlike AC systems with currents in the amps, the current through the grounding resistor for this DC system is only on the order of 10 mA.

During the steady-state ground fault, the current and voltage waveforms are determined by CM characteristics. A transient is not observed when the ground fault clears. Instead, the neutral voltage decays back to zero in about one second.

Compared to an ungrounded AC system, the threat of an intermittent ground fault is much reduced in a DC system. This is primarily due of the lack of a transient when the ground fault clears.

7.2 Grounding Methods

7.2.1 Line-to-Ground Resistors

Figure 7.6 depicts the grounding of a DC system using line-to-ground resistors. These line-to-ground resistors reduce the neutral-to-ground voltage dependency on temperature and humidity. To be effective, the line-to-ground resistance R_{LG} should be an order of magnitude less than the line-to-ground resistance of the

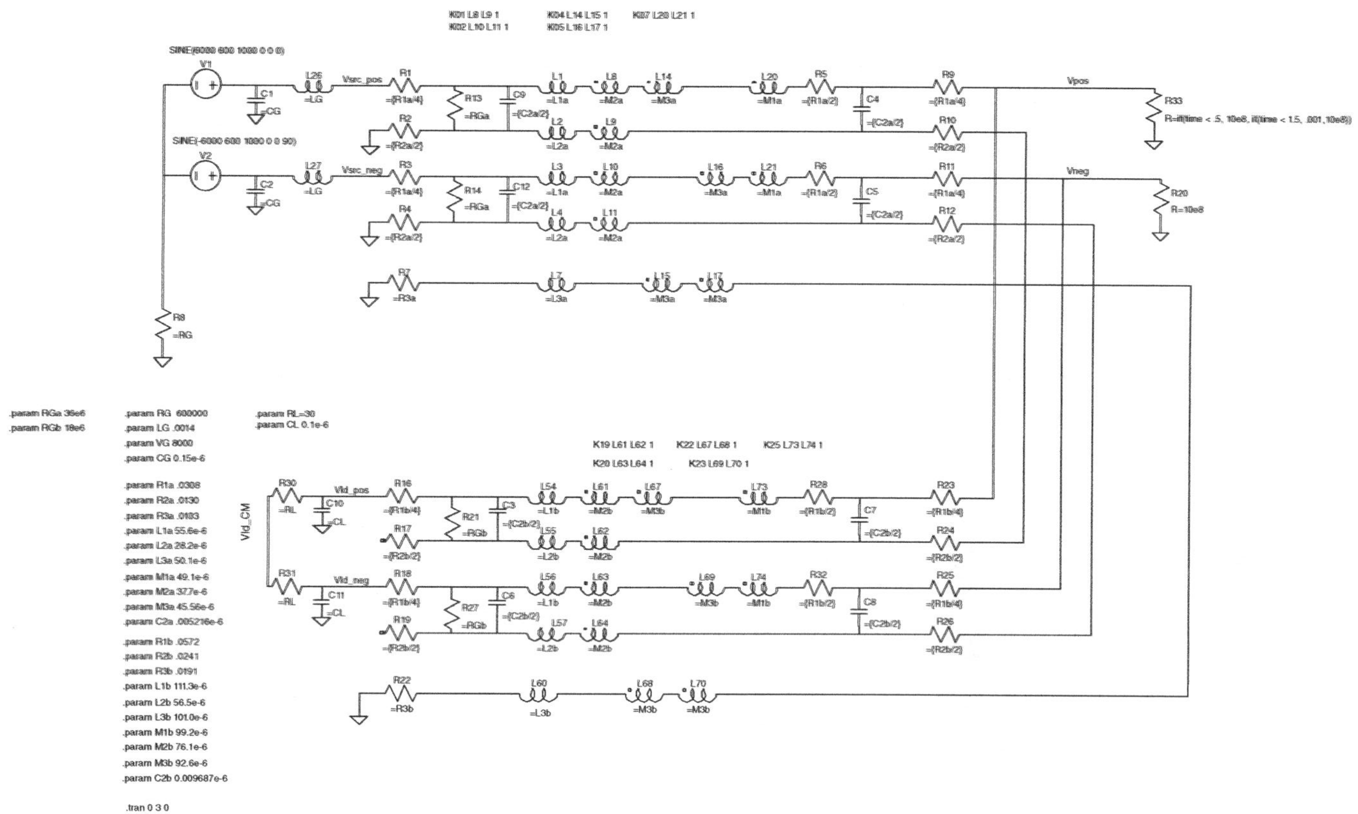

Figure 7.2 Schematic of HRG system for DC.

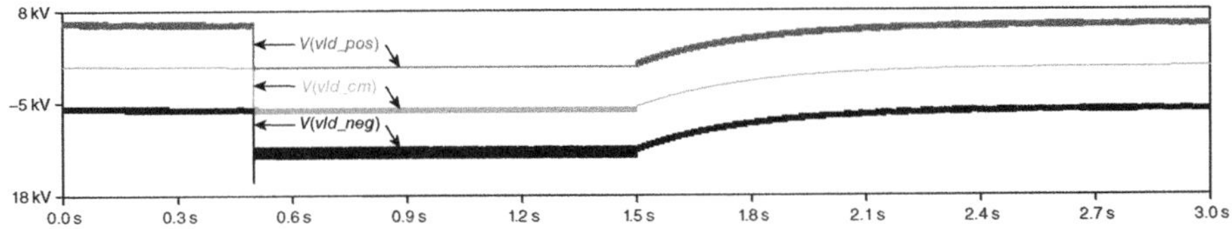

Figure 7.3 Line-to-ground and neutral-to-ground voltages at the load.

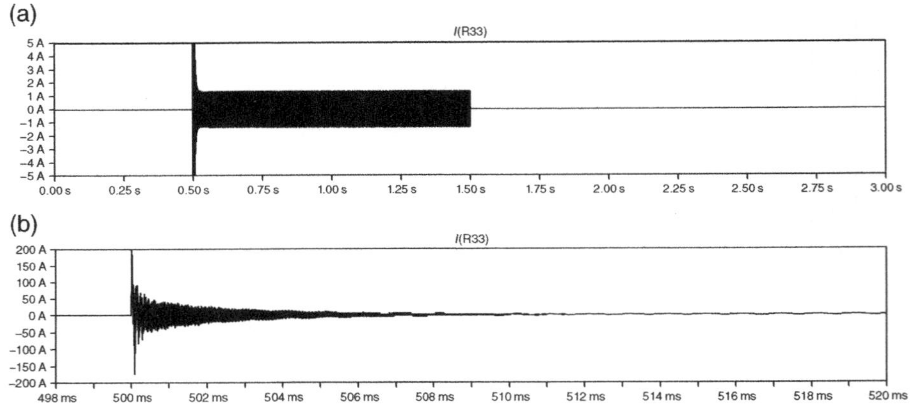

Figure 7.4 Current through the ground fault. (a) Complete waveform; (b) initial transient.

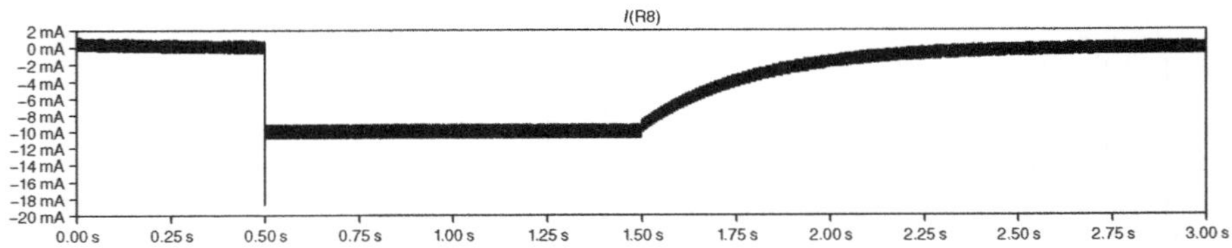

Figure 7.5 Current through the grounding resistor.

insulation. Thus, the maximum value of R_{LG} is recommended to be 100 kΩ/kV of the nominal system voltage with a 100 kΩ minimum. At this level, the ground fault current is roughly 10 mA for nominal system voltages above 1 kV, which reduces the risk of injury to personnel and damage to equipment. Without the grounding resistor, the ground fault current will flow through the insulation where it is the weakest. The current flow can further heat the insulation resulting in further degradation and eventually insulation failure.

The steady-state current through the resistors is about 5 mA for nominal system voltages above 1 kV. A current of 5 mA at 12 kV ($R_{LG} = 1.2$ MΩ) results in 60 W for

Figure 7.6 DC line-to-ground resistors.

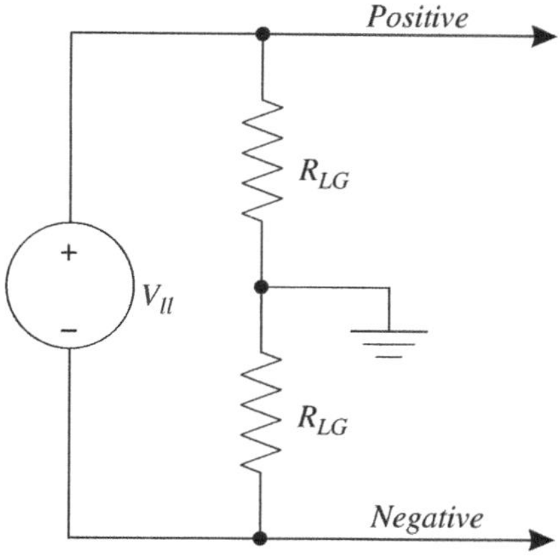

the combined dissipation of the two resistors (30 W each). Under ground fault conditions, the 1.2 MΩ line-to-ground resistor that is not shorted will carry 10 mA and dissipate 120 W, while the shorted 1.2 MΩ line-to-ground resistor will not dissipate any power. At these low power levels, the grounding resistors can be mounted directly within switchboards.

Each generator set rectifier switchboard is recommended to have a set of line-to-ground resistors. These switchboards may serve one or multiple generator sets.

One of the concerns with using line-to-ground resistors is the possibility of ground currents among multiple sets of line-to-ground resistors. This can happen if the grounding resistors do not have equal resistances. If one of the resistances increases by 10%, then the current through the hull (or grounding bus if one is employed) will be roughly 0.5 mA for nominal system voltages greater than 1 kV. For this reason, the design of the grounding resistors should be such that their resistances do not vary from each other significantly. In particular, efforts should be made to ensure both grounding resistors are at the same temperature and protected from becoming wet.

7.2.2 Split Power Supply

As depicted in Figure 7.7, a split power supply consists of effectively two power supplies connected in series, with their common connection connected to ground via the grounding resistor. This resistor has value equal to R_G and should have a value equal to 50 kΩ/kV of the nominal system voltage, with a minimum of 50 kΩ. Under normal balanced operation, where the system neutral is at ground potential, no current will flow through the grounding resistor. Under a ground fault, the grounding resistor will carry about 10 mA for nominal system voltages greater

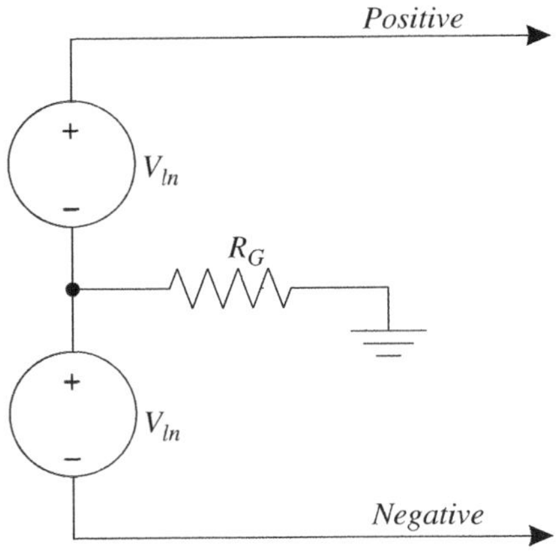

Figure 7.7 Split power supply.

than 1 kV. For a 12 kV system, the grounding resistor will dissipate about 60 W during a ground fault.

7.2.3 Grounding Bus

A grounding bus connecting all the DC switchboards is recommended to provide a ground reference at each switchboard and a path for ground currents. Each switchboard should have a grounding switch and a current sensor to measure the current through the grounding switch and a voltage sensor across the grounding switch. To comply with regulations and class rules, under normal operation, only one grounding switch would be closed to provide a single-point ground. In addition to the grounding resistors, all EMI and CM filters that require a ground connection would connect to this grounding bus.

Ideally, DC ground currents caused by tolerances in grounding resistors and voltage regulation would flow primarily through the grounding bus instead of the ship's hull. As discussed in Chapter 15, DC hull currents may result in unacceptable levels of corrosion.

7.2.4 Current Sensors

Unlike AC systems, DC systems cannot use current transformers for measuring current. Instead, most current sensors are either of the DC shunt resistor variety or of the Hall effect variety.

A DC shunt resistor usually has a calibrated resistance between 0.0001 and 0.01 Ω. The resistance is chosen so that the rated current results in a voltage across the shunt resistor to be between 50 and 100 mV. The current through the shunt resistor is calculated by measuring the voltage across the shunt and then dividing by the resistance. Since the resistance of materials changes with temperature, the current measurement accuracy will also depend on the temperature.

Hall effect sensors produce a voltage that is proportional to the strength of a magnetic field. In a Hall effect current sensor, a magnetic core surrounds the conductor being measured; the Hall effect sensor is located in a gap in the magnetic core; the strength of the measured magnetic field is used to determine the measured current. A signal conditioning circuit is used to provide the desired form of the measured current.

7.3 Modeling

7.3.1 Four-Conductor Shielded Cable

DC power systems with nominal system voltages above 1 kV can be expected to have both conductor shields and cable shields, as depicted in Figure 7.8. The conductor shields are typically grounded on one end to provide a constant electric field gradient through the conductor insulation. The cable shield encircles all of the conductors and is normally grounded on both ends. This shield provides a path for CM currents such that hull currents are minimized. At rated current levels above 100 A (and perhaps at lower current levels as well), four-conductor cable is recommended to prevent high DC fields and to prevent permanently magnetizing portions of the hull structure. In the four-conductor cables, two sets of conductors are connected in parallel, as depicted in Figure 7.8. A transient model of this cable is depicted in Figure 7.9.

The self and mutual inductances (μH) with respect to the conductors are given by Eqs. (6.1) and (6.2).

Figure 7.8 Four-conductor shielded DC cable.

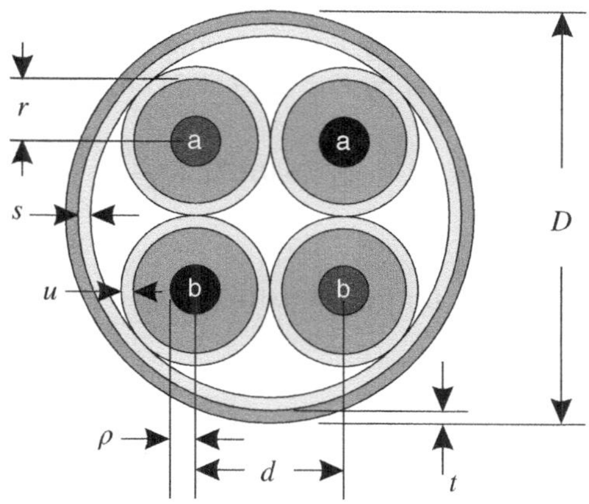

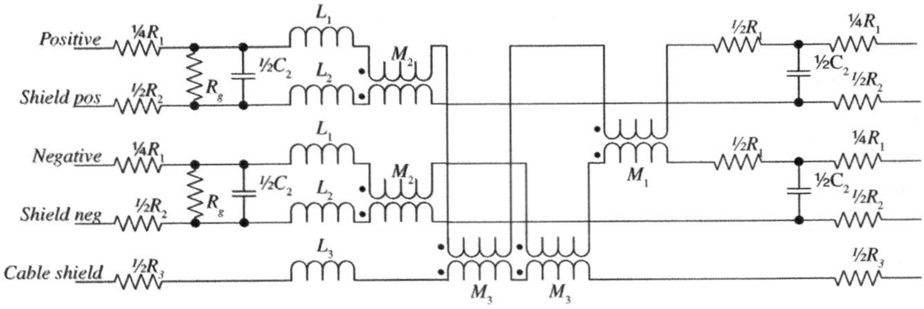

Figure 7.9 Shielded DC cable transient model.

$$L_1 = 0.2l\left[\ln\left(\sqrt{2}\right) + \ln(l) - \frac{1}{2}\ln(\rho) - \frac{1}{2}\ln(d) - \frac{7}{8}\right]$$

$$M_1 = M_{ab} \approx 0.2l\left[\ln\left(\sqrt{2}\right) + \ln(l) - \ln(d) - 1\right]$$

The self and mutual inductances with respect to the shields are those derived in Chapter 5.

The self-inductances (μH) of the shields are given by

$$L_2 = 0.1l\left[\ln\left(\frac{2l}{r+u}\right) + \ln(\xi) - 1\right] \quad (7.1)$$

$$L_3 = 0.2l\left[\ln\left(\frac{2l}{\frac{D}{2} - t}\right) + \ln(\xi) - 1\right] \quad (7.2)$$

L_2 is halved because two shields are in parallel.

If the thickness of the shield is much smaller than its radius, then $\ln(\xi)$ can be approximated by Eqs. (7.3) and (7.4). The mutual inductances M_2 and M_3 are approximated by Eqs. (7.5) and (7.6).

$$\text{For } L_2\text{: } \ln(\xi) \approx 0.0167\left(20\frac{u}{r+u}\right) \quad (7.3)$$

$$\text{For } L_3\text{: } \ln(\xi) \approx 0.0167\left(20\frac{s}{\frac{D}{2} - t}\right) \quad (7.4)$$

$$M_2 \approx \frac{1}{2}\left(L_1 + L_2 - 0.2l\left[\ln\left(\frac{r+u}{\rho}\right) + 0.25 + \ln(\xi)\right]\right) \quad (7.5)$$

$$M_3 \approx \frac{1}{2}\left(L_1 + L_3 - 0.2l\left[\ln\left(\frac{\frac{D}{2}-t}{\rho}\right) + 0.25 + \ln(\xi)\right]\right) \tag{7.6}$$

The resistance R_1 is usually provided in cable data sheets. IEEE Std 1580-2021 (2021), MIL-HDBK-299 (1991), and ASTM B8-11 (2017) may also be consulted for values.

The shield resistances, R_2 and R_3, can be approximated by the resistivity of the shield material. If we assume the shields are cylinders, then the resistances of the shields are given by

$$R_2 = R_{vol}\frac{l}{2\pi(2r + u)u} \tag{7.7}$$

$$R_3 = R_{vol}\frac{l}{\pi(D - 2t - s)s} \tag{7.8}$$

where R_{vol} is the resistivity of the shield material in ohm meters (Ωm). Note that for R_2, the equation accounts for the conductor shields of both conductors.

The impact of temperature on resistivity is given by

$$R_{vol}(T) = R_{vol}(T_0) \times (1 + \alpha_0(T - T_0)) \tag{7.9}$$

where

$R_{vol}(T)$, resistivity (Ωm) at a given temperature T

$R_{vol}(T_0)$, resistivity (Ωm) at a reference temperature T_0

α_0, temperature coefficient (1/°C).

R_{vol} for solid annealed copper is about 1.72×10^{-8} Ωm at 20 °C. The temperature coefficient (α_0) of annealed copper is 0.00393/°C. Thus, at 40 °C, R_{vol} for annealed copper is about 1.86×10^{-8} Ωm. The anticipated operating temperature of the cable should be used.

Because a braid is not a solid material, the resistance of braided shields will be greater; lacking additional information, doubling, or even tripling the calculated resistance value for a braided shield is not unreasonable.

The conductor-to-conductor shield capacitance (pF) is given by

$$C_2 = l\frac{2\pi\varepsilon_0\kappa}{\ln\left(\frac{r}{\rho}\right)} \tag{7.10}$$

where

ε_0, electric permittivity of free space

$$\varepsilon_0 = 8.85\ \text{pF/m}$$

κ, dielectric constant.

Chapter 2 provides nominal dielectric constants for some insulators.

7.4 Ground Fault Detection

Ground faults are easily detected by measuring the line-to-ground voltages or the neutral-to-ground voltage. The current through the grounding resistors can also be used to detect ground faults. Grounding lights may be used at voltages less than 280 V but are typically not used for higher voltages. Insulation monitoring devices are generally not compatible with resistance grounding systems.

7.5 Ground Fault Localization

Ground faults may be localized by comparing the CM current of the different feeders. Figures 7.10 through 7.13 depict waveforms for a DC split power supply HRG system. Figure 7.14 depicts the schematic used to generate the waveforms; it consists of two balanced sources supplying three-feeder cables: Cable 1, Cable 3, and Cable 4. Cables 3 and 4 are directly connected to loads, while Cable 1 connects to Cable 2, which is connected to a load. The ground fault is placed on the positive conductor at the junction of Cables 1 and 2. The sources have a small 6 V, 1 kHz AC waveform superimposed on each conductor voltage to represent a small voltage ripple. The voltage-to-ground waveforms (Figure 7.10) are dominated by the DC shift during the ground fault and the exponential decay back to pre-fault values after the fault clears. As expected, the current through the grounding resistor (Figure 7.11) follows the same shape as the neutral voltage and, during the ground

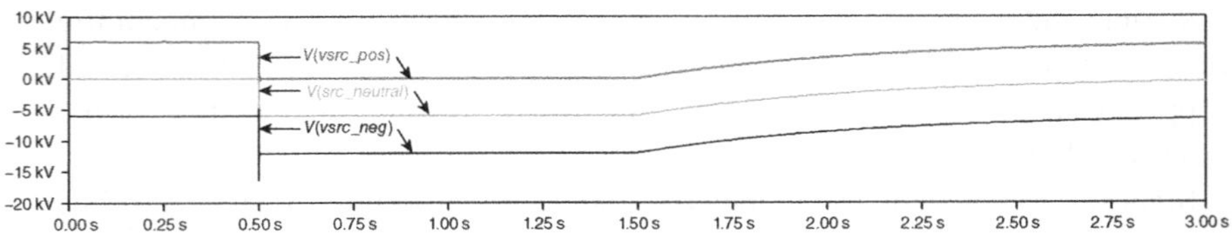

Figure 7.10 Voltages of positive and negative conductors, and neutral voltage of three-feeder system with ground fault between 0.5 and 1.5 seconds.

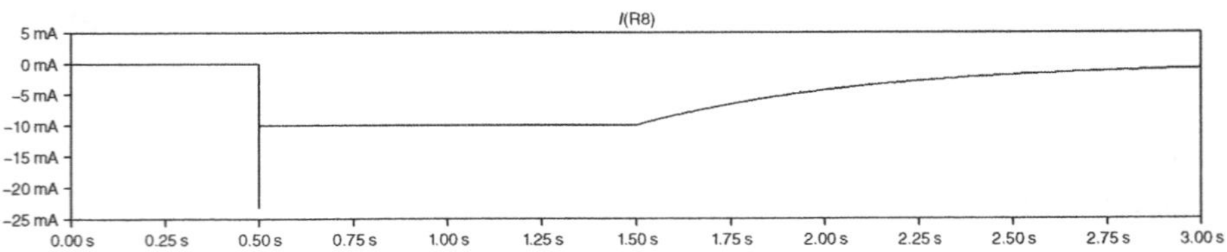

Figure 7.11 Grounding resistor current of three-feeder system with ground fault between 0.5 and 1.5 seconds.

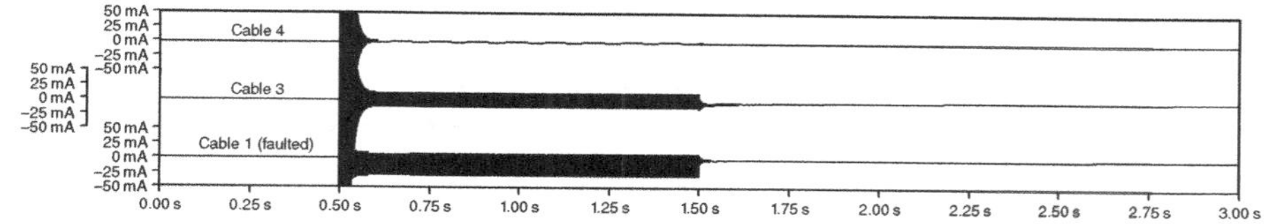

Figure 7.12 CM current of each of the three feeders.

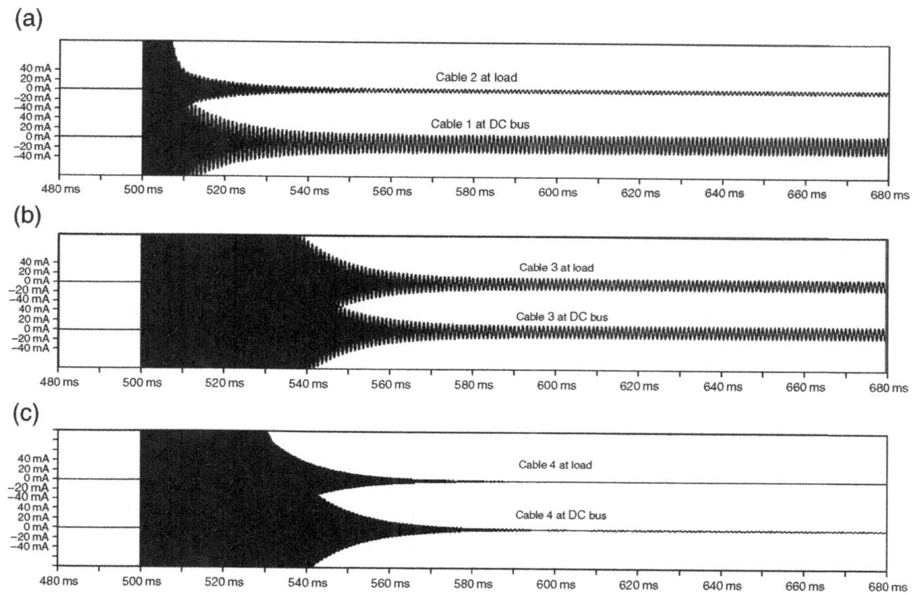

Figure 7.13 CM current at the source and load on each of the feeders. (a) Cable 1 and cable 2 (faulted); (b) Cable 3; (c) Cable 4.

fault, has a steady-state DC magnitude of 10 mA. As shown in Figure 7.12, this same 10 mA DC offset is present in the CM current of the faulted feeder. The CM current of the other un-faulted feeders does not show this DC offset. Being able to detect this DC offset is one method of localizing the faulted feeder. As shown in Figure 7.13, the DC offset in the CM current of the faulted feeder disappears between the ground fault and the load.

The voltage ripple of the DC supply directly influences the AC component of the fault currents of the feeder cables. If the ripple is of sufficient magnitude, the resulting AC currents can also be used to help locate the ground fault. If necessary, the ripple can be artificially increased. Figures 7.15 through 7.18 depict the waveforms for a system with a ripple voltage magnitude 100 times that for Figures 7.10

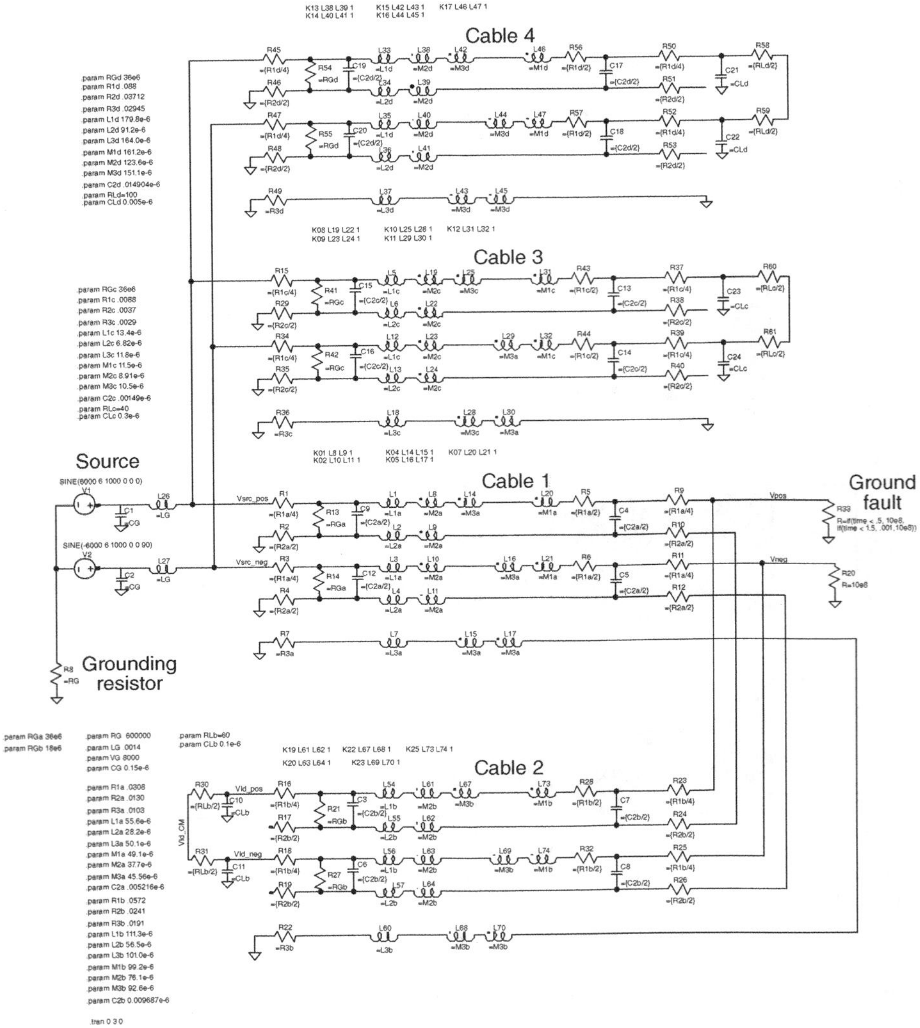

Figure 7.14 Schematic of DC circuit with three feeders.

through 7.13. The AC component of the CM current dwarfs the 10 mA DC offset on the faulted feeder. The magnitude of the CM current ripple drops substantially between the ground fault and the load. This drop in CM current ripple can also be used to localize the fault to a particular location on the faulted feeder.

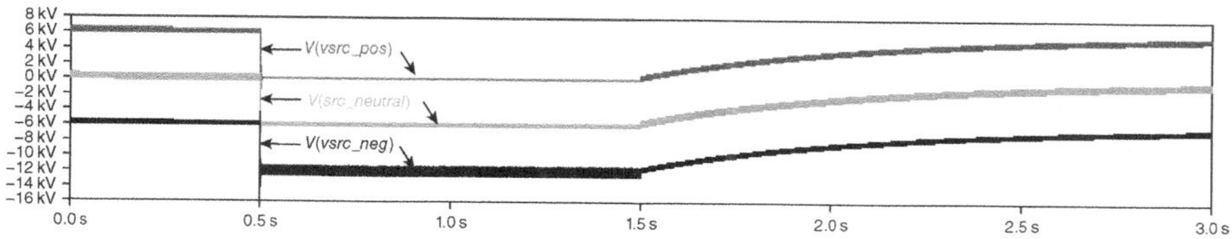

Figure 7.15 Voltages of positive and negative conductors, and neutral voltage of three-feeder system with high ripple voltage.

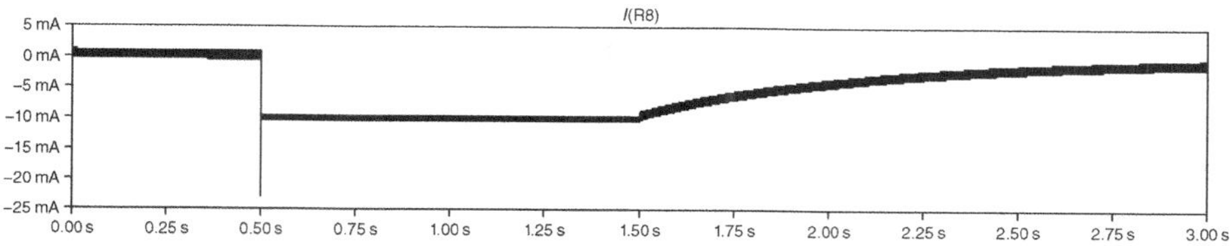

Figure 7.16 Grounding resistor current of three-feeder system with high voltage ripple.

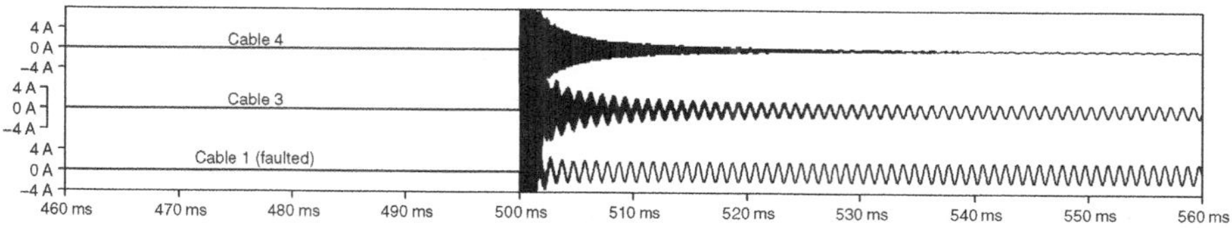

Figure 7.17 CM current of each of the three feeders for a system with high voltage ripple.

The faulted feeder may not always have the largest CM current, but a good strategy is to check the feeders starting with the one with the largest CM current at the DC bus and working down to the feeder with the smallest CM current. The CM current at the DC bus of each feeder should be compared to the CM current at the load end of the feeder: un-faulted feeder cables will not show a significant drop in CM current magnitude while the faulted feeder cable will show a significant drop. Once the faulted feeder has been identified, one can measure the CM current at different locations on the feeder cable; the ground fault will be near where the measurements show a large change. When making the CM current measurements, one should try to only include the conductor currents and not the shields; current in the shields may impact the CM current measurements.

The pulsing system described in Chapter 3 may also be employed to localize the ground fault.

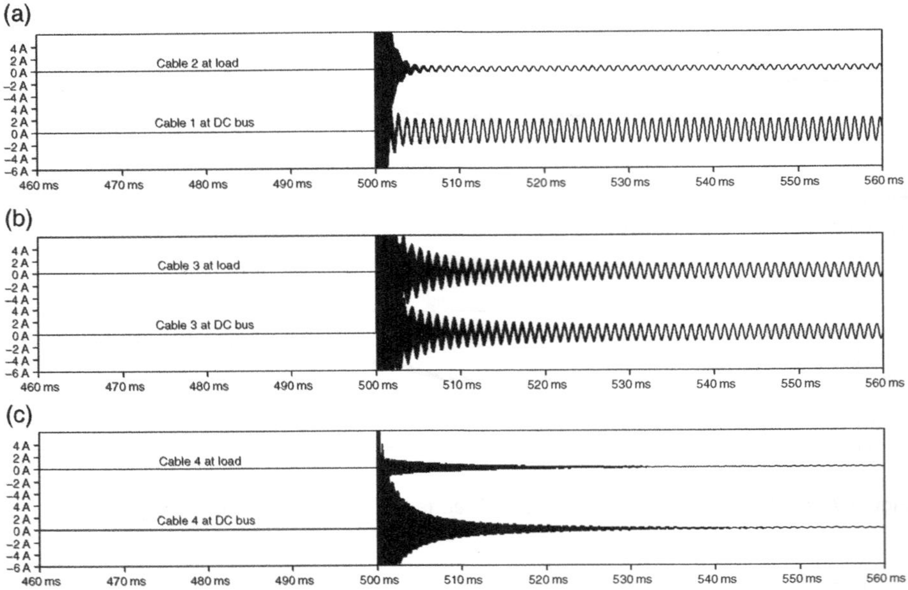

Figure 7.18 CM current on the feeder at the DC bus and at the load with high voltage ripple. (a) Cable 1 and cable 2 (faulted); (b) Cable 3; (c) Cable 4.

7.6 Electrical Insulation Impacts

7.6.1 Insulation Voltage Rating

The insulation for each conductor should be rated to take the nominal system voltage (line-to-line) as well as the superimposed CM voltage. This CM voltage will depend on the details of the rectifier design, as well as the equipment and system feeding the rectifier. A rule of thumb is to multiply the nominal system voltage by 1.2 and then use the next higher standard DC insulation rating. Cables designed for AC use generally have a DC line-to-line voltage rating of approximately 150% of the AC rating.

7.6.2 Partial Discharge

As with cables for AC primary distribution systems, cables for DC primary distribution systems are susceptible to partial discharge (PD). Because the electric field strength is relatively constant under DC, PD measurements for DC cable often have a lot of noise that must be filtered out. Hence, a PD monitoring system designed for AC cable may not be satisfactory for DC cables; one should ensure that the PD monitoring equipment is designed for DC cable applications.

7.6.3 Space Charge

Space charge (SC) is another mechanism for cable insulation deterioration in DC primary distribution systems. SC is an accumulation of electric charge within the cable insulation; this distribution of charge changes the electric field within the insulation to be nonuniform. The nonuniformity of the electric field, in turn, can lead to more SC accumulation, in addition to localized areas where the electric field is large enough to cause insulation breakdown. The accumulation of SC is a function of the insulation material, field strength, temperature, and exposure of the insulation to the environment. The rate of SC accumulation can be controlled by

- Cable design, including insulation material selection; high-quality cables are important for a long service life.
- Cable installation practices to avoid sharp bends and properly ground shields.
- Minimizing exposure of the cable to high temperatures.
- Keeping current levels below ampacity rating of the cable to limit the temperature gradient in the insulation.
- Minimizing exposure of the cable to liquids and corrosive gases.

At present, instrumentation is not commercially available to determine the level of SC in a DC cable onboard a ship.

References

ASTM B8-11 (2017). *Standard Specification for Concentric-Lay-Stranded Copper Conductors, Hard, Medium-Hard, or Soft.* Philadelphia, PA: ASTM International. Available from http://webstore.ansi.org.

IEEE Std 1580-2021 (2021). *IEEE Recommended Practice for Marine Cable for Use on Shipboard and Fixed or Floating Facilities.* New York: Institute of Electrical and Electronics Engineers. Available from http://ieeexplore.ieee.org.

MIL-HDBK-299 (1991). *Cable Comparison Handbook, Data Pertaining to Electric Shipboard Cable.* Washington, DC: Department of Defense. Available from http://quicksearch.dla.mil.

8

System Grounding: Shipboard Solidly Grounded DC Systems (No Greater than 1 kV)

8.1 Characteristics

As shown in Figure 8.1, DC power systems with nominal system voltages below 1 kV serving noncritical loads may be solidly grounded. Critical loads are specified by ship requirements, regulations, or class rules. Low-voltage special circuits may have requirements met by a solidly grounded DC system.

Solidly grounded systems are typically grounded at the negative conductor (Figure 8.2) or at the point of common connection for a split power supply (Figures 8.3 and 8.4). Solidly grounding the positive conductor is also possible, but it is not usually encountered.

A line-to-ground fault of an ungrounded conductor in a solidly grounded system will result in large fault currents. The magnitude of the fault current will depend on the characteristics of the source. Some sources do not have the ability to limit current, and the fault current will only be limited by the internal resistance of the source and the line resistance. Other sources, typically power electronic active rectifiers, may have the capability of limiting the ground current to a value somewhat greater than the full rated current of the fault. In either case, the fault protection system is relied upon to selectively trip interruption devices, such as circuit breakers or fuses.

Solidly grounded DC systems do not inherently have the ability to continue operation without isolating a ground fault of a non-grounded conductor. For this reason, solidly grounded systems should not be used to supply critical loads.

Solidly grounding the negative conductor (Figure 8.2) should only be done if the ripple voltage is small or if there is very little line-to-ground capacitance. Otherwise, the ripple voltage can cause a substantial hull current through the line-to-ground parasitic capacitances. Systems where batteries are directly connected to the DC bus often have very little voltage ripple and, therefore, grounding the

Design of Shipboard Power System Grounding/Earthing, First Edition. Norbert Doerry, Mohammed M. Islam, and John Prousalidis.
© 2025 The Institute of Electrical and Electronics Engineers, Inc.
Published 2025 by John Wiley & Sons, Inc.

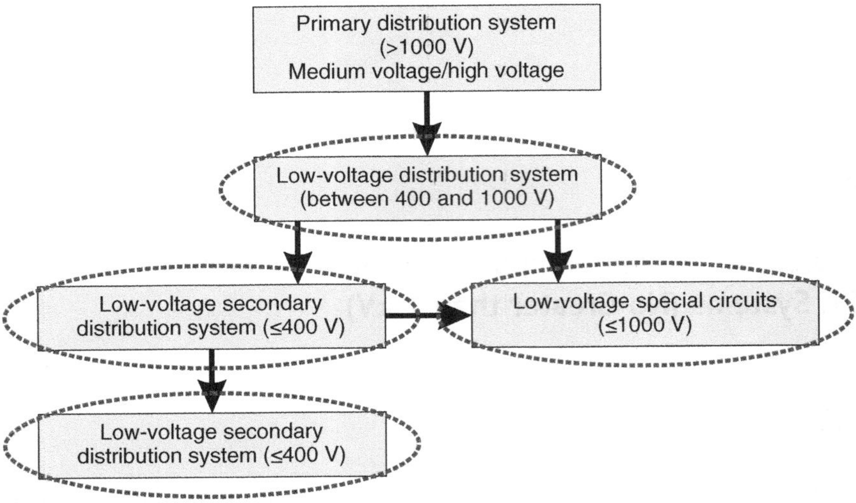

Figure 8.1 Applicability of solidly grounded DC systems.

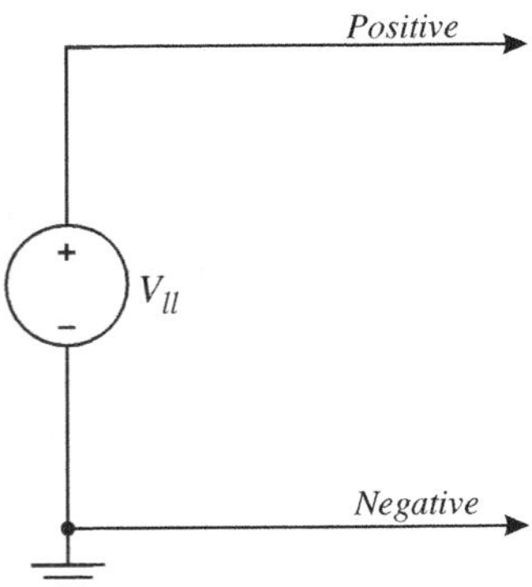

Figure 8.2 Solidly grounded DC power system with grounded negative conductor.

negative conductor may prove beneficial. The hull should never be used as a current-carrying conductor.

In systems with the negative conductor grounded, circuit protection is only needed on the positive conductor; single-pole breakers are adequate. The negative conductor is a grounded conductor, and the positive conductor is often called a hot conductor.

Solidly grounded systems with line-to-line voltages above about 150 V, or those below 150 V with significant voltage ripple and line-to-ground capacitance, should use a split power supply with the neutral point grounded, as depicted in Figure 8.3 or Figure 8.4. In these configurations, line-to-line voltage ripple will not result in significant CM currents flowing through the hull. For systems with a split power supply, circuit protection should be provided on both the positive and negative conductors; two-pole breakers are needed. A two-wire system, as depicted in Figure 8.3, provides a single voltage level to all loads. A three-wire system allows loads to connect between the positive and negative conductors, or between either the positive or negative conductor and the neutral conductor. In this manner, loads may be provided either the full line-to-line voltage or half the line-to-line

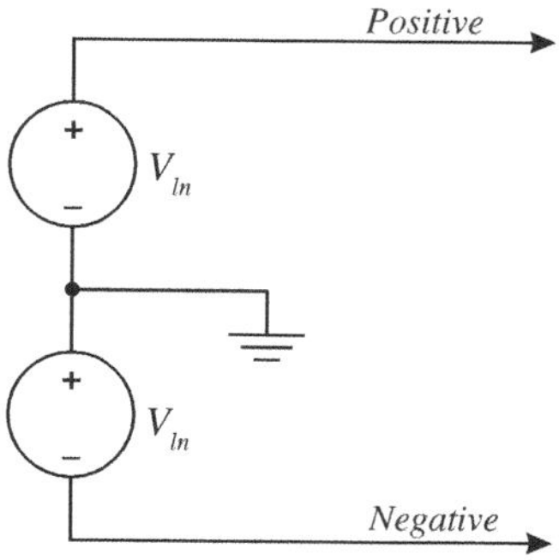

Figure 8.3 Solidly grounded two-wire DC power system with split power supply and grounded neutral.

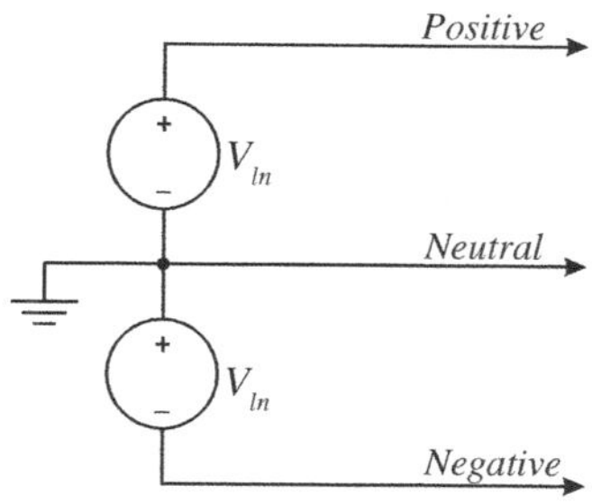

Figure 8.4 Solidly grounded three-wire DC power system with split power supply and grounded neutral.

voltage. Loads should never be powered by connecting between the positive or negative conductor and the ship's hull.

In general, solidly grounded systems should have a deliberate connection to the ship's hull at only one location, typically the source; parasitic capacitances and resistances within the system will provide a distributed connection to the ship's hull. A current meter should be provided on the deliberate connection to ground to enable the operator to monitor the current flow between the power system and the ship's hull. In some cases, ground faults of the grounded conductor may be detected in this manner.

8.2 Corrosion

Chapter 15 details how a DC current in the hull of the ship can result in a voltage difference between different regions of the hull. This voltage difference, in the presence of the saltwater electrolyte, can result in galvanic corrosion of the region that is more positive. A fraction of a volt difference between areas of the ship's hull is sufficient to cause a corrosion current. For this reason, the ship's hull should never be used as a current-carrying conductor for a section of the power system.

8.3 Modeling Shipboard Solidly Grounded DC Systems

Properly modeling the fault behavior of the source is very important in understanding the ground fault characteristics of a solidly grounded AC system. Ideally, the fault behavior of the source would be measured via testing of actual hardware and a dynamic model derived from this test data. If testing is not possible, then other methods include querying the manufacturer for the fault behavior or modeling published specifications for the source.

The modeling of the CM performance of a solidly grounded DC system is possible; Chapter 13 demonstrates the impact of solidly grounding the negative conductor on CM currents versus solidly grounding the neutral conductor in a split power supply.

8.4 Ground Fault Detection and Localization

The same protection system for line-to-line short circuits can be relied upon to detect and clear ground faults on ungrounded (hot) conductors. The circuit protection should measure and trip based on overcurrent on any ungrounded conductor.

A DC-rated GFCI/RCD/ALCI may also be employed on all current-carrying conductors to detect and isolate ground faults where the line-to-ground impedance is too high to trip the overcurrent protection.

Figure 8.5 depicts a solidly grounded DC system with a resistive load and with each of the conductors of a two-conductor cable having a resistance of 10 mΩ. At time 5 ms, a ground fault through a 10 Ω resistor is inserted near the load on the negative conductor which is solidly grounded at the source. As depicted in Figure 8.6, before the ground fault, the CM current through the two-conductor cable is zero. With the ground fault, the CM current magnitude is 10 mA, which is sufficient to trip a GFCI. If a GFCI is not present, the CM current would flow through the ship's hull back to the ground connection.

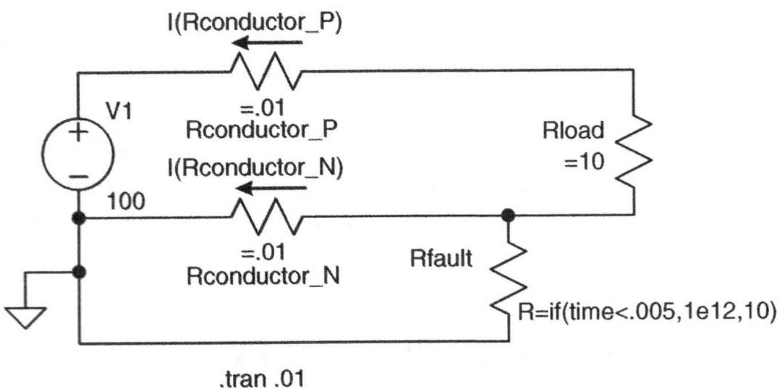

Figure 8.5 LTspice model of a ground fault on the grounded conductor of a two-wire system with grounded negative conductor.

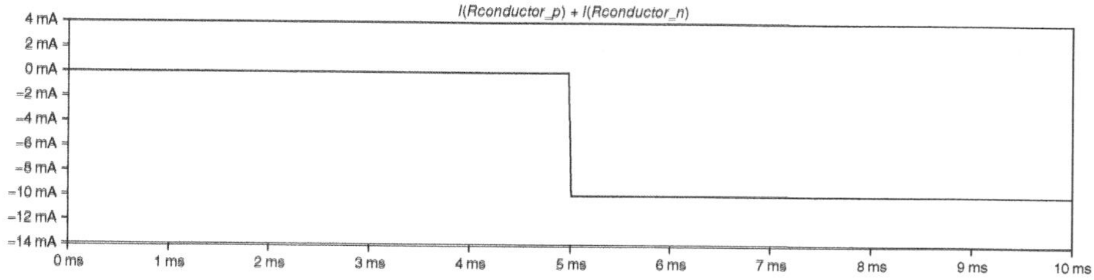

Figure 8.6 CM current due to ground fault on the grounded conductor of a two-wire system.

To reduce the rate of hull corrosion, operating with a ground fault for extended periods of time should be avoided.

Figure 8.7 depicts the circuit of Figure 8.5 except the ground fault is applied to the positive conductor near the load. At time 4 ms, a high-resistance (10 kΩ) ground fault is applied. As shown in Figure 8.8, the ground fault results in a CM current of 10 mA, enough to trip the GFCI. At time 7 ms, the resistance of the ground fault is lowered to 10 Ω. As shown in Figure 8.9, the current through

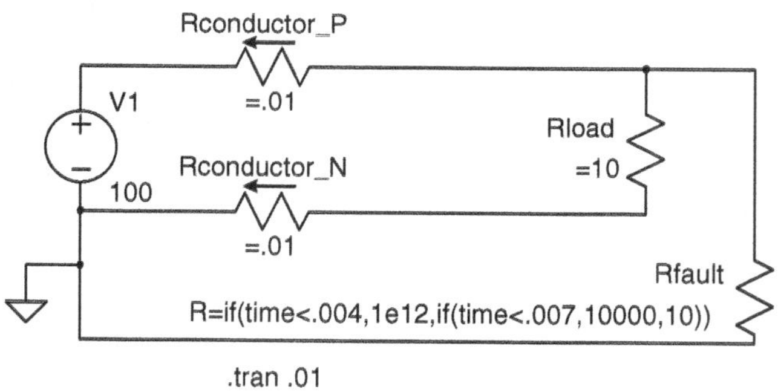

Figure 8.7 LTspice model of a ground fault on the "hot" conductor of a two-wire system with grounded negative conductor.

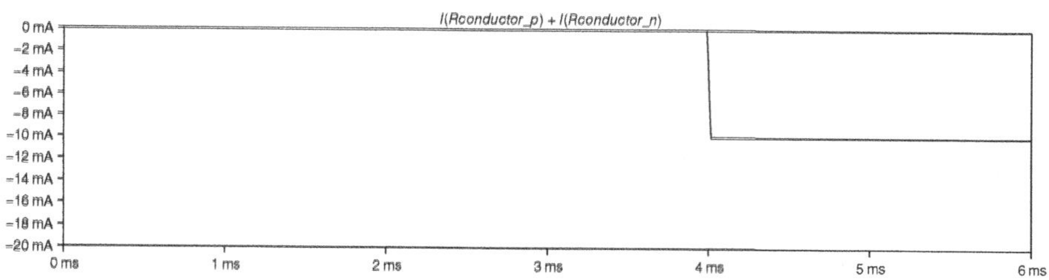

Figure 8.8 CM current due to high-resistance ground fault on "hot" conductor at 4 ms.

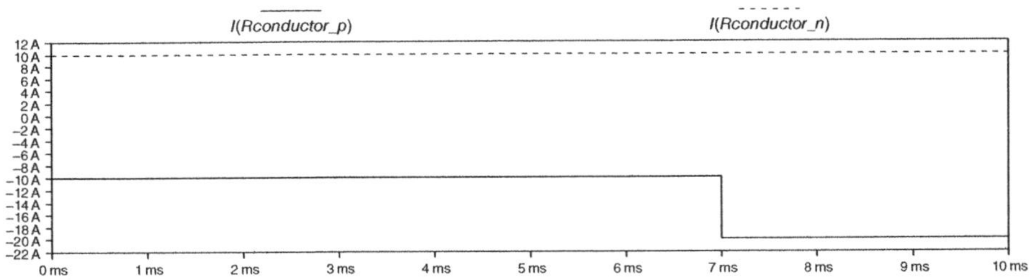

Figure 8.9 Phase currents due to low-resistance ground fault on "hot" conductor at 7 ms.

the positive conductor doubles, likely resulting in either the overcurrent protection tripping or, since the CM current has a magnitude of 10 A, the GFCI tripping.

8.5 Electrical Insulation Impacts

If one of the conductors is grounded, the insulation around each conductor should be rated for at least the maximum line-to-line voltage, including peak ripple voltage. For split power supplies, the insulation around each conductor should be rated for at least half the maximum line-to-line voltage, including peak ripple voltage. To ensure sufficient insulation thickness in cases of abrasion of the insulation during installation and operation, cable with a minimum AC voltage rating of 300 V is typically used. Standard wire and insulation AC voltage ratings include 300, 600, and 1000 V. The equivalent DC rating is 1.4 times the AC rating and reflects that the peak value of an AC waveform is $\sqrt{2}$ (approximately 1.4) times its rms value.

8.6 Cable Insulation Colors

DC cable insulation for solidly grounded systems will typically use the following insulation color convention:

- Positive conductor: red or brown.
- Negative conductor: black, gray, white, light blue, or blue.
- Neutral conductor (for split power supply): white or blue.
- Protective ground: green, green with yellow stripe, or bare.

The insulation color should never be considered absolute; the voltage of a conductor should be verified through measurement.

9

Designing Shipboard Power System Grounding/Earthing Systems

9.1 Introduction

The previous chapters described the different grounding/earthing methods that are applicable to shipboard power systems. This chapter provides guidance on deciding which method to use for a particular application. This guidance is in alignment with IEEE Std 45.1-2023 (2023) and is summarized in Table 9.1. The grounding methods in this chapter are applicable for ground fault currents at the fundamental frequency or DC; the control of CM currents at nonfundamental frequencies is covered in Chapter 13.

A ship will usually have multiple power distribution systems, each defined by a nominal system voltage and a nominal system frequency (or DC) interconnected via transformers or power converters. The power converters are typically power electronics-based but may also include a motor-generator set. The transformers and power converters can be used to connect power distribution systems of different voltages and, in the case of power converters, different frequencies (including DC). The power converters should be designed to enable each power distribution system to which it connects to operate as a separately derived system, as defined in IEEE Std 3003.1-2019 (2019). In a separately derived system, a ground fault on one power distribution system will not result in a change to the neutral-to-ground voltage on another power distribution system; the two systems are considered to be isolated. A transformer, operating either at the nominal system frequency or, more likely, at a frequency between 1 kHz and 40 kHz, is often used within electronic power converters to provide the isolation. There are other technologies available to provide isolation (such as motor-generator sets), but they are not widely used.

If two power distribution systems with different nominal system voltages are connected via power conversion equipment, but they are each part of the same separately derived system (not isolated), then a ground fault on the higher voltage

Design of Shipboard Power System Grounding/Earthing, First Edition. Norbert Doerry, Mohammed M. Islam, and John Prousalidis.
© 2025 The Institute of Electrical and Electronics Engineers, Inc.
Published 2025 by John Wiley & Sons, Inc.

Table 9.1 Recommended grounding methods.

Application	Ungrounded	High-resistance grounded	Solidly grounded
AC primary distribution system	No	Yes	No
AC low-voltage distribution system	Yes	Yes	No
AC low-voltage secondary distribution system	No	No	Yes
AC low-voltage special circuits	See text	See text	See text
DC primary distribution system	No	Yes	No
DC low-voltage distribution system	Yes	Yes	No
DC low-voltage secondary distribution system	Yes	No	Yes
DC low-voltage special circuits	See text	See text	See text

system could result in a line-to-ground voltage on the lower voltage system equaling the line-to-neutral voltage of the higher voltage system plus the line-to-neutral voltage of the lower voltage system. As a consequence, the cable and equipment insulation for the lower voltage system should be rated to a much higher voltage than if the lower voltage system were a separately derived system.

For example, Figure 9.1 depicts the line-to-ground voltages for two DC distribution systems, one with a 12 kV nominal system voltage and the other with a 1 kV nominal system voltage. A ground fault of the negative conductor of the 12 kV system occurs at time 5 (transients are not shown). If the power distribution systems are not isolated, the line-to-ground voltages of the 1 kV conductors will rise from −500 and 500 V to 5.5 and 6.5 kV. If isolated, the line-to-ground voltages of the 1 kV conductors will not change (neglecting transients). If the power distribution systems are not isolated, cable and equipment on the 1 kV system should be rated in excess of 6.5 kV in contrast to a rating of 1 kV needed for isolated power distribution systems.

9.2 AC Primary Distribution Systems

AC primary distribution systems are three-phase systems with a nominal system voltage greater than 1 kV; the power is usually directly produced by generator sets. Normally, AC primary distribution systems are employed when the total installed power generation capacity results in the steady-state current at low

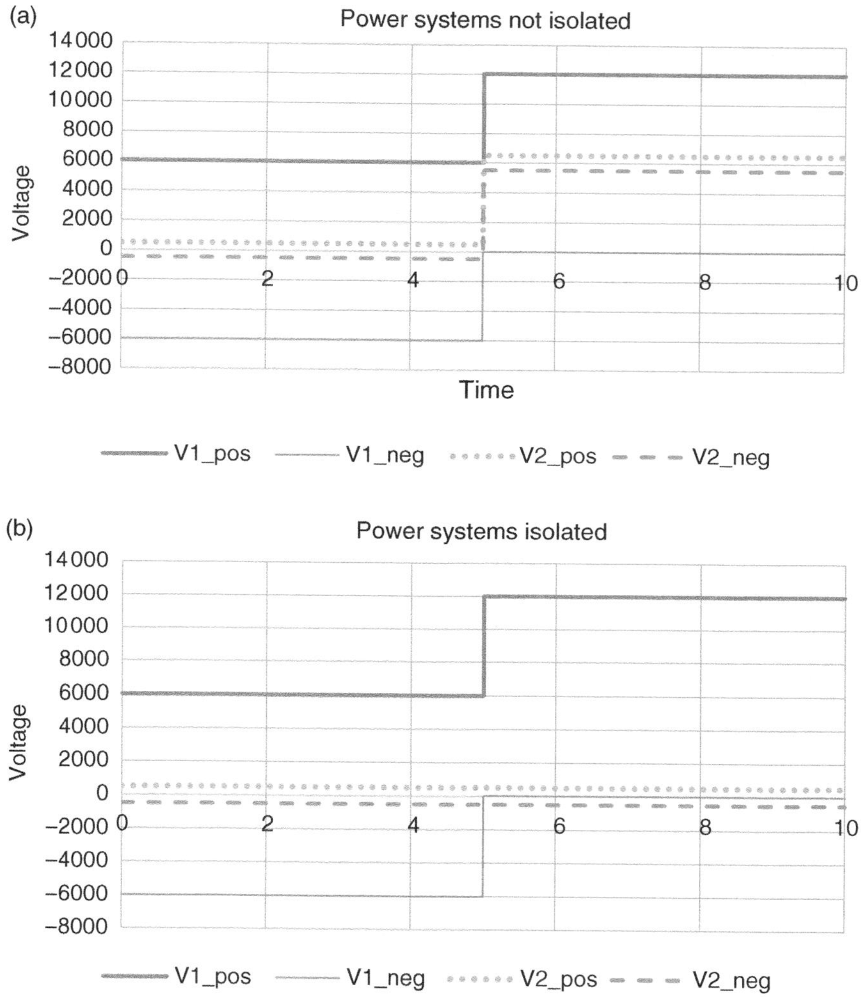

Figure 9.1 Line-to-ground voltages for 1 and 12 kV HRG systems that are (a) not isolated; (b) isolated.

voltage exceeding the current rating of available low-voltage circuit breakers. Most of the time, the transition from the use of AC low-voltage distribution systems to AC primary distribution systems occurs when the total generation capacity is between 6 and 12 MW.

Typically, individual loads on an AC primary distribution system will be three-phase and exceed 100 kW; smaller loads are served by low-voltage distribution systems powered via transformers or power converters from the primary power distribution system or other low-voltage distribution systems. Because the total number of sources and loads on an AC primary distribution system is limited (compared to the low-voltage distribution system), the total length of cable associated with an AC primary distribution system is also limited. The shorter length of cable tends to reduce the parasitic line-to-ground capacitance, but the insulation materials and conductor shields of cable rated for more than 1 kV, compared to low-voltage cable, tend to have higher line-to-ground parasitic capacitance.

Shipboard power systems with an AC primary distribution system should employ high-resistance grounding. As described in Chapter 5, high-resistance grounding limits ground fault currents and transient line-to-ground voltages during a ground fault. Continued operation for a short period of time with a ground fault provides the operator with options for reconfiguring the electrical system and supported load systems to avoid placing the ship in danger. An unexpected loss of power to critical equipment, such as propulsion, can place the ship in danger; in the case of a single line-to-ground fault, a high-resistance grounding system can help avoid this outcome. Shipboard operators should have the capability of quickly identifying when a line-to-ground fault occurs (even if it is an intermittent ground fault) and localizing the ground fault to take corrective action when such action can be safely taken.

An ungrounded system is not recommended due to the potential of excessive voltage stress on cable and equipment insulation from voltage transients associated with a ground fault.

A solidly grounded system is not recommended due to the high ground fault currents and inability for continued operation with one line-to-ground fault. The high ground fault currents pose a personnel safety hazard (high localized voltage differences in the ship structure, as well as localized heating) and can damage equipment (generally due to localized heating) until the fault is cleared. The inability to support continued operation with a single line-to-ground fault can also place the ship in danger due to an unexpected loss of power to critical equipment.

9.3 AC Low-Voltage Distribution Systems

AC low-voltage (less than 1 kV) distribution systems are generally three-phase systems with no line-to-neutral loads. Power for these systems is typically provided directly from generator sets or via transformers or power converters from an AC primary distribution system. AC low-voltage distribution systems can span the entire ship or only a single electrical zone; the amount of power cable can be very large, resulting in high line-to-ground parasitic capacitance. As discussed in Section 9.2, AC low-voltage distribution systems are typically appropriate for systems with a total electrical power generation capacity below a value between 6 and 12 MW.

Shipboard power systems with an AC low-voltage distribution system should either employ high-resistance grounding or be ungrounded. High-resistance grounding is appropriate for AC low-voltage distribution systems that experience significant stresses to cable and equipment insulation. These stresses include frequent voltage transients, frequent voltage spikes, high temperatures, and high levels of waveform distortion. High-resistance grounding is also appropriate if a large number of loads are connected to it and thereby increasing the probability of an intermittent ground fault. If these conditions do not exist, then either a high-resistance grounded or an ungrounded system is suitable.

9.4 AC Low-Voltage Secondary Distribution Systems

AC low-voltage secondary distribution systems are generally either single-phase systems, possibly center-tapped, with a nominal system voltage between 120 and 240 V, or three-phase systems having line-to-neutral loads and with a nominal system voltage between 208 and 400 V. AC low-voltage secondary distribution systems typically are used to provide power to lighting circuits, power receptacles, and relatively low-power equipment. AC low-voltage secondary distribution systems generally do not have a capacity exceeding 100 kW and, for large ships, normally serve a relatively small region of the ship. For most ships, AC low-voltage secondary distribution systems receive power via a transformer or power converter from an AC low-voltage distribution system.

Very small vessels (less than about 100 kW total installed power) may generate and distribute only single-phase AC power with a nominal system voltage of either 120 or 240 V.

AC low-voltage secondary distribution systems should be solidly grounded. Much of the equipment connected to this power system is commercial or industrial and designed for terrestrial power systems that are solidly grounded. Commercial or industrial equipment may not have been tested for use on an ungrounded or high-resistance grounded system and may malfunction if operated with a power system that is not solidly grounded. In some cases, equipment connected to a power system that is not solidly grounded will report a "building wiring error" and immediately shut down.

9.5 AC Low-Voltage Special Circuits

AC low-voltage special circuits serve loads that otherwise would be served by an AC low-voltage secondary distribution system but require continued operation with a single ground fault. For example, special circuits may be used to serve some or all applicable emergency loads, vital loads, and medical facility loads. A special circuit typically only serves equipment in a few spaces. For most ships, AC low-voltage special circuits receive power via a transformer or power converter from an AC low-voltage distribution system.

AC low-voltage special circuits should be either ungrounded or high-resistance grounded. An ungrounded system is normally used because of lower cost, and the risks of an ungrounded system are mitigated by the limited scope of the special circuit.

9.6 DC Primary Distribution Systems

DC primary distribution systems are two conductor systems with a nominal system voltage about 1 kV; the power is produced by an AC generator and then rectified. DC primary distribution systems are suitable when the total installed power generation capacity would result in the steady-state current at low-voltage exceeding the current rating of available low-voltage circuit breakers. Most of the time, the transition from the use of low-voltage DC power generation and distribution to DC primary distribution systems occurs when the total generation capacity is between 6 and 12 MW.

Typically, individual loads on a DC primary distribution system will exceed 100 kW; smaller loads are served by low-voltage systems powered via power converters from the DC primary distribution systems and other low-voltage systems.

DC primary distribution systems should employ high-resistance grounding of the neutral. As described in Chapter 7, high-resistance grounding limits transient line-to-ground voltages during a ground fault. High-resistance grounding also provides a means for balancing line-to-ground voltages on the parasitic capacitances upon removal of a ground fault. Continued operation for a short period of time with a ground fault provides the operator with options for reconfiguring the electrical system and supported load systems to avoid placing the ship in danger. An unexpected loss of power to critical equipment, such as propulsion, can place the ship in danger; in the case of a single line-to-ground fault, a high-resistance grounding system can help avoid this outcome. Shipboard operators should have the capability of quickly identifying when a line-to-ground occurs (even if an intermittent ground fault) and localizing the ground fault to take corrective action when such action can be safely taken.

An ungrounded system is not recommended due to the potential for voltage stress on insulation systems due to large neutral-to-ground voltages resulting from an imbalance in line-to-ground insulation resistance and the inability to quickly balance line-to-ground voltages upon removal of a ground fault.

A solidly grounded system is not recommended due to the high ground fault currents and inability for continued operation with one line-to-ground fault. The high ground fault currents pose a personnel safety hazard (high localized voltage differences in the ship structure, as well as localized heating) and can damage equipment (generally due to localized heating) until the fault is cleared. The inability to support continued operation with a single line-to-ground fault can also place the ship in danger due to an unexpected loss of power to critical equipment.

9.7 DC Low-Voltage Distribution Systems

Low-voltage DC distribution systems are two conductor systems with nominal system voltages greater than 400 V and less than or equal to 1 kV.

Power for these systems is typically provided from generator sets via rectifiers or via power converters from an AC or DC primary distribution system. In some cases, energy storage may be the source of power. These power systems can span the entire ship or only a single electrical zone. As discussed in Section 9.6, low-voltage DC distribution systems are typically appropriate for systems with a total electrical power generation capacity below a value between 6 and 12 MW.

Low-voltage DC distribution systems should employ either high-resistance grounding or be ungrounded. High-resistance grounding is appropriate for low-voltage DC distribution systems, which experience significant stressors to cable and equipment insulation. These stressors include frequent voltage transients, frequent voltage spikes, high temperatures, and high levels of ripple. If these conditions do not exist, then either a high-resistance grounded or ungrounded system is suitable.

9.8 DC Low-Voltage Secondary Distribution Systems

DC low-voltage secondary distribution systems are generally two conductor systems with nominal system voltages no greater than 300 V. DC low-voltage secondary distribution systems typically are used to provide power to lighting circuits and relatively low-power equipment. DC low-voltage secondary distribution systems normally serve a relatively small region of the ship. For most ships, DC low-voltage secondary distribution systems receive power via a power converter from an AC or DC low-voltage distribution system or from an AC low-voltage secondary distribution system.

DC low-voltage secondary distribution systems with a nominal system voltage above 50 V should have the negative conductor solidly grounded to keep the shipboard practice aligned with terrestrial practices. DC low-voltage secondary distribution systems with a nominal system voltage no greater than 50 V should be ungrounded to prevent potential DC currents through the hull that can lead to corrosion.

9.9 DC Low-Voltage Special Circuits

Some DC loads with a nominal system voltage no greater than 1 kV require a grounding system that differs from the recommendations in Sections 9.7 and 9.8. These loads should be served from a low-voltage special circuit powered from a power converter that, in turn, is powered by a separately derived system. This low-voltage special circuit should be provided with the grounding system required by the load.

9.10 Examples

9.10.1 AC Low-Voltage Distribution System for Commercial Mechanical Drive Ship

Figure 9.2 depicts a shipboard power system typical of commercial ships with a mechanical drive and a relatively low ship service electrical load. Power is generated and distributed at 450 V AC through a low-voltage distribution system. Emergency loads are powered via an emergency switchboard that is normally powered by one of the two ship service generator switchboards; the emergency generator is employed when the main generators are not available. The two ship service generator switchboards may be paralleled or operated independently in a split plant configuration. The 440 V loads are powered directly from the generator and emergency switchboards. Lower voltage loads (120 and 208 V loads) are powered via transformers through a secondary distribution system.

As depicted in Figure 9.3, the 450 V AC distribution system employs a high-resistance ground using a zigzag transformer on each switchboard. The grounding resistor should have a resistance approximately equal to the magnitude of the CM capacitive impedance to ground due to the loads, distribution cable, and

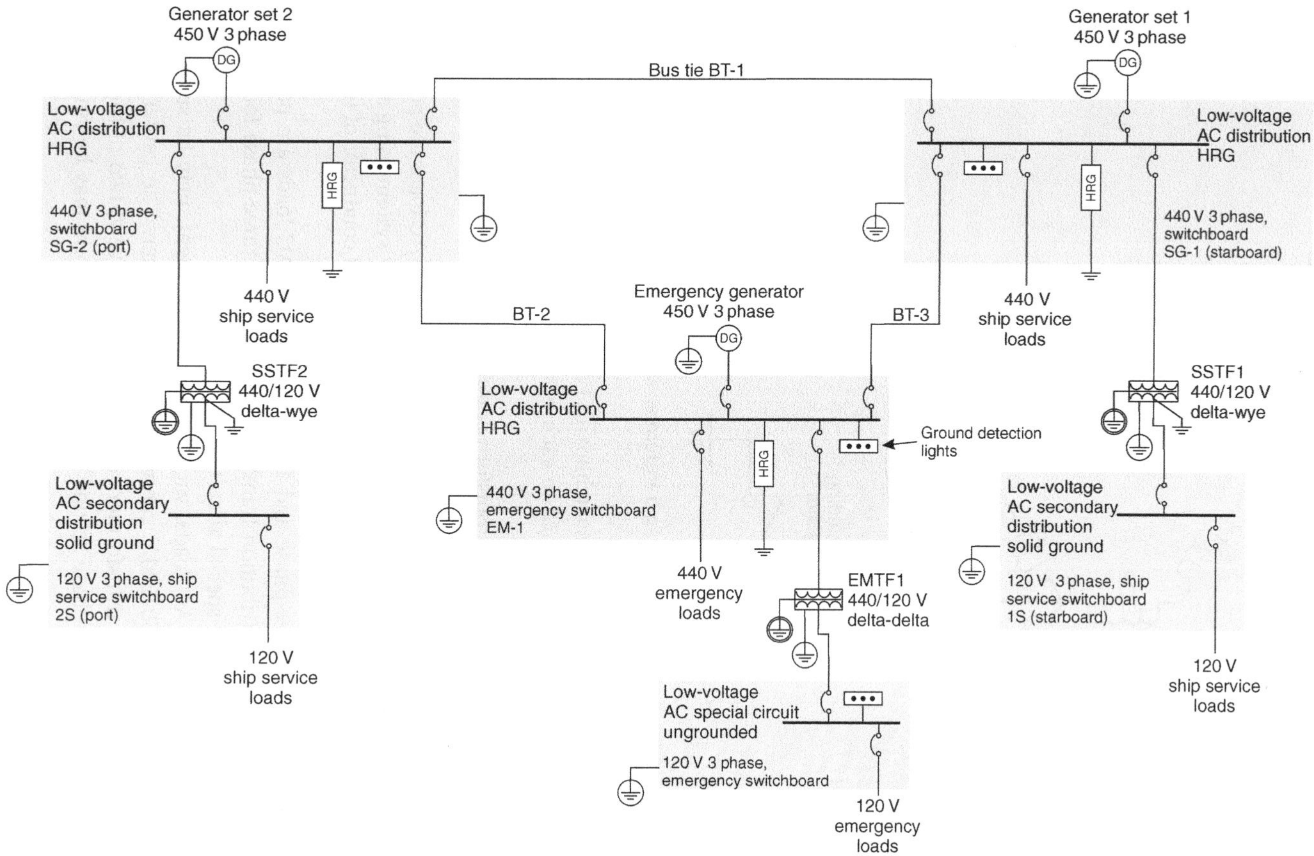

Figure 9.2 Shipboard power system for a mechanical drive ship with low ship service loads and HRG circuit.

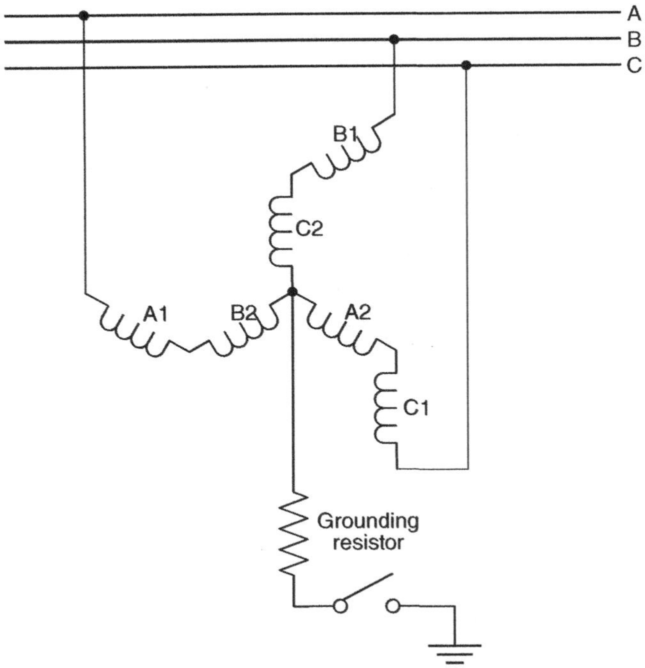

Figure 9.3 HRG circuit using a zigzag transformer with grounding switch.

generators associated with each switchboard. For bus ties, half of the CM capacitive impedance would be assigned to each of the two switchboards the bus tie connects.

Wye-delta or wye-broken delta transformer grounding circuits may be employed instead of the zigzag transformer.

Nonemergency 208 V three-phase and 120 V single-phase AC loads are powered through delta-wye distribution transformers from the 450 V system using a three-phase four-wire secondary distribution system. The neutral conductor of the wye secondary is solidly grounded.

Emergency 208 V three-phase and 120 V single-phase AC loads are powered through a delta-wye distribution transformer as part of a three-phase four-wire ungrounded low-voltage special circuit.

Nonemergency 240 V AC loads, such as galley equipment, may be supplied using a single-phase 450–240 V center-tapped transformer. The center tap is solidly grounded at the transformer secondary. Alternatively, two of the three phases of a 208 V three-phase secondary distribution system may be employed for these loads.

Ships may use voltages other than those depicted in Figure 9.3. Instead of 450 V, common low-voltage distribution system voltages include 400 and 690 V. Likewise, 240 V may be employed in the AC low-voltage secondary distribution system instead of 120 V.

Class rules or regulations may require single-point grounding of a power system. As discussed in Chapter 3, the technical basis for single-point grounding is questionable. If allowed, each grounding resistor grounding circuit should be grounded to the ship's hull. If not allowed, the grounding resistors should have a resistance roughly equal to the magnitude of the CM capacitive impedance to ground due to the loads, distribution cable, and generators associated with the entire ship, not just the switchboard. A grounding switch between the grounding resistor is also provided, as shown in Figure 9.3; controls ensure only one grounding resistor is connected to ground for each set of connected switchboards. If a switchboard is not connected to another switchboard, its grounding switch should be closed.

If the generator stators are wye configured, then an NGR and a grounding switch may be connected to the neutral conductor, as depicted in Figure 9.4, to eliminate the need for the grounding transformer. The NGR should have a resistance sized for the entire ship and the grounding switches controlled by the machinery control system. The NGR may be installed in a dedicated equipment cabinet, as depicted in Figure 9.4a, or may be integrated within the overall generator switchboard enclosure, as depicted in Figure 9.4b.

A disadvantage to using switched grounding resistors is that the ground fault current may become larger than desired when one switchboard is isolated from the others; the grounding resistor should have a resistance roughly equal to the magnitude of the CM capacitive impedance to ground for the entire ship, not just for the switchboard. The requirement for a control system interface also adds complexity.

Another approach, depicted in Figure 9.5, is to connect the grounding resistor grounding circuits to a grounding bus; the grounding bus is then connected to the ship's hull in one location via a grounding switch. The grounding bus should have sufficient ampacity to conduct the maximum fault current for the time required to clear the fault. It may be desirable to provide a grounding switch to connect the grounding bus to ground at each switchboard; only one of the switchboard ground switches would be closed at any time to provide the single-point ground; which grounding switch is closed is independent of whether all the switchboards are connected into a single separately derived system or not. The multiple ground switches provide a means of providing a ground connection at every switchboard should the grounding bus conductor break for some reason. A control system interface is not required. Any system ground connection, such as for EMI filters, should connect to the grounding bus and not to the hull directly.

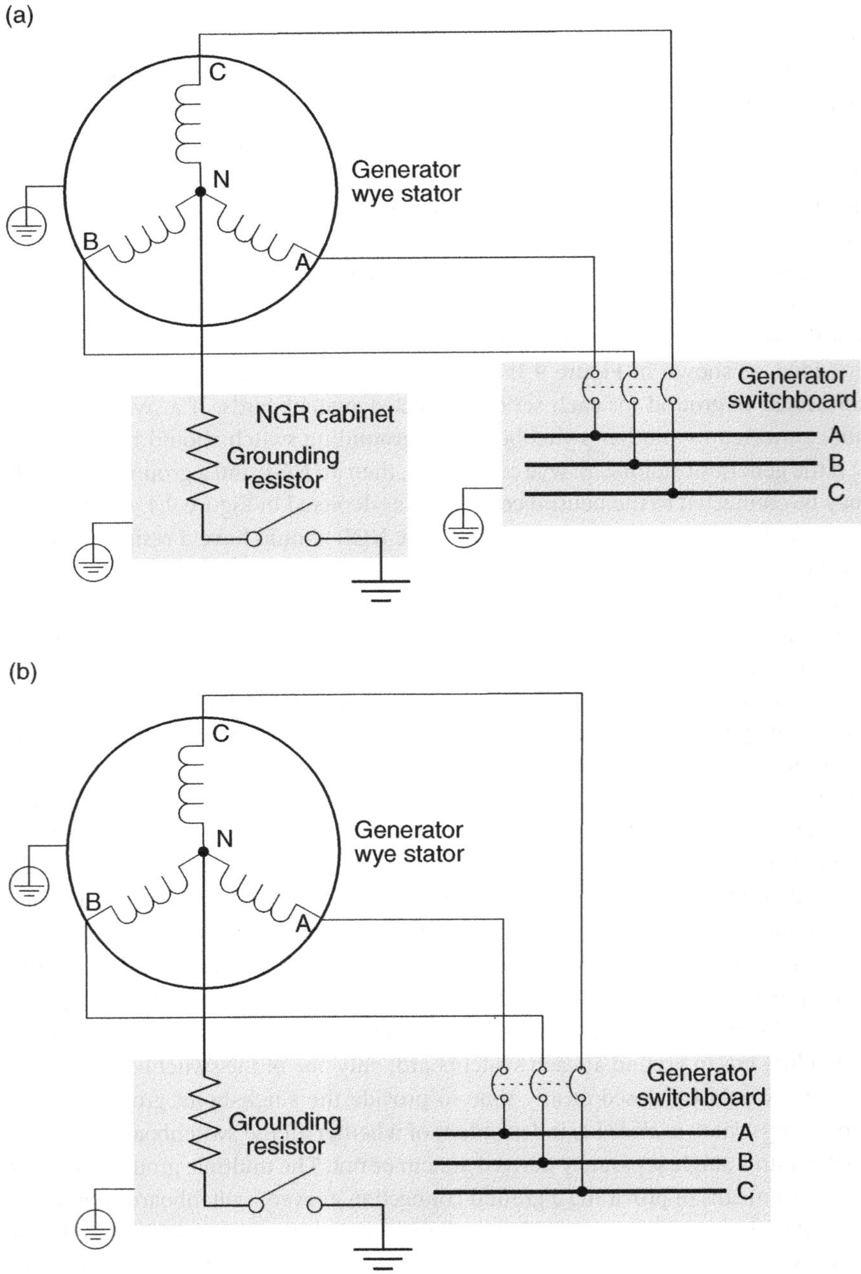

Figure 9.4 Generator NGR with grounding switch: (a) separate NGR cabinet; (b) integrated into generator switchboard.

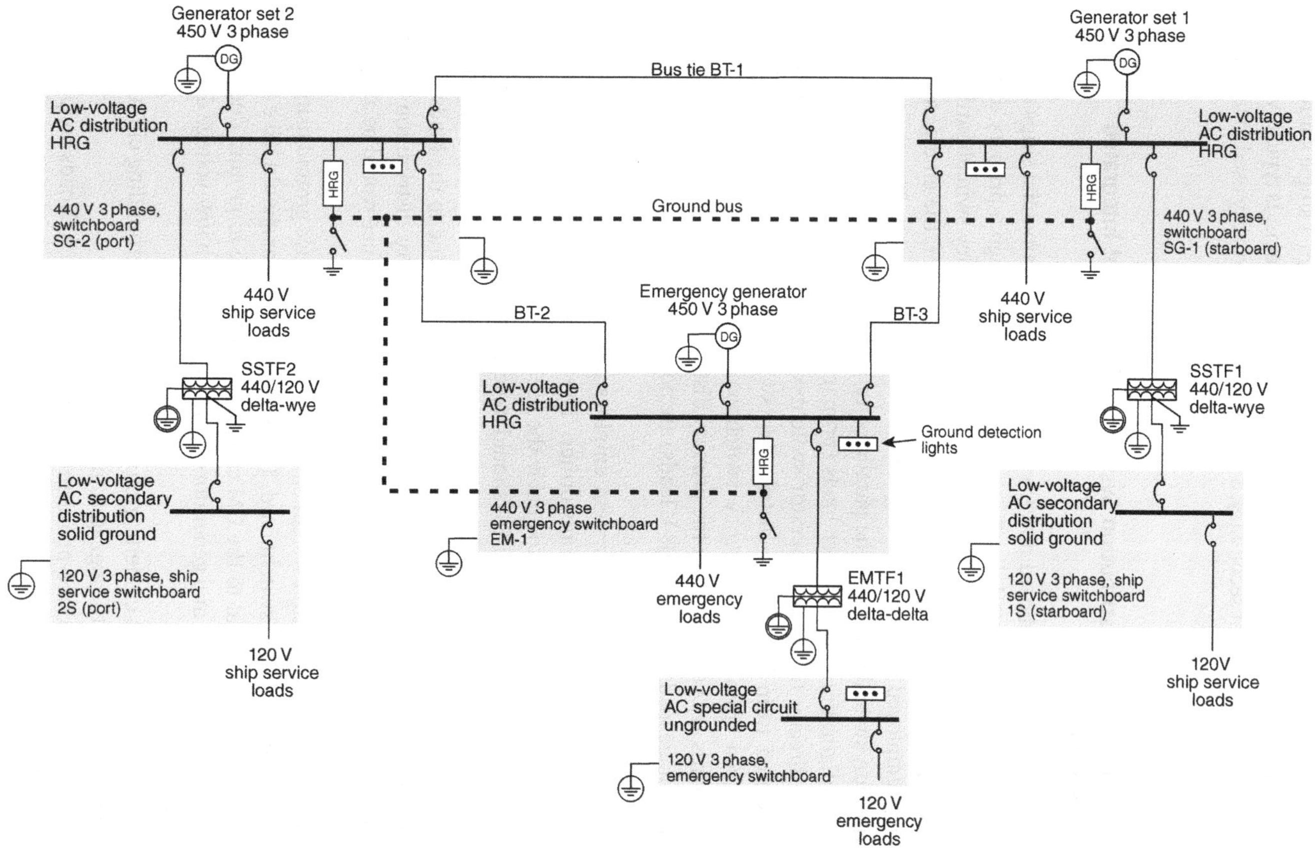

Figure 9.5 Shipboard power system for a mechanical drive ship with low ship service loads and a grounding bus.

Distribution systems providing 120/208 V three-phase power and 120/240 V single-phase power typically employ radial distribution from a single source such as a transformer. Single-point grounding, for grounded systems, is typically implemented at the transformer secondary; a wye secondary is normally used to facilitate grounding.

9.10.2 AC Primary Distribution System for Commercial Integrated Power System Ship

Figure 9.6 depicts a shipboard power system typical of commercial ships with an IPS. Power is generated and distributed at 6.6 kV AC through a primary distribution system. Emergency loads are powered via an emergency switchboard that is normally powered from the low-voltage distribution system; the emergency generator is employed when the main generators are not available. The two primary distribution switchboards may be paralleled or operated independently in a split plant configuration. High-power loads (primarily propulsion and thrusters) are powered directly from the generator switchboards. Nonemergency 440 V loads are powered from low-voltage distribution switchboards and emergency loads are powered from the emergency switchboard. The low-voltage distribution switchboards are powered from the primary distribution system via transformers. Lower voltage loads (120 and 208 V loads) are powered via transformers through a secondary distribution system.

The 6.6 kV AC distribution system employs a high-resistance ground using a wye-broken delta grounding transformer. The secondary winding of the transformer operates at low voltage to enable the use of low-voltage sensors for voltage and current. The grounding resistor should have a resistance such that the ground current on the primary connection to ground is roughly equal in magnitude to the current through the CM capacitive impedance to ground due to the loads, distribution cable, and generators associated with each primary distribution system switchboard. For bus ties, half of the CM capacitive impedance would be assigned to each of the two switchboards the bus tie connects.

The 440 V AC distribution system employs a high-resistance ground using a zigzag transformer on each switchboard. The grounding resistor should have a resistance approximately equal to the CM capacitive impedance to ground due to the loads, distribution cable, and generators associated with each low-voltage distribution switchboard.

Wye-delta or wye-broken delta grounding transformer grounding circuits may be employed instead of the zigzag transformer.

Nonemergency 208 V three-phase and 120 V single-phase AC loads are powered through delta-wye distribution transformers from the 440 V system using a

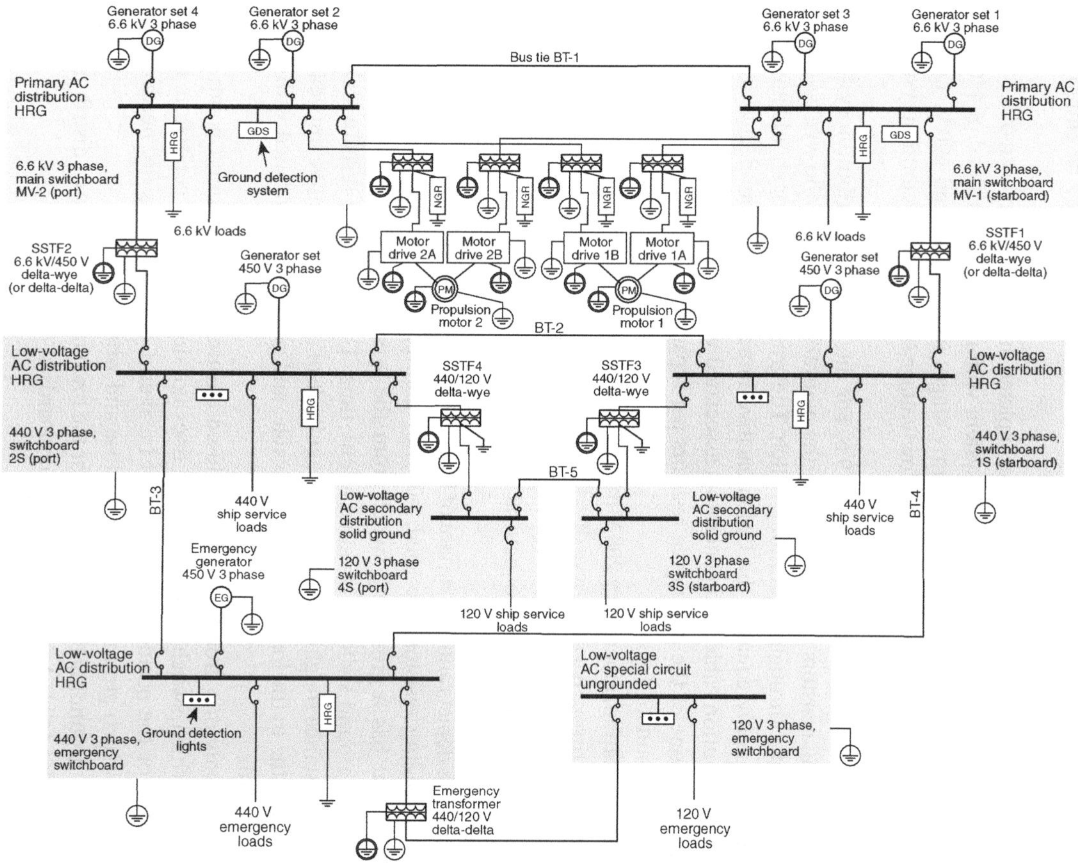

Figure 9.6 Shipboard power system for an IPS ship with HRG circuits.

three-phase four-wire secondary distribution system. The neutral conductor of the wye secondary is solidly grounded.

Emergency 208 V three-phase and 120 V single-phase AC loads are powered through a delta-wye distribution transformer as part of a three-phase four-wire ungrounded low-voltage special circuit.

Nonemergency 240 V AC loads, such as galley equipment, may be supplied using a single-phase 450–240 V center-tapped transformer. The center tap is solidly grounded at the transformer secondary. Alternatively, two of the three phases of a 208 V three-phase secondary distribution system may be employed for these loads.

Ships may use voltages other than those depicted in Figure 9.6. Instead of 6.6 kV, another common primary distribution system voltage is 11 kV. Instead of 450 V, common low-voltage distribution system voltages include 400 and 690 V. Likewise, 240 V may be employed in the AC low-voltage secondary distribution system instead of 120 V.

Figure 9.7 depicts a variant of Figure 9.6. For the primary distribution system, the grounding resistors are connected at the neutral conductor of the wye wound generator stator through a grounding switch. The grounding resistor should have a resistance roughly equal to the CM capacitive impedance to ground due to the loads, distribution cable, and generators associated with the entire ship. Controls ensure only one of the grounding resistors per separately derived system is connected to ground.

In this variation, emergency 208 V three-phase and 120 V single-phase AC loads are powered through a delta-wye distribution transformer as part of a three-phase four-wire high impedance grounded low-voltage special circuit. The grounding resistor is connected to the neutral conductor of the wye secondary of the transformer.

Figure 9.8 depicts another variant of Figure 9.6. For the primary distribution system, grounding resistors are used, but they connect to a grounding bus. Each switchboard has a grounding switch from the grounding bus to ground; only one of the ground switches is closed at any time. The grounding resistor should have a resistance roughly equal to the magnitude of the CM capacitive impedance to ground due to the loads, distribution cable, and generators associated with the switchboard.

Having only one grounding switch closed ensures a single ground point; the choice of which switch ground to close does not depend on the generators that are online or the position of any circuit breaker. Control logic is not needed to determine which grounding switch to close and which to open.

A similar grounding bus is employed for the low-voltage (440 V AC) distribution including the 440 V AC emergency switchboard.

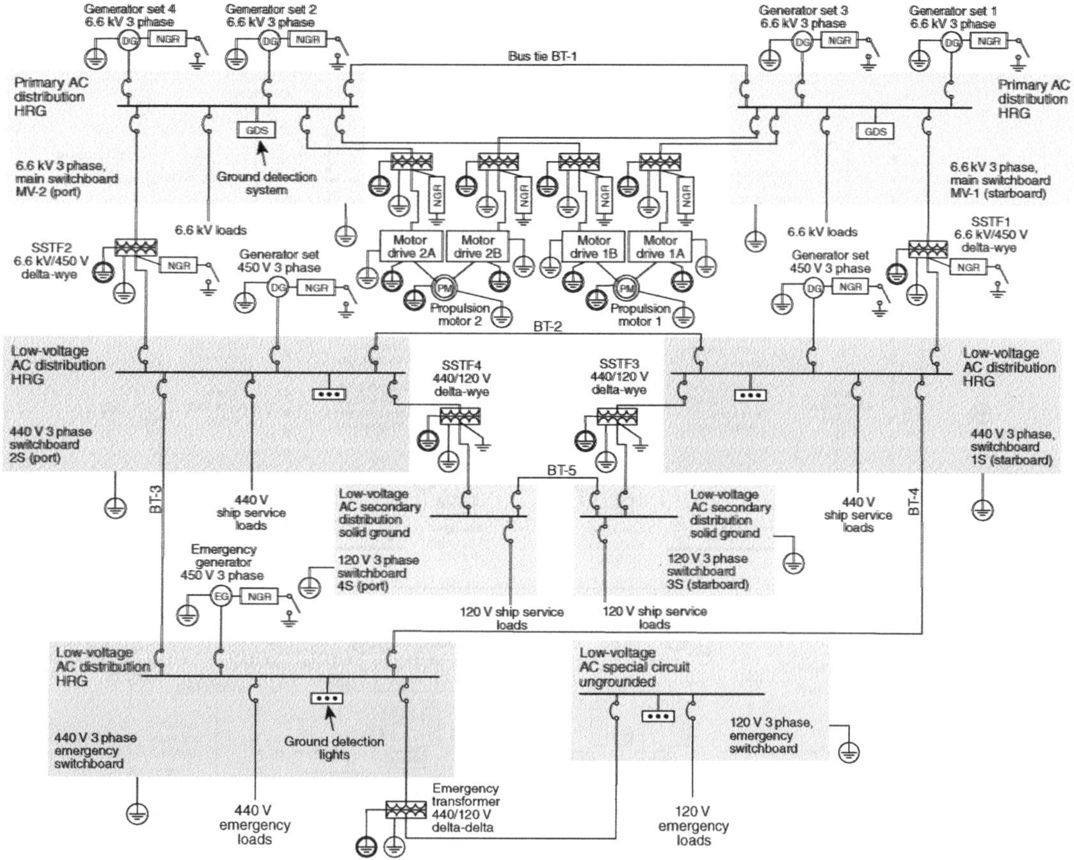

Figure 9.7 Shipboard power system for an IPS ship with NGRs.

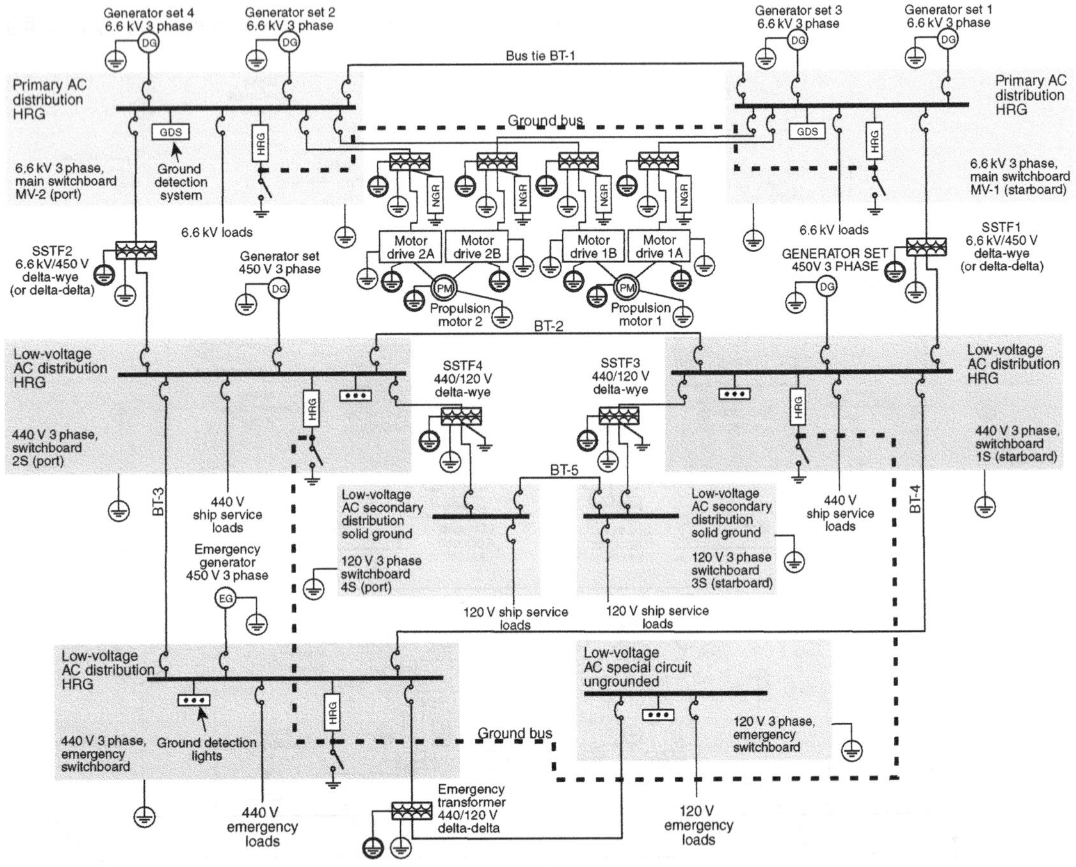

Figure 9.8 Shipboard power system for an IPS ship with grounding buses.

9.10.3 Low-Voltage AC Zonal Distribution

Figure 9.9 depicts a low-voltage AC zonal distribution system representative of those found on some mechanical drive naval ships. Many loads on these ships require normal and alternate supplies of power; providing these alternate supplies from port and starboard switchboards rather than from forward and aft switchboards usually results in the total amount of cable required for zonal systems being less than that required by a system depicted in Section 9.10.1.

In this zonal distribution system, three 450 V, three-phase generator sets (typically gas turbine-based or diesel-based) are connected to generator switchboards. These switchboards power several large local loads and longitudinal bus switchboards. The ship has two longitudinal buses: port and starboard. The longitudinal bus switchboards segment the longitudinal buses and provide power to load centers; the load centers provide power to 440 V AC loads and to transformers for secondary distribution systems and low-voltage special circuits.

Zigzag grounding transformers and grounding resistors are provided at each generator switchboard. Alternatively, the generators are provided with a wye configured stator and the grounding resistor is connected to the generator neutral conductor.

Under normal operation, two of the generator sets are online; one is powering each of the two longitudinal buses. Consequently, the grounding resistors should have a resistance roughly equal to half the magnitude of the CM capacitive impedance to ground due to the loads, distribution cable, and generators associated with the entire ship. The grounding transformer for the offline generator set is isolated by opening the bus-tie breakers of the generator switchboard to the longitudinal bus switchboards.

AC low-voltage secondary distribution systems are powered from transformers connected to a load center and are solidly grounded. The transformers are typically delta-wye with the neutral conductor of the secondary solidly grounded. Low-voltage special circuits are powered from delta-delta transformers connected to a load center and are ungrounded.

9.10.4 Zonal AC Primary Distribution

Figure 9.10 depicts a zonal AC primary distribution system representative of those found on larger naval vessels, possibly with electric propulsion. Three-phase power with a nominal system voltage of 4.16 kV is generated and distributed zonally via generator switchboards and longitudinal bus switchboards. Most loads are powered via the longitudinal bus switchboards. Some large loads, however, could be powered from the generator switchboards. Normally, the port and starboard longitudinal buses are operated independently; the number of online generator sets connected to each bus depends on the total load of the respective buses.

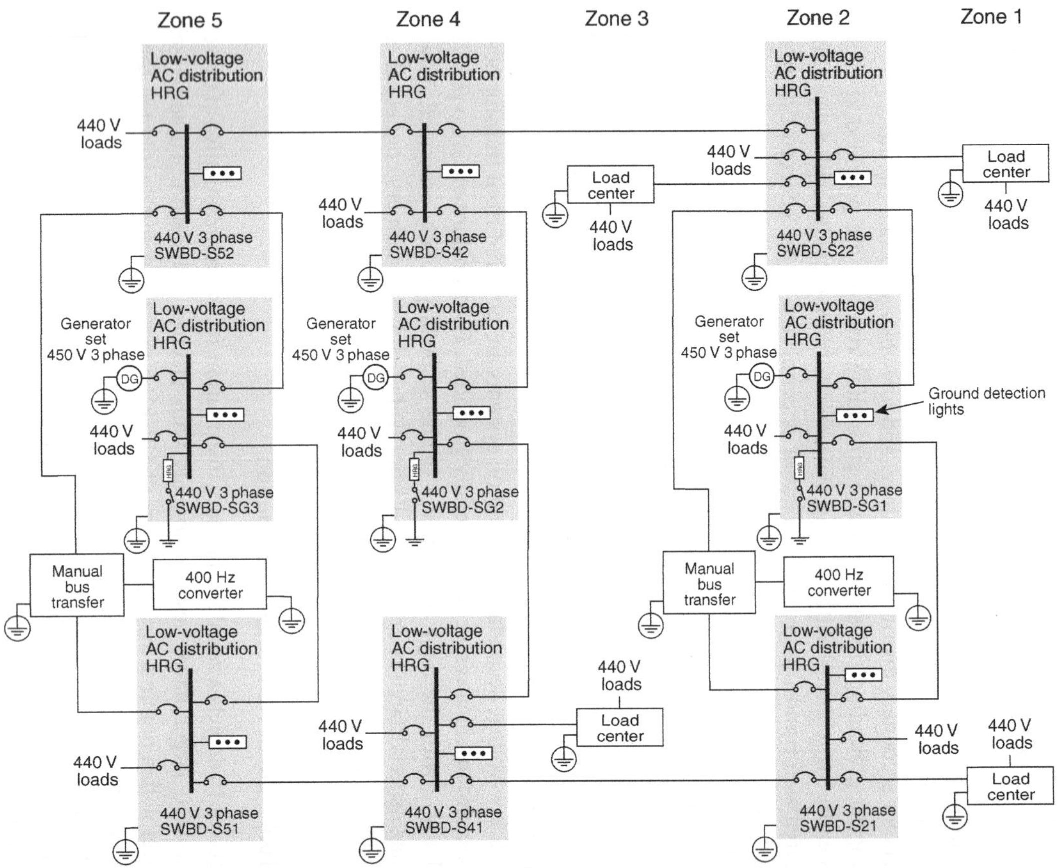

Figure 9.9 Low-voltage AC zonal distribution system.

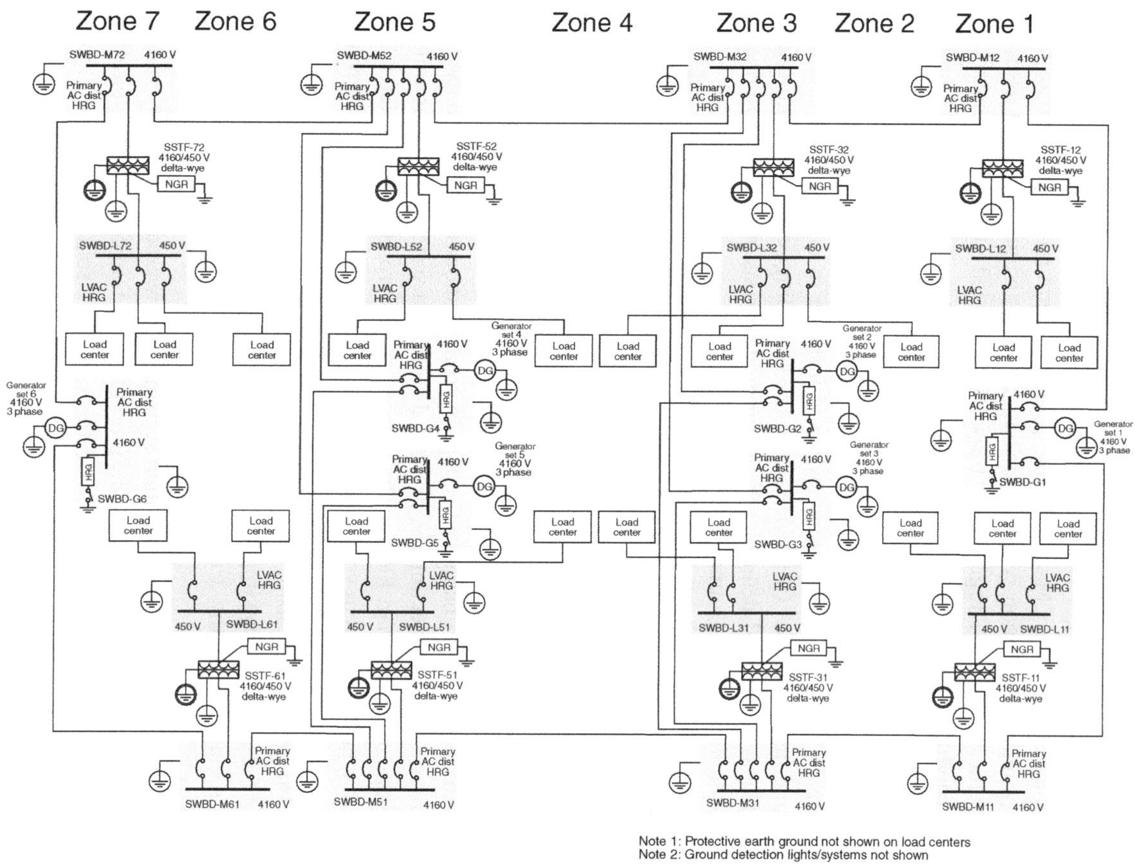

Figure 9.10 Zonal AC primary distribution system with HRG in generator switchboards.

Within each zone, multiple AC low-voltage distribution systems are powered via transformers from the longitudinal bus switchboards. The 440 V loads connect to load centers that, in turn, are connected to transformer switchboards. Secondary AC distribution and low-voltage special circuits are powered via transformers from the load centers.

For the primary distribution system, the grounding resistor may be installed in either the generator switchboards or in the longitudinal bus switchboards. If installed in the generator switchboards, the grounding resistor should have a resistance such that the HRG ground current is roughly equal to half the magnitude of the ground current of the CM capacitive impedance to ground due to the loads on the AC primary distribution system, distribution cable, and generator sets associated with the entire ship. The grounding resistor is connected either to the generator neutral conductor (if the generator stator is a wye winding configuration) or to a grounding transformer. The grounding transformer would be configured as a wye broken delta to enable low-voltage instrumentation on the secondary. If the ground fault current would be excessive when multiple grounding resistors are on the same bus, then a grounding switch is installed to enable disconnecting the grounding transformer (if provided) and the grounding resistor from ground; the control system would determine which grounding resistor switches are open or closed to manage the ground fault current.

If the grounding resistor is installed in the longitudinal bus switchboards, as depicted in Figure 9.11, the grounding resistor should have a resistance such that the HRG ground current is roughly equal to the magnitude of the ground current of the CM capacitive impedance to ground due to the primary distribution system loads, distribution cable, and generators associated with the switchboard. For bus ties, half of the CM capacitive impedance would be assigned to each of the two switchboards the bus tie connects. The grounding transformer would be configured as a wye-broken delta to enable low-voltage instrumentation on the secondary.

The transformers connecting the primary distribution system (with a nominal system voltage greater than 1 kV) to the in-zone AC low-voltage distribution system (with a nominal system voltage no greater than 1 kV) should be a delta-wye, with the grounding resistor connected to the transformer secondary neutral conductor. The grounding resistor should have a resistance roughly equal to the magnitude of the CM capacitive impedance to ground due to the loads and distribution cable associated with the transformer's AC low-voltage distribution system switchboard.

AC low-voltage secondary distribution systems are powered from transformers connected to a load center and are solidly grounded. The transformers are typically delta-wye, with the neutral conductor of the secondary solidly grounded. Low-voltage special circuits are powered from delta-delta transformers connected to a load center and are ungrounded.

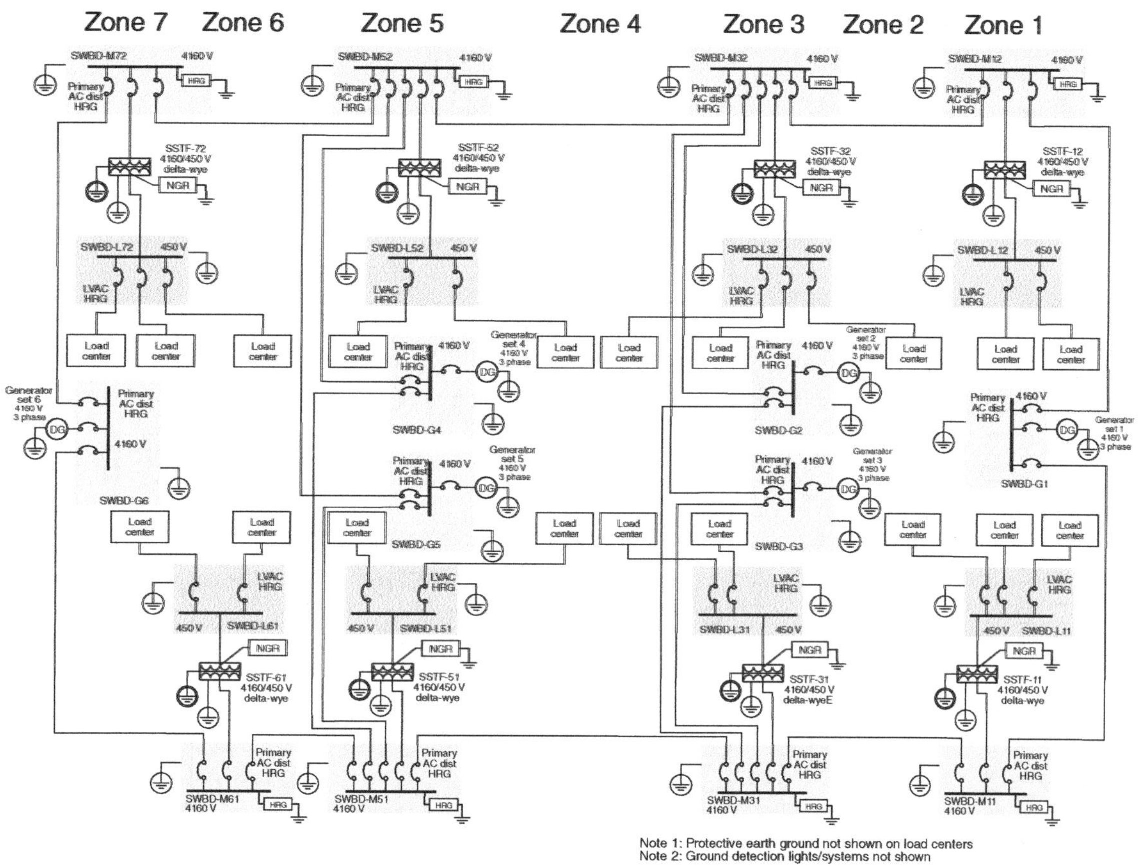

Figure 9.11 Zonal AC primary distribution system with HRG in zonal switchboards.

9.10.5 Commercial Ship DC Distribution

Figure 9.12 depicts a shipboard power system employing a DC distribution system that is becoming common on smaller vessels, such as platform supply vessels. Contemporary systems are limited to a nominal system voltage of 1 kV; they rely upon fuses for circuit protection and fuses are only available for DC operation at or below 1 kV. At low voltages, integrating the power electronics into the switchboard, as depicted in Figure 9.12, is common.

Advanced circuit protection technologies are under development, and systems employing voltages above 1 kV are anticipated in the near future. At higher voltages and power ratings, the power conversion equipment may be housed within their own enclosures, as depicted in Figure 9.13.

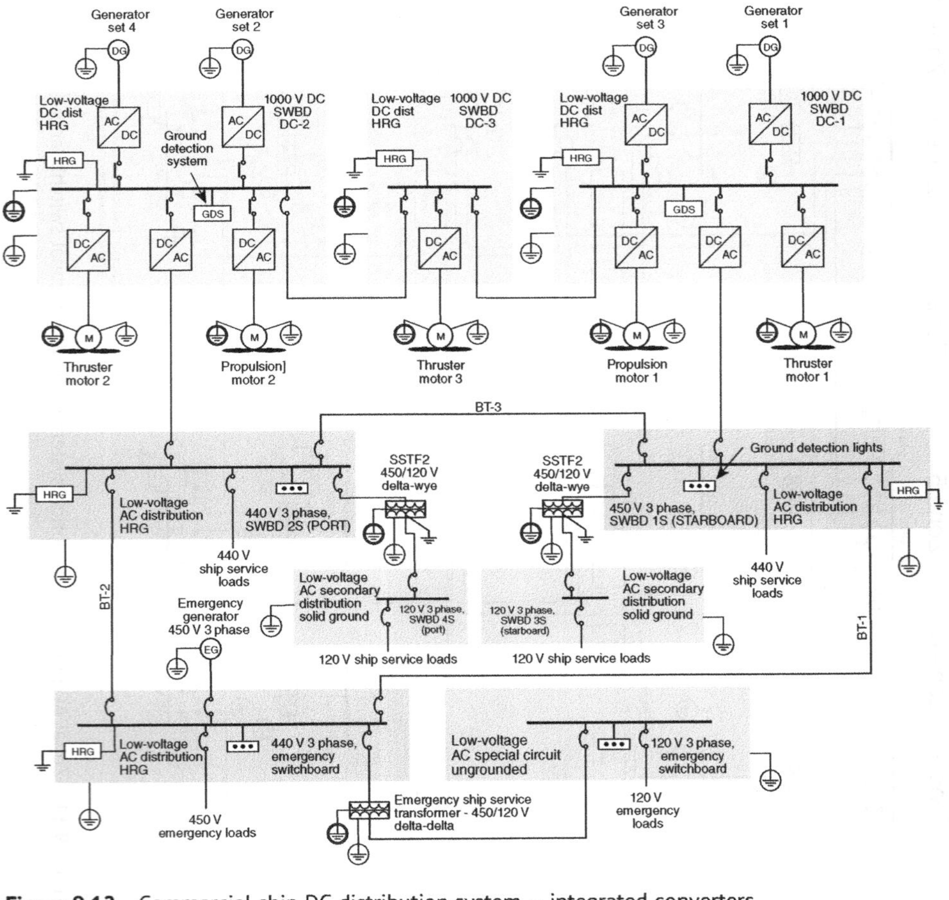

Figure 9.12 Commercial ship DC distribution system – integrated converters.

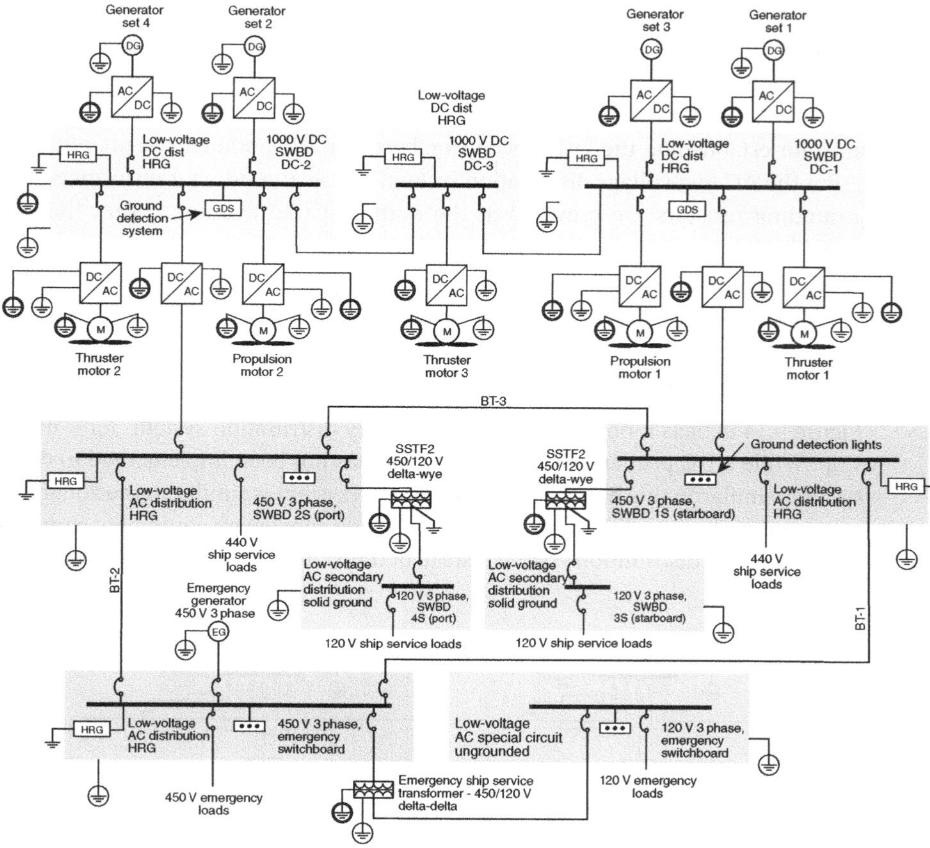

Figure 9.13 Commercial ship DC distribution system – separate converters.

DC distribution enables diesel generator sets to operate at the optimal speed/frequency to minimize fuel consumption for a given power output. DC distribution also facilitates the integration of energy storage; energy storage can further reduce fuel consumption under certain operational conditions.

Inverters are used to convert the DC power to AC for supplying the AC low-voltage distribution systems. These inverters should be isolated power conversion equipment; ground faults on the DC bus should not impact the system neutral on the AC bus, and vice versa. As with the other systems, the secondary AC distribution and low-voltage special circuits are generally powered via transformers from the AC low-voltage distribution system.

Each of the generator switchboards is high-resistance grounded. As discussed in Chapter 7, two different methods may be employed to implement the high-resistance ground: two line-to-ground resistors or a split power supply with the grounding resistor connected to the neutral conductor. Chapter 7 provides guidance for establishing the resistance value based on the nominal system voltage.

For the AC low-voltage distribution system, zigzag grounding transformers and grounding resistors are provided at the output of the inverters within the AC switchboard. As with the other systems described in this chapter, the secondary AC distribution systems are solidly grounded, and the low-voltage special circuits are ungrounded.

9.10.6 Zonal DC Primary Distribution

Figure 9.14 depicts a notional zonal DC primary distribution system for a naval ship. While a DC primary distribution system has not been implemented to date, systems similar to this have been proposed. This system is similar to the zonal AC primary distribution with the exception that power conversion equipment supplies the in-zone AC distribution system instead of transformers. These inverters should

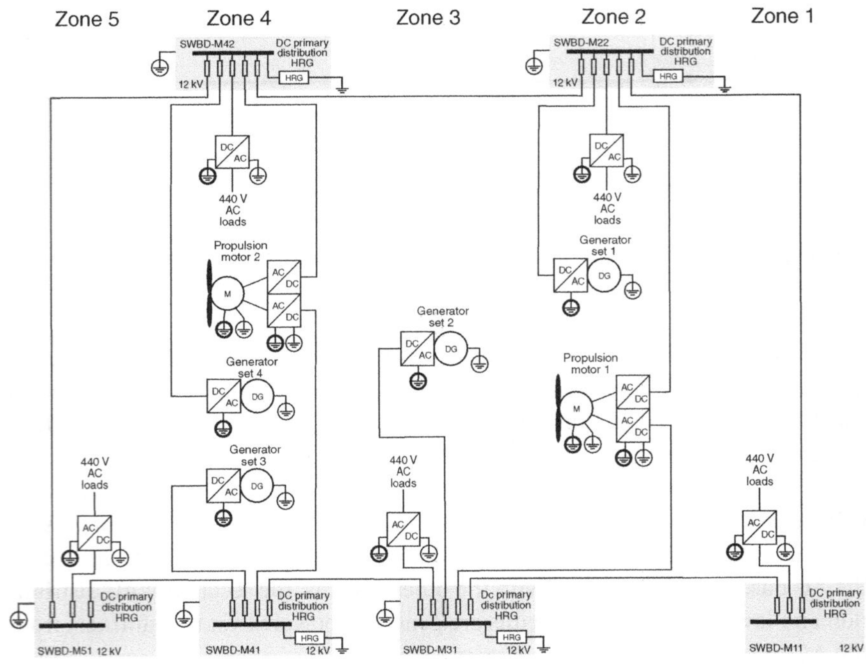

Figure 9.14 Zonal DC primary distribution system.

be isolated power conversion equipment; ground faults on the DC bus should not impact the system neutral on the AC bus, and vice versa.

As with the other systems, the AC secondary distribution system and low-voltage special circuits are generally powered via transformers from an AC low-voltage distribution system.

Each of the switchboards with a connected generator set is high-resistance grounded. As discussed in Chapter 7, two different methods may be employed to implement the high-resistance ground: two line-to-ground resistors or a split power supply with the grounding resistor connected to the neutral conductor. Chapter 7 provides guidance for establishing the resistance value based on the nominal system voltage.

For the AC low-voltage distribution system, zigzag grounding transformers and grounding resistors are provided at the output of the inverters. As with the other systems described in this chapter, the secondary AC distribution systems are solidly grounded, and the low-voltage special circuits are ungrounded.

References

IEEE Std 45.1-2023 (2023). *IEEE Recommended Practice for Electrical Installations on Shipboard – Design*. New York: Institute of Electrical and Electronics Engineers. Available from http://ieeexplore.ieee.org.

IEEE Std 3003.1-2019 (2019). *IEEE Recommended Practice for System Grounding of Industrial and Commercial Power Systems*. New York: Institute of Electrical and Electronics Engineers. Available from http://ieeexplore.ieee.org.

10

Power Conversion Equipment Grounding

10.1 Introduction

Power conversion equipment includes transformers, variable frequency drives (VFDs), and power electronics that convert power of one type (voltage, frequency, and number of phases) to another type. In general, these equipment have two or more power interfaces. Each power interface may be either unidirectional or bidirectional. At least one power interface can act as an input (load) while at least one input can act as an output (source). Two interfaces are isolated if a ground fault on one interface does not result in a change in the line-to-ground impedance on the other interface; otherwise, the two interfaces are non-isolated.

The integration of power conversion equipment with a power system includes managing the following characteristics:

- DM current harmonics and DM components of EMI conducted emissions.
- Ground fault currents.
- CM voltages and currents under normal and faulted conditions.

While the first item is important, its relationship to shipboard grounding is primarily through the coupling of DM currents and CM voltages/currents via circuit asymmetry. This chapter focuses on the second and third items.

10.2 Transformers

While transformers may provide isolation for the purpose of ground faults, they do not inherently provide isolation for CM voltages and currents. Primary winding to secondary winding capacitance can provide a CM path between the power systems

Design of Shipboard Power System Grounding/Earthing, First Edition. Norbert Doerry, Mohammed M. Islam, and John Prousalidis.
© 2025 The Institute of Electrical and Electronics Engineers, Inc.
Published 2025 by John Wiley & Sons, Inc.

connected to the primary and secondary windings. Transformers intended to provide CM isolation have a shield between the primary and secondary windings; this shield is connected to a CM ground.

As depicted in Figure 10.1, there are four configurations of three-phase transformers: delta-delta, delta-wye, wye-delta, and wye-wye. Three-phase transformers may be constructed from three individual single-phase transformers or all six windings may be wound on the same transformer core. See Doerry (2022) for additional details on the characteristics of three-phase transformers.

Multi-winding transformers may also be used to reduce DM current harmonics on the power system feed to the primary. One common configuration used to supply VFD rectifiers incorporates a delta primary (although a wye primary is also permissible) and delta and wye secondaries to provide two three-phase sets that

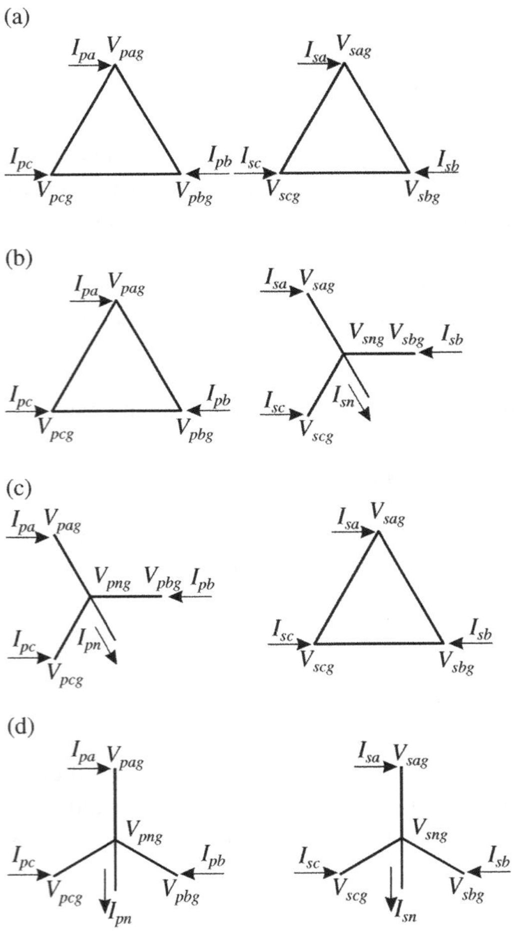

Figure 10.1 Three-phase transformer configurations: (a) delta-delta; (b) delta-wye; (c) wye-delta; (d) wye-wye.

are 30 electrical degrees apart. These two three-phase sets enable partial cancellation of DM current harmonics; thereby reducing the DM current harmonics through the primary. Additional multi-winding transformer configurations exist that enable even greater DM current harmonic cancellation. See IEEE Std C57.18.10-2021 (2021) for additional details.

If we neglect the capacitive coupling of the windings to the shield, a delta winding cannot conduct a CM current to the three-phase power system to which it connects; without a connection to ground, the sum of the currents entering the transformer must be zero. Similarly, a wye winding in a wye-wye transformer cannot conduct CM currents to the three-phase power systems to which it connects unless the neutral conductors of both the primary and secondary are grounded either solidly or via an impedance.

In a delta-wye or wye-delta transformer with the neutral conductor of the wye grounded either directly or via an impedance, a CM current may flow from the power system connected to the wye winding through the wye winding and then to ground. This CM current is linked to the delta winding in the form of a circulating current through the three delta windings; the CM current does not flow through the power system connected to the delta winding. For a delta-wye transformer, the circulating current in the delta winding will be equal to $\frac{1}{3n}$ times the CM current through the wye winding; n is the turns ratio from the primary to the secondary. The delta windings should be rated to carry this additional circulating current. If the neutral conductor of the wye winding is not grounded, no CM current will flow in the wye winding or the delta winding.

Wye-wye transformers have interesting CM characteristics. If the neutral conductors of the primary and secondary windings are not both grounded (either solidly or through an impedance), no CM current will flow. If both neutral conductors are grounded, then the CM currents of the primary and secondary will be related by the turns ratio; the transformer will not isolate the CM current.

For a wye-wye transformer, if the neutral conductor on the primary is ungrounded, but the neutral conductor on the secondary is solidly grounded or impedance grounded, no current will flow from the neutral conductor to ground, even during a ground fault. This transformer configuration has the following properties:

- The grounding impedance (or solid ground) on the secondary has no impact on the ground fault current.
- The system behaves as if it were ungrounded; the primary line currents, secondary line currents, and load currents are not impacted by the ground fault.
- The voltage of the neutral conductor of the primary is almost equal to the faulted phase of the primary voltage, resulting in very little voltage across the faulted phase transformer.

- Under a ground fault, the wye-wye transformer essentially behaves as an open-delta transformer; the faulted phase transformer is essentially replaced with a short on both the primary and secondary. If the transformer is implemented through the use of three single-phase transformers, the un-faulted phase transformers are at risk of being overloaded due to the faulted phase (in this case, Phase A) transformer not contributing any power. The voltage across each of the un-faulted phase transformers will increase from line-to-neutral to line-to-line. The current through the faulted phase winding will equal the negative sum of the current through the two un-faulted phases; a large fault current does not exist to quickly activate the fault protection system.

Capacitive coupling of the windings to the transformer shield provides a path for CM currents to flow in both wye and delta windings. Since the circulating current in a delta winding is of lower impedance than the winding-to-ground capacitance, coupling of CM currents between the power systems connected to the primary and secondary of a transformer with a delta winding is likely to be small. On the other hand, the capacitive coupling of the primary and secondary windings to the ground of a transformer with only wye windings provides a mechanism for linking the CM currents of the primary and secondary windings.

The use of wye-wye transformers is not recommended. Should one be employed, the neutral conductors on the primary and secondary should both be left ungrounded. A wye-wye transformer should not be used if the secondary neutral conductor is intended to connect to a solid ground or impedance ground; use a delta-wye instead.

Large multimegawatt transformers often use pre-magnetization circuitry on the transformer secondary when energizing the transformer to reduce the inrush current on the primary. These circuits typically do not have a significant impact on the grounding system.

10.3 Isolated Power Conversion Equipment

Isolated power conversion equipment almost always incorporates transformers. Motor-generator sets are also isolated power conversion equipment, but they are rarely employed in modern systems. VFDs often employ transformers, including multi-winding transformers, to reduce current waveform distortion and the propagation of CM currents. Other isolated power conversion equipment employ high-frequency transformers to reduce transformer size and weight. Figures 10.2 and 10.3 depict typical configurations of power conversion stages used in power conversion equipment. In some cases, energy storage is incorporated into the DC links of the conversion stages.

Figure 10.2 Isolated power conversion equipment architectures for AC loads: (a) line-frequency transformer; (b) high-frequency transformer; (c) DC bus supplied.

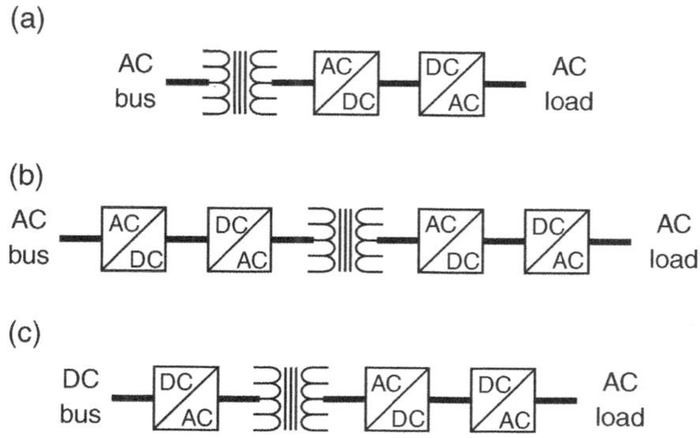

Figure 10.3 Isolated power conversion equipment architectures for DC loads: (a) line-frequency transformer; (b) high-frequency transformer; (c) DC bus supplied.

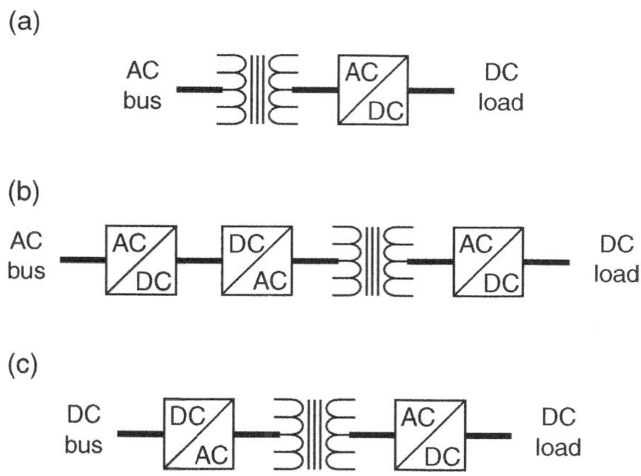

Every power electronic conversion stage depicted in Figures 10.2 and 10.3 is a potential CM voltage source. The transformers should be chosen to limit the linkage of CM currents between the primary and secondary windings. Ground faults on the primary side should not impact the secondary side; ground faults on the secondary side should not impact the primary side.

Power conversion stages on the bus side of the transformer should be analyzed or tested to characterize the CM voltage source they present to the system. Similarly, power conversion stages on the load side of the transformer should be analyzed or tested to characterize the CM voltage source they present to the loads.

10.4 Non-Isolated Power Conversion Equipment

Non-isolated power conversion equipment typically do not include transformers. Rectifiers as well as many VFDs are examples of non-isolated power conversion equipment. Figures 10.4 and 10.5 depict typical configurations of power

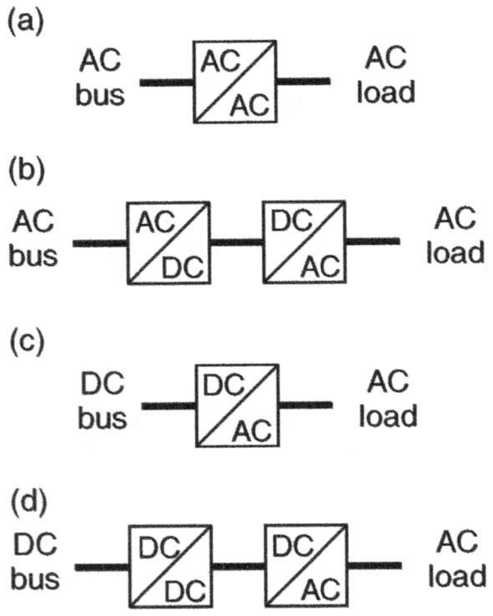

Figure 10.4 Non-isolated power conversion equipment architectures for AC loads: (a) direct conversion; (b) two-step conversion; (c) DC bus direct conversion; (d) DC bus two-step conversion.

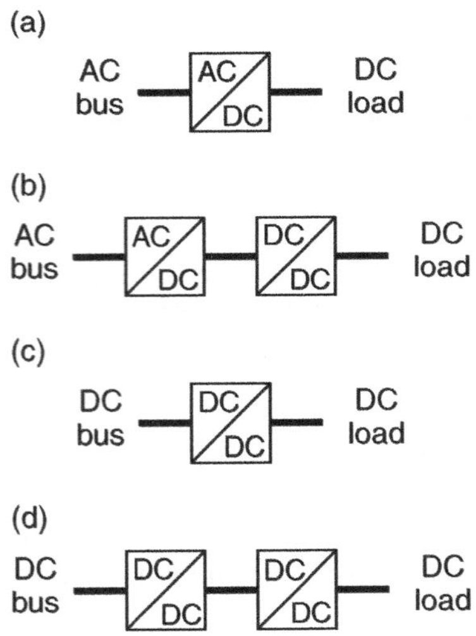

Figure 10.5 Non-isolated power conversion equipment architectures for DC loads: (a) direct conversion; (b) two-step conversion; (c) DC bus direct conversion; (d) DC bus two-step conversion.

conversion stages used in power conversion equipment. In some cases, energy storage is incorporated into the DC links of the conversion stages. For non-isolated power conversion equipment, the impact of a ground fault on one interface on another interface should be well understood and taken into account to ensure safety and to ensure equipment and system reliability. If the ratio of the peak voltages on any two interfaces exceeds roughly 2.5, consideration should be given to using isolated power conversion equipment; a ground fault on one interface could lead to excessive line-to-ground voltages on a different interface.

As was the case with the isolated power conversion equipment, every power electronic conversion stage depicted in Figures 10.4 and 10.5 is a potential CM voltage source. Power conversion should be analyzed or tested to characterize the CM voltage source they present to the system and the load.

10.5 CM Voltage and Current Control

The preferred method of controlling CM voltages and currents is to eliminate or minimize CM voltage sources. Without CM voltages, there can be no CM currents. Techniques to reduce CM voltages include

- Choosing power conversion equipment with topologies and switching algorithms that avoid circuit asymmetry.
- Choosing a power conversion equipment–load interface that enables eliminating circuit asymmetry. For example, a load interface with two three-phase circuits 60 electrical degrees apart enables operating the load supplying inverter in a manner that does not result in a CM voltage source. Note that the rectifier stage may still represent a CM voltage source.
- Choosing an isolation transformer–power conversion interface that enables eliminating circuit asymmetry. The CM voltage between the DC output of the rectifier and the combined phases of the AC input should be minimized.

If these techniques are not practicable, then CM mitigation efforts may be needed. The goal of CM mitigation is to control CM voltages and currents in a manner that reduces the risk of personnel injury, equipment degradation, and structural degradation. CM mitigation is accomplished by deliberately designing a CM circuit instead of allowing the CM circuit to be dictated by parasitic capacitances. The typical goals of the CM circuit are

- To reduce the CM voltage seen by the loads. This is particularly important for motor loads where high CM voltages can lead to bearing pitting and potentially insulation failure.
- To reduce the CM currents in the feeder cable from the power bus to reduce the path CM currents take through the ship's structure. CM currents should be localized as much as possible to the geographical vicinity of the CM voltage sources that create them.

The design components used to construct the CM circuit include

- CM chokes that present a high impedance to CM currents but a low impedance to DM currents. CM chokes are used inline on either or both the bus side and load side of the power conversion equipment to limit the magnitude of the CM current.

- CM shunts that present a low impedance to CM currents but a high impedance to DM currents. CM shunts are used line-to-ground (CM ground) on either or both the bus side and load side of the power conversion equipment to reduce the CM voltage at the bus and load interfaces.
- VFD cable, to provide a path for CM currents to flow instead of through the ship's hull.
- Design implementation details for motor loads, to include shaft brushes, insulated bearings, and insulated shaft coupling. The type of grease used in the bearings can also influence the occurrence of pitting in bearing surfaces.

Line reactors used to manage DM current harmonics also play a role in controlling CM currents; their impedance, however, may not be sufficient by itself to adequately limit CM currents.

Chapter 13 provides details on using these components within a CM circuit design.

10.6 VFD Cable

10.6.1 VFD Cable Description and Use

As depicted in Figure 10.6, VFD cable includes extra conductors, commonly called drain wires, to provide a lower impedance path for CM currents than the ship's hull; normally, three drain wires are provided and are located between the main phase conductors. These drain wires are in addition to one or more overall cable shields, as shown in Figure 10.7. VFD cable intended for applications above 1 kV nominal system voltage usually have additional conductor shields. The normal practice for the grounding of the drain wires and shields is shown in Figure 10.8. To avoid electric field stresses on the conductor insulation, cables for applications above 1 kV normally employ termination kits, as detailed in Section 5.9.

The purpose of the conductor shield (along with semiconducting insulation layers), when provided, is to ensure a uniform electric field surrounds the conductor insulation. This shield, however, has a parasitic capacitance between it and the conductor; hence, an AC current will flow through this shield from the conductor to ground. Under balanced operation, the three conductor shield currents will cancel; no current will actually flow in the ship's hull. If a ground fault occurs, then the conductor shield currents will not cancel and a hull current will flow between the ground fault connection to the ship's hull and the conductor shield's connection to the ship's hull.

Figure 10.6 VFD cable showing main conductors (without conductor shields) and smaller drain wires. *Source:* Courtesy of Marmon Industrial Energy & Infrastructure.

Figure 10.7 VFD cable showing cable shields, main conductors, and drain wires for applications with a nominal system voltage below 1 kV. *Source:* Courtesy of Marmon Industrial Energy & Infrastructure.

If a ground fault occurs between a conductor shield and its conductor on one cable, then the ground fault current will flow from where the grounding resistor is grounded and from where conductor shields of other cables are grounded, through the hull, to the location where the faulted cable conductor shields are grounded. To minimize the distance that the ground fault current flows through the ship's hull for this case, the grounding points for all the shields should be close together and close to where the grounding resistor is located. For most architectures, this translates into grounding the conductor shields at the source (generator)

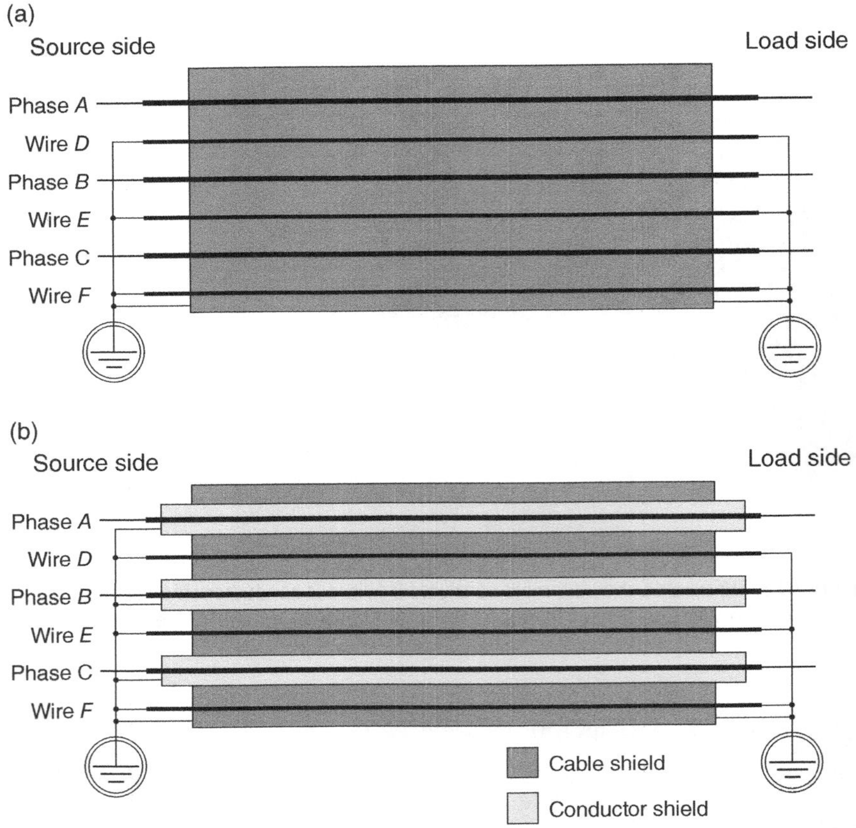

Figure 10.8 Grounding of VFD cable drain wires and shields: (a) without conductor shields; (b) with conductor shields.

end of the cable. For bus ties between generators, the choice of which end to connect the conductor shield to ground is arbitrary.

The purpose of the cable shield and drain wires is to divert some of the CM current from the hull; hence, the cable shield and drain wires should be grounded at both ends. While the resistances of the cable shield and drain wires are likely much greater than the ship's hull, their mutual inductance with the power conductors is also much greater than that between the conductors and the ship's hull. The net result is that the CM impedance of the current path through the cable shield and drain wires is likely to be about the same magnitude as the impedance of the ship's hull.

The connection boxes or compartments should have sufficient volume for routing and terminating the main power conductors, as well as the drain wires, conductor shields, and cable shields. Care should be taken to ensure creepage and clearance distance requirements are met. With covers off or doors open, workers should have clear access to all screw terminals and cable terminations.

See IEEE 45.8-2016 (2016) and IEEE 1580-2021 (2021) for additional details on VFD cable.

10.6.2 VFD Cable Modeling (Without Conductor Shields)

Figure 10.9 depicts the geometry of a VFD cable without conductor shields. Figure 10.10 depicts a detailed and a simplified model of the cable. The detailed model includes the three drain wires and the cable shield as separate items. The detailed model does ignore the mutual inductances and capacitances among the drain wires and shield; these conductors are normally grounded on both ends, and thus their electrical interactions are small. This detailed model is appropriate if one is modeling a ground fault in a cable between its two ends. The simplified model assumes the drain wires are paralleled and can be combined into a single equivalent combined conductor.

The parameters for the detailed model are

R_1 = resistance of the primary current-carrying conductors
R_D = resistance of the drain wires
R_3 = resistance of the cable shield
L_1 = self-inductance of the primary current-carrying conductors
L_D = self-inductance of the drain wires
L_3 = self-inductance of the cable shield
M_{AD} = mutual inductance between conductor A and drain wire D
M_{AF} = mutual inductance between conductor A and drain wire F
M_{DE} = mutual inductance between drain wires

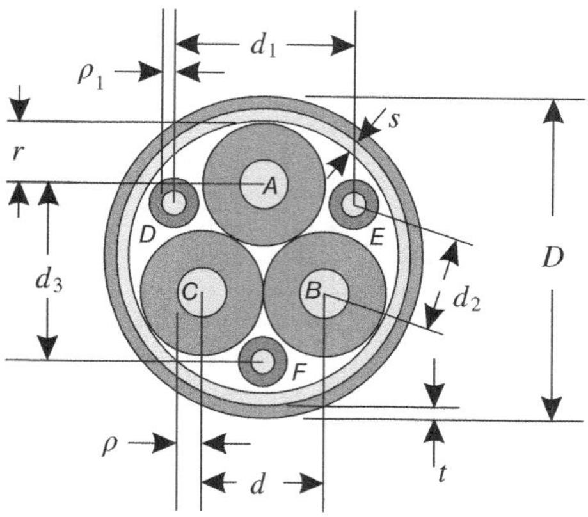

Figure 10.9 Geometry of VFD cable without conductor shields.

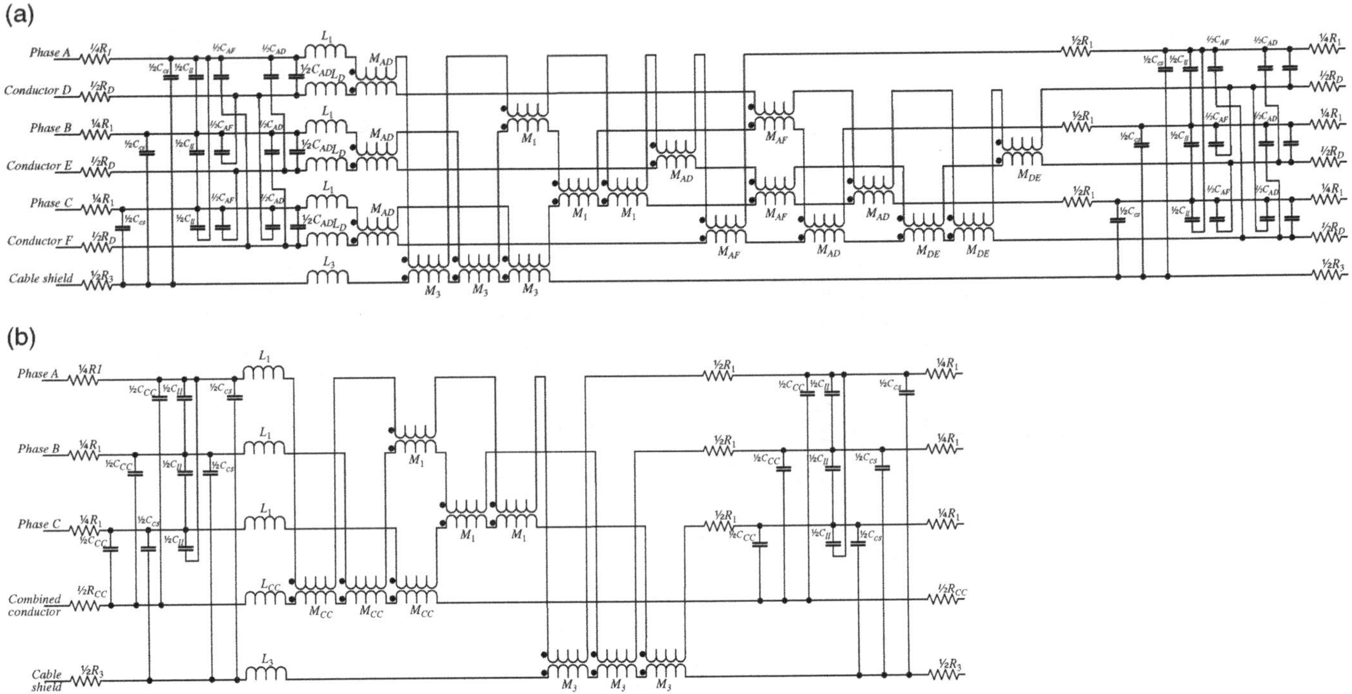

Figure 10.10 Cable transient model: (a) detailed; (b) simplified.

M_1 = mutual inductance between primary current-carrying conductors

M_3 = mutual inductance between a primary current-carrying conductor and the cable shield

C_{cs} = capacitance between a primary current-carrying conductor and the cable shield

C_{ll} = capacitance between primary current-carrying conductors

C_{AD} = capacitance between a primary current-carrying conductor A and drain wire D

C_{AF} = capacitance between a primary current-carrying conductor A and drain wire F.

The additional parameters for the simplified model are

R_{CC} = combined resistance of the drain wires and shield (R_D)

C_{CC} = combined capacitances of a primary current-carrying conductor to the drain wires (C_{AD} and C_{AF})

L_{CC} = combined self-inductance of the drain wires and shield (L_D and M_{DE})

M_{CC} = combined mutual inductance of the drain wires and shield with the primary current-carrying conductors (M_{AD} and M_{AF}).

See Chapter 5 for estimating R_1, R_3, L_1, L_3, M_1, and M_3.

R_D can be determined from cable data sheets, IEEE Std 1580-2021 (2021), MIL-HDBK-299 (1991), and ASTM B8-11 (2017).

For a cable of length l m, the self-inductance and mutual inductances (μH) including the drain wires are given by

$$L_D = 0.2l\left[\ln\left(\frac{2l}{\rho_1}\right) - \frac{3}{4}\right]$$

$$M_{AD} \approx 0.2l\left[\ln\left(\frac{l\sqrt{2}}{d_2}\right) - 1\right]$$

$$M_{AF} \approx 0.2l\left[\ln\left(\frac{l\sqrt{2}}{d_3}\right) - 1\right]$$

$$M_{DE} \approx 0.2l\left[\ln\left(\frac{l\sqrt{2}}{d_1}\right) - 1\right]$$

The capacitances (pF) can be estimated by

$$C_{cs} \approx \frac{2\pi\varepsilon_0\kappa l}{\ln\left(\frac{\sqrt{\left(\frac{D}{2} - s - t\right)r}}{\rho}\right)}$$

$$C_{ll} \approx \frac{\pi\varepsilon_0 \kappa l}{\operatorname{acosh}\left(\frac{d}{2\rho}\right)}$$

$$C_{AD} \approx \frac{2\pi\varepsilon_0 \kappa l}{\operatorname{acosh}\left(\frac{d_2^{\,2} - \rho_1^{\,2} - \rho^2}{2\rho_1\rho}\right)}$$

$$C_{AF} \approx \frac{2\pi\varepsilon_0 \kappa l}{\operatorname{acosh}\left(\frac{d_3^{\,2} - \rho_1^{\,2} - \rho^2}{2\rho_1\rho}\right)}$$

The simplified model parameters are given by

$$R_{CC} = \frac{R_D}{3}$$

$$C_{CC} = 2C_{AD} + C_{AF}$$

$$L_{CC} = \frac{L_D}{3} + \frac{2M_{DE}}{3}$$

$$M_{CC} = \frac{(2M_{AD} + M_{AF})}{3}$$

10.6.3 VFD Cable Modeling (with Conductor Shields)

Figure 10.11 depicts the geometry of a VFD cable with conductor shields. If the conductor shields are grounded at one end and the cable shield and drain wires are grounded at both ends, the following assumptions can be made.

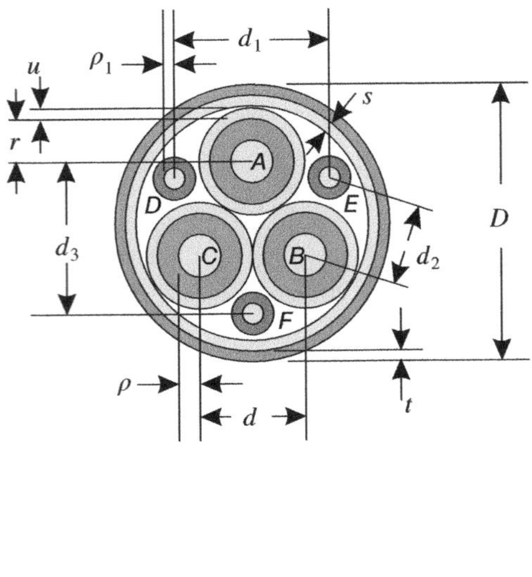

Figure 10.11 Geometry of VFD cable with conductor shields.

- The line-to-ground capacitance is dominated by the conductor-to-conductor shield capacitance.
- The mutual inductance between the conductor shield and the conductor can be ignored.
- The mutual inductances among the drain wires, conductor shields, and cable shield can be ignored.

With these assumptions, the simplified model of Figure 10.12 may be used. The parameters in addition to those defined in Section 10.6.2 can be estimated by

$$R_{CS} = \frac{R_2}{3}$$

$$L_{CS} = \frac{L_2}{3}$$

$$C_{CS} = C_2$$

where R_2, L_2, and C_2 are defined in Chapter 5.

10.7 Examples

10.7.1 Directly Connected VFD and Motor – No More than 1 kV

Figure 10.13 depicts a motor on a skid powered via a VFD. If the VFD does not represent a large CM voltage source, and if the CM impedance is sufficiently large, then the circuit depicted in Figure 10.13 will likely be successful. The line reactors, while installed to reduce the DM harmonic distortion, also help to a lesser extent to increase the CM impedance. The shaft brush on the motor shunts safely to ground the CM current coupled to the motor rotor from the stator; the shaft brush thus protects the shaft bearings from CM current-induced pitting.

One of the challenges with the circuit of 10.13 is that CM currents are largely determined by parasitic capacitances; one typically will not be able to predict if the proposed installation will be successful until the equipment is installed in the ship and commissioning tests are completed. If a problem at this stage of vessel construction is identified, the cost to remedy it may be large to avoid late delivery of the ship. If the CM voltage properties of the VFD are known ahead of time, either through conversations with the vendor, or through laboratory testing, then one can specify the requirements for one or two CM chokes to keep the CM current below a given value. If the VFD does not have a significant parasitic capacitance to ground, then a CM choke, as depicted in Figure 10.14 may be sufficient; the CM impedance of the choke is set to limit the CM current assuming the only

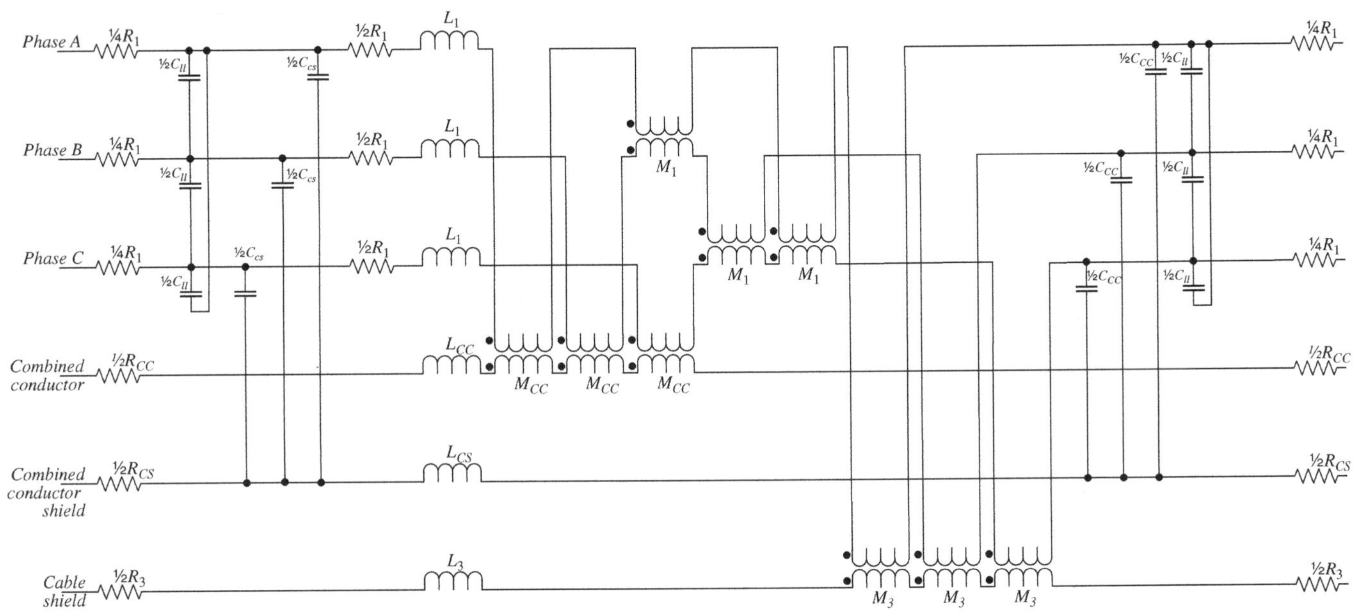

Figure 10.12 Simplified model of VFD cable with conductor shields.

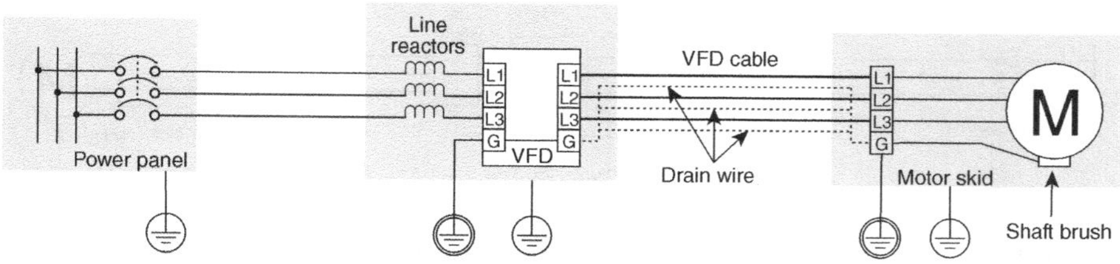

Figure 10.13 Low-voltage VFD and motor.

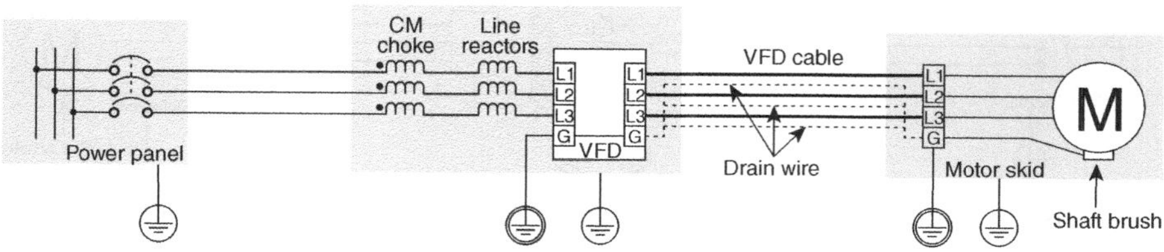

Figure 10.14 Low-voltage VFD and motor with supply-side CM choke.

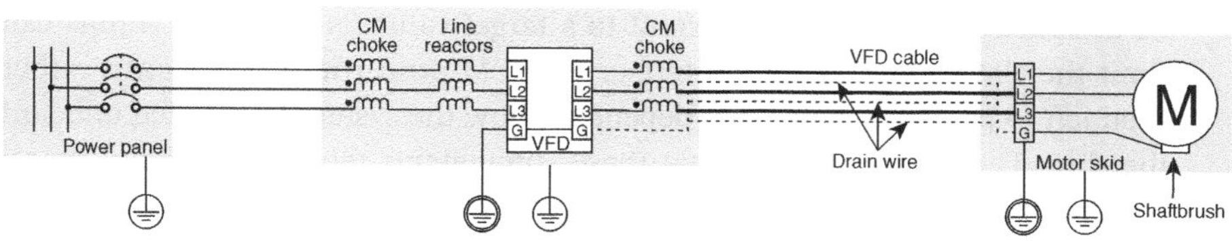

Figure 10.15 Low-voltage VFD with supply-side and load-side CM chokes.

CM impedances to the CM voltage are the CM impedance of the choke and the line reactors.

In some cases, the VFD may internally have a low enough CM impedance to ground to effectively create two CM paths, one on the source side and one on the load side. The VFD can be modeled as two CM voltage sources in series with a parasitic capacitance between the two sources to ground. As depicted in Figure 10.15, a CM choke would be required for each CM voltage source (associated with the source side and the load side) to control the CM currents.

CM shunts (see Chapter 13) can also be used in addition to the CM choke(s) to limit the CM voltage at both the input and output interfaces of the VFD. CM shunts have a high impedance for DM currents, but a low impedance for CM currents. If the CM shunt on the output interface is sufficiently effective, then the cable to the motor need not be a VFD cable. Figure 10.16 depicts a

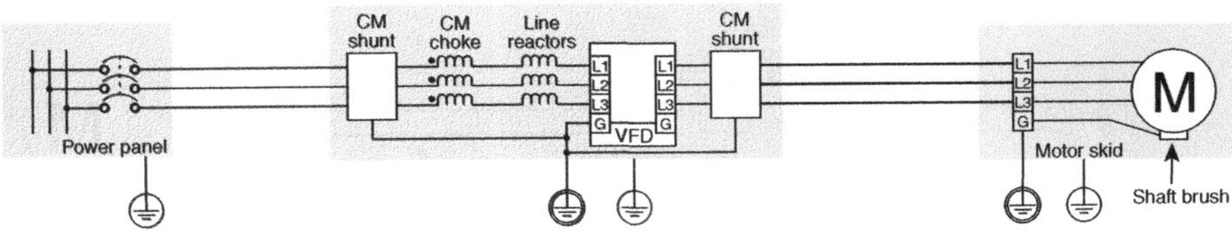

Figure 10.16 Low-voltage VFD and motor with supply-side and load-side CM shunts and supply-side CM choke.

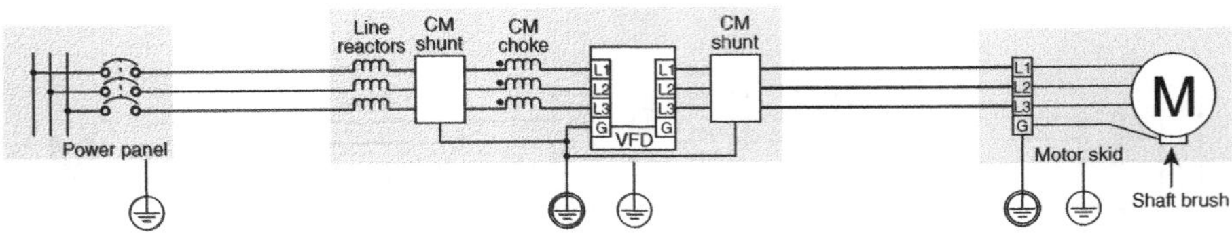

Figure 10.17 Low-voltage VFD and motor using line reactors to limit CM currents from the power system.

low-voltage VFD with supply-side and load-side CM shunts in addition to a supply-side CM choke. This configuration has the advantage that the CM choke can be designed to limit the CM current to a target value, and the CM shunts can limit the CM voltage at each interface; the CM properties are designed rather then left to parasitic values that depend on how the system is installed onboard the ship. The disadvantage is that these components require additional space, weight, and cost.

If the VFD has an internal isolation transformer, then Figure 10.16 may be modified by inserting a second CM choke between the VFD and load-side CM shunt. Each side of the isolation transformer has its own CM circuit; each CM circuit should be managed via CM shunts and CM chokes.

If one is concerned about the supply-side shunt having to handle both CM currents from its VFD as well as external CM currents from the power system, the configuration in Figure 10.17 may be beneficial. Although designed to reduce DM current harmonics, the line reactors may provide sufficient CM impedance to limit the impact of power system CM currents. If the line reactors are not sufficient, then an additional CM choke may be needed on the supply side of the line reactors.

10.7.2 Transformer-Supplied VFD and Motor – No Greater than 1 kV

Isolating the VFD and motor from the power supply via a power transformer has a number of benefits. First, the voltage of the drive and motor need not be the same as the voltage of the power distribution system. Second, the use of a transformer with a delta winding prevents CM currents from magnetically coupling between

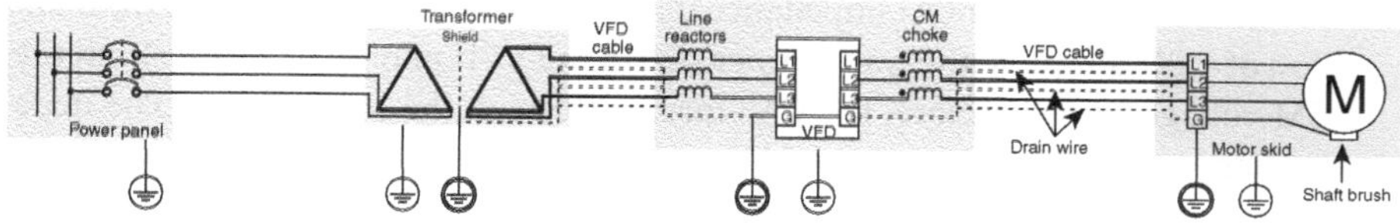

Figure 10.18 Transformer-isolated low-voltage VFD and motor.

the primary and secondary sides. A shield with a CM ground between the primary and secondary prevents electric field coupling (capacitive coupling) between the primary and secondary windings. And finally, the windings of a transformer are less susceptible to CM voltages than the motor. In many cases, the circuit depicted in Figure 10.18 will be sufficient to manage the CM currents. In fact, if the line reactors provide sufficient CM impedance, the CM choke may not be required. Because the grounded transformer shield is parasitically coupled capacitively to the transformer secondary windings, the cable between the transformer and the VFD should be a VFD cable to reduce the amount of CM current flowing through the hull.

If a CM shunt is used, it may be possible to remove the requirement for VFD cable for the connection to the motor; the shunt should reduce the CM voltage to a low enough level at the motor to obviate the need for the VFD cable. Figure 10.19 shows the CM shunt on the load supply of the VFD; the transformer side still requires VFD cable to provide a path for the CM current from the transformer shield.

If the distances between the transformer, VFD, and motor are not large, the system on the secondary of the transformer may be ungrounded from a system grounding perspective. In some cases, where the risk of operating an ungrounded system is considered unacceptable, a delta-wye transformer with an NGR may be desired, as depicted in Figure 10.20. The delta-wye transformer with NGR

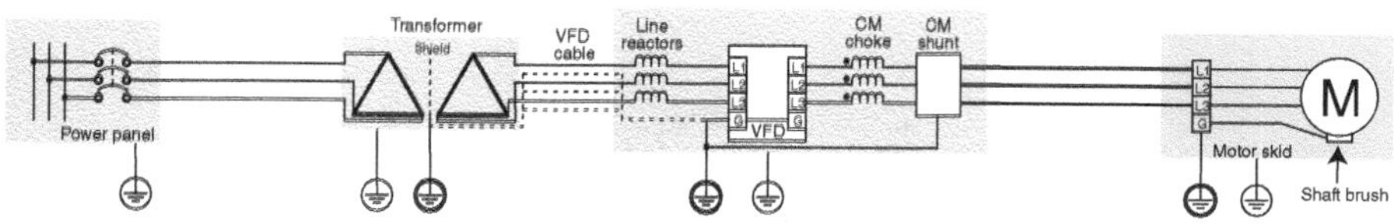

Figure 10.19 Transformer-isolated low-voltage VFD and motor with CM shunt.

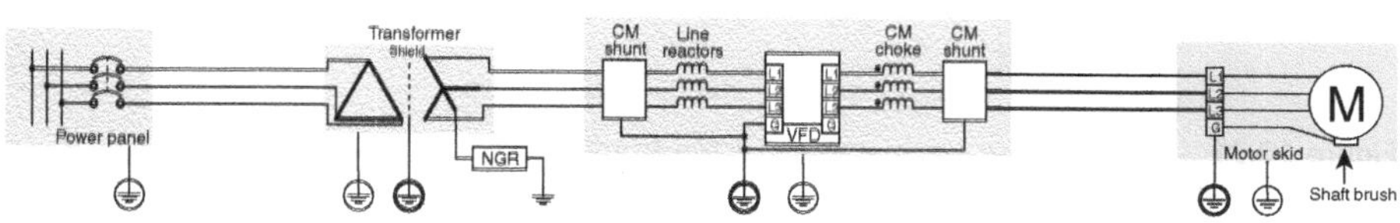

Figure 10.20 Transformer-isolated low-voltage VFD and motor with NGR.

provides a much lower CM impedance than the delta-delta transformer; the addition of a CM shunt on the transformer side of the VFD may be needed to limit the CM voltage and, therefore, the CM current through the transformer. The delta primary and the transformer shield between the primary and secondary windings prevent the CM currents on the secondary from coupling with the power from the power panel. A VFD cable is not required between the power panel and transformer primary.

10.7.3 Propulsion Motor Drives – Greater than 1 kV

Propulsion motors are typically one of the largest, if not the largest, loads onboard a ship. Consequently, the motor drives for the propulsion motors typically use a propulsion transformer with multiple sets of secondary windings; these multiple windings reduce DM current harmonics through the primary windings. If the propulsion transformer has two sets of secondary windings with a 30° offset between them, such as the transformer depicted in Figure 10.21, the resulting DC waveform on the output of the rectifier stage will have 12 pulses per fundamental frequency period. Hence, these rectifier stages are called 12-pulse rectifiers. The 12-pulse rectifiers are constructed from two three-phase rectifiers. In many cases the rectifiers are connected in parallel through the use of balancing reactors on their outputs, as depicted in Figure 10.21; the balancing reactors ensure the diodes in each three-phase, six-pulse, rectifier commutate properly and do not interfere with the commutation of rectifiers in the other three-phase rectifier.

In Figure 10.21, the system ground is often established by connecting an NGR to the neutral conductor of the wye secondary winding. Because the rectifier stage and the inverter stage of the propulsion motor drive are likely CM voltage sources, there will likely be CM voltages on the DC link at the output of the rectifier stage, as well as the AC output of the inverter. The parasitic capacitance in the motor, as well as the parasitic capacitance in the transformer secondary windings to the

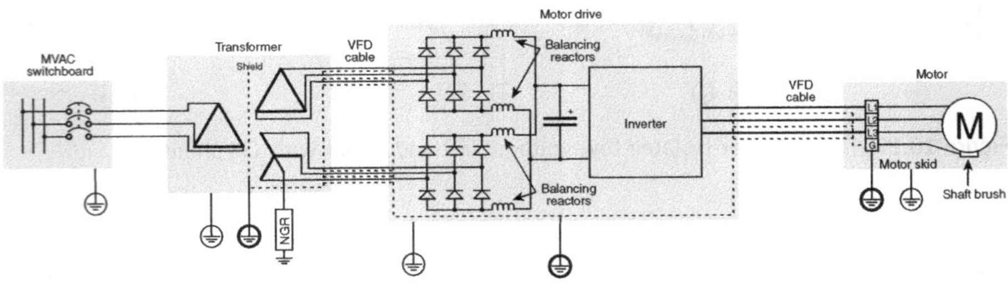

Figure 10.21 Twelve-pulse rectifier constructed from two six-pulse rectifiers in parallel (greater than 1 kV).

grounded transformer shield, provide a path for CM currents. As such, the cables between the propulsion transformer and propulsion motor drive, and the cable between the propulsion motor drive and the propulsion motor should be VFD cables where the drain wires are grounded on both ends. The drain wires reduce the amount of CM current flowing through the hull. The conductor and cable shields of all the cables are not shown in Figure 10.21; see Section 10.6.1 for the termination of the conductor and cable shields.

For redundancy, many propulsion motors incorporate two independent stator windings that are each powered by independent propulsion motor drives, as depicted in Figure 10.22. Each independent stator winding is normally part of a separately derived system; the two drive circuits are each grounded through their own NGR. The conductor and cable shields of all the cables are not shown in Figure 10.22; see Section 10.6.1 for the termination of the conductor and cable shields.

While the previous examples used two three-phase, six-pulse rectifiers in parallel, it is also possible to connect the two six-pulse rectifiers in series and insert the NGR on the connection between the two rectifiers, as depicted in Figure 10.23. Connecting the rectifiers in series does require the conductor-to-ground insulation of the propulsion transformer secondary windings to be greater than normal since a CM voltage exists between the conductors connected to each propulsion transformer and ground. The conductor and cable shields of all the cables are not shown in Figure 10.23; see Section 10.6.1 for the termination of the conductor and cable shields.

Figure 10.24 depicts an alternate grounding method; an HRG circuit is connected to the DC bus at the output of the rectifier. Figure 10.25 depicts an example of an HRG circuit. The line-to-ground capacitances and the capacitor-to-ground resistor represent a CM shunt. The ground resistor has a low resistance based on the need to limit CM current at a particular frequency. The resistors in parallel with the capacitors have a very high resistance, perhaps as high as 1 MΩ; their purpose is to balance the voltages across the capacitors. An RC time constant on the order of tens of seconds is likely acceptable. The conductor and cable shields of all the cables are not shown in Figure 10.24; see Section 10.6.1 for the termination of the conductor and cable shields.

In some cases, propulsion transformers with three or four sets of secondary windings may be encountered. These extra sets of windings enable 18-pulse or 24-pulse rectifiers that further reduce DM current harmonics. As with the 12-pulse rectifiers, the three-phase rectifiers may either be connected in parallel or in series. If connected in parallel with balancing reactors, the high-resistance ground may be established with either an NGR on the neutral conductor of a wye secondary or through an HRG grounding circuit on the DC bus. These same options are available for an 18-pulse rectifier as long as the rectifier for the wye winding is placed in the middle. For a 24-pulse rectifier, one can insert the grounding resistor in the middle as was done in Figure 10.23 or an HRG grounding circuit on the DC bus may be employed.

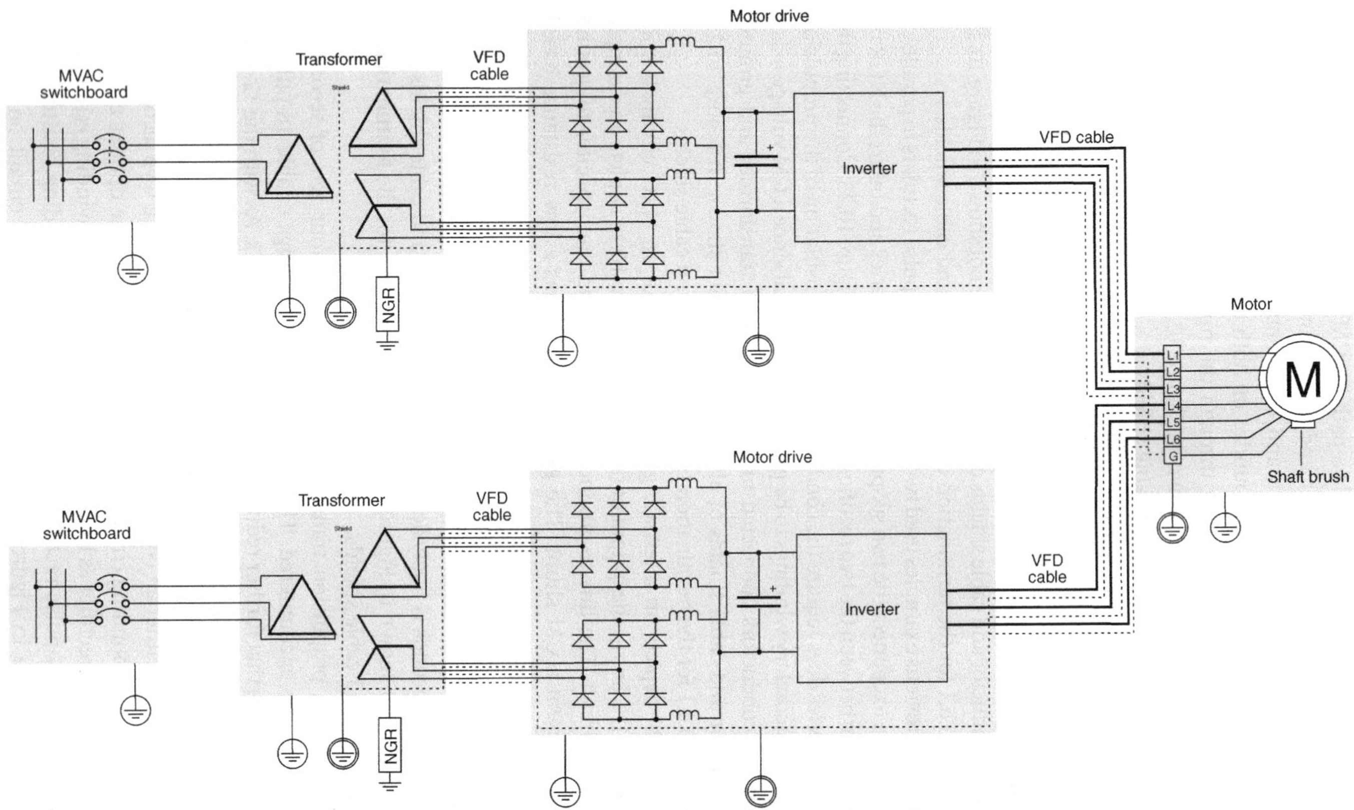

Figure 10.22 Twelve-pulse rectifiers for double stator winding motor in parallel (greater than 1 kV).

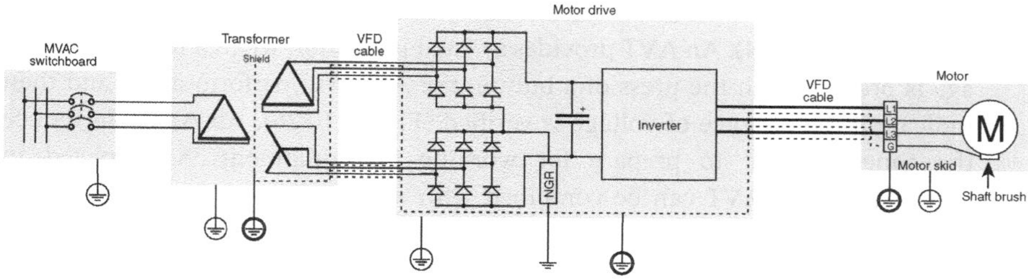

Figure 10.23 Twelve-pulse rectifier constructed from two six-pulse rectifiers in series (greater than 1 kV).

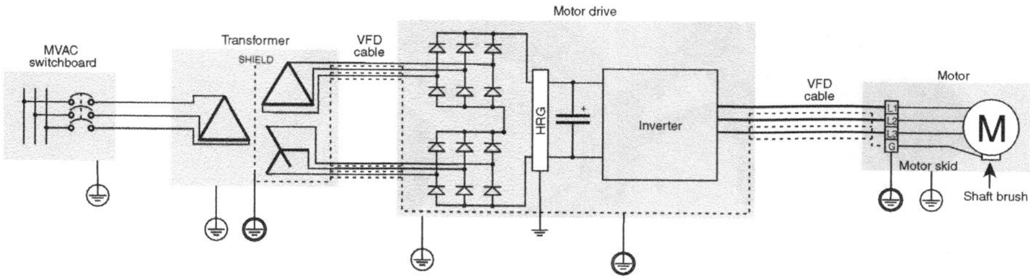

Figure 10.24 Twelve-pulse rectifier with HRG circuit in parallel (greater than 1 kV).

Figure 10.25 Example HRG circuit.

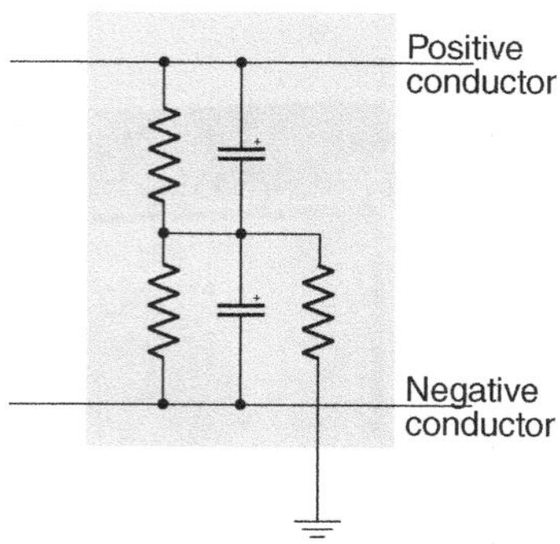

10.8 Maintenance Considerations

When equipment is developed or specified, consideration should be given to ensuring the electrical equipment can be easily verified to be safe to work on. One way is to provide easy access to power conductor and circuit test points to verify with a portable voltmeter that there is an absence of voltage between the conductors and between the conductors and ground. Another way is to have an absence of voltage tester (AVT) permanently installed in the hardware, as detailed in NFPA 70E (2024). An AVT provides a visual indication when a hazardous voltage is present; with the press of a button, the AVT will perform a test and then indicate if the absence of voltage is verified (Figure 10.26). An AVT can reduce the time required to prepare for working on equipment. As depicted in Figure 10.27, an AVT can be combined with a voltage test indicator to enable maintenance personnel to additionally perform voltage measurements without opening equipment cabinet doors.

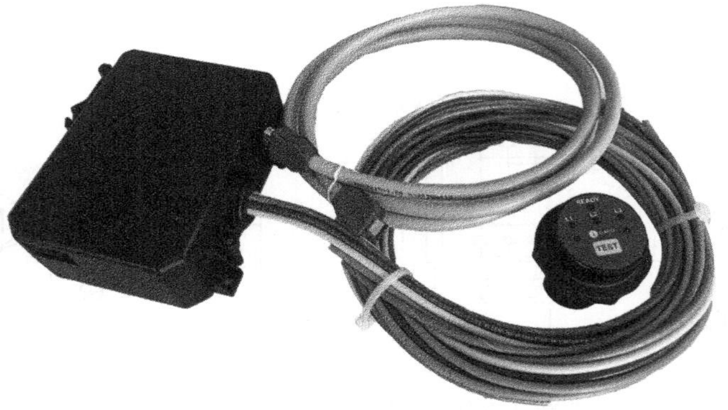

Figure 10.26 AVT control unit, cable, and display. *Source:* Courtesy of I-Gard.

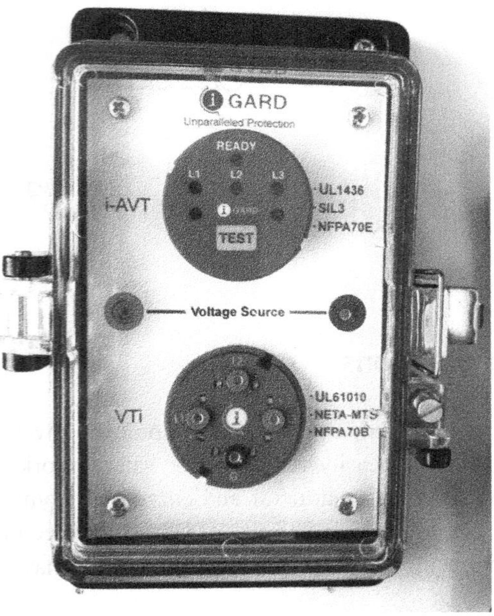

Figure 10.27 Combination AVT and voltage test indicator. *Source:* Courtesy of I-Gard.

References

ASTM B8-11 (2017). *Standard Specification for Concentric-Lay-Stranded Copper Conductors, Hard, Medium-Hard, or Soft*. Philadelphia, PA: ASTM International. Available from http://webstore.ansi.org.

Doerry, Norbert (2022). Shipboard Three-Phase Power Transformer Analysis, Burke, VA. 10 December 2022. Available from http://doerry.org/norbert/papers/papers.htm.

IEEE Std C57.18.10-2021 (2021). *IEEE Standard Practices and Requirements for Semiconductor Power Rectifier Transformers*. New York: Institute of Electrical and Electronics Engineers. Available from http://ieeexplore.ieee.org.

IEEE Std 45.8-2016 (2016). *IEEE Recommended Practice for Electrical Installations on Shipboard – Cable Systems*. New York: Institute of Electrical and Electronics Engineers. Available from http://ieeexplore.ieee.org.

IEEE Std 1580-2021 (2021). *IEEE Recommended Practice for Marine Cable for Use on Shipboard and Fixed or Floating Facilities*. New York: Institute of Electrical and Electronics Engineers. Available from http://ieeexplore.ieee.org.

MIL-HDBK-299 (1991). *Cable Comparison Handbook, Data Pertaining to Electric Shipboard Cable*. Washington, DC: Department of Defense. Available from http://quicksearch.dla.mil.

NFPA 70E (2024). *Standard for Electrical Safety in the Workplace*. Quincy, MA: National Fire Protection Association. Available from www.nfpa.org.

11

Shore Power (Cold Ironing) Connection Grounding

11.1 Introduction

While the US Navy and other navies of the world have long used shore power while in port to enable shutting down the shipboard electrical power generator sets, it has only been in the last decade that this practice has become common in commercial ships. Outside the United States, the term "cold ironing" is often employed rather than "shore power"; the two terms have identical meanings and may be used interchangeably. The trend toward greater use of cold ironing in the commercial sector is largely driven by the need to reduce emissions in some ports. A second possible advantage is that the cost of electricity from the shore power connection may be less than the cost of the electricity if it were generated onboard.

As depicted in Figure 11.1, many ships use one or more shore power cables to connect a shore power station on the pier and the shore power terminals onboard the ship. The cables are positioned manually, or through the use of a general-purpose crane. As shown in Figure 11.2, this can be a very labor-intensive process. More recently, some ports have installed cable handling systems to reduce the amount of labor needed to connect and disconnect shore power (Figure 11.3).

In the past few years, electric ferries using battery energy storage have been placed in service. These ferries typically charge using DC power with a nominal system voltage of about 1 kV; the shore power connection may be rated up to 5 MW. For short-distance ferries, charging time can be as short as six minutes; the shore power connection is made robotically. The grounding interface is proprietary.

Smaller vessels may charge using electric vehicle charging interfaces such as CHAdeMO and the Combined Charging System (CCS). These charging interfaces

Design of Shipboard Power System Grounding/Earthing, First Edition. Norbert Doerry, Mohammed M. Islam, and John Prousalidis.
© 2025 The Institute of Electrical and Electronics Engineers, Inc.
Published 2025 by John Wiley & Sons, Inc.

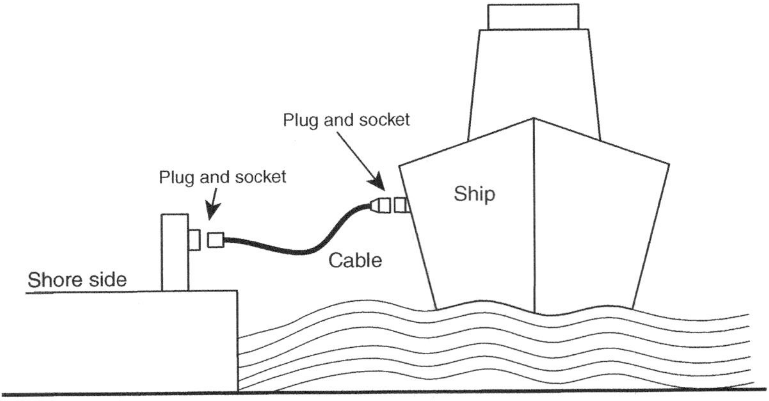

Figure 11.1 Shore-to-ship connection using cables.

Figure 11.2 Sailors manually disconnecting shore power cables. *Source:* MC2 Michael Sanchez/United States Navy/Public domain.

can currently support charging rates up to 350 kW. Interfaces under development are anticipated to achieve 1 MW or more. Some vessels use multiple electric vehicle interfaces to achieve a sufficiently high charging rate. In general, the electric vehicle charging interfaces incorporate a ground connection, a ground integrity connection that prevents charging should the ground connection be broken, and ground fault circuit interrupters (GFCIs).

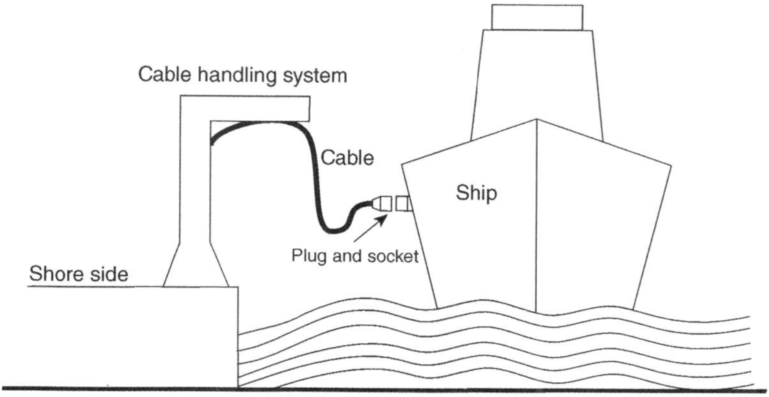

Figure 11.3 Shore-to-ship connection using a cable handling system.

11.2 Low-Voltage Shore Connections

While an international standard does not yet exist for low-voltage shore connection (LVSC) systems, IEC and IEEE are currently developing the IEC/IEEE DIS 80005-3; a preliminary version of the standard, IEC PAS 80005-3:2014 (2014), was published, and a revision is currently under development. Figures 11.4 and 11.5 depict a shore power connection, as described in the current draft revision. Figure 11.4 does not depict the circuits for the pilot contacts. Characteristics of the low-voltage shore power connection include

- The LVSC is intended for providing up to 1 MVA using three-phase AC power with a nominal system voltage of 400, 440, or 690 V AC.
- The frequency provided by the LVSC system and the ship's nominal frequency must match; a frequency converter may be required.
- The shoreside supply is normally via a delta-wye transformer with an NGR; a grounding switch is provided to disconnect the NGR when necessary.
- Where an isolated power electronic converter is employed (such as when a change in frequency is required), a grounding transformer with NGR and an associated grounding switch may be employed.
- If the ship's power system employs NGRs, these NGRs should be disconnected when receiving power from shore; this will happen automatically if the ship's NGRs are implemented by connecting to the generator neutrals.
- If the ship's power system is normally ungrounded, preference should be given to operating with the shore supply NGR. Both the person in charge of the shore

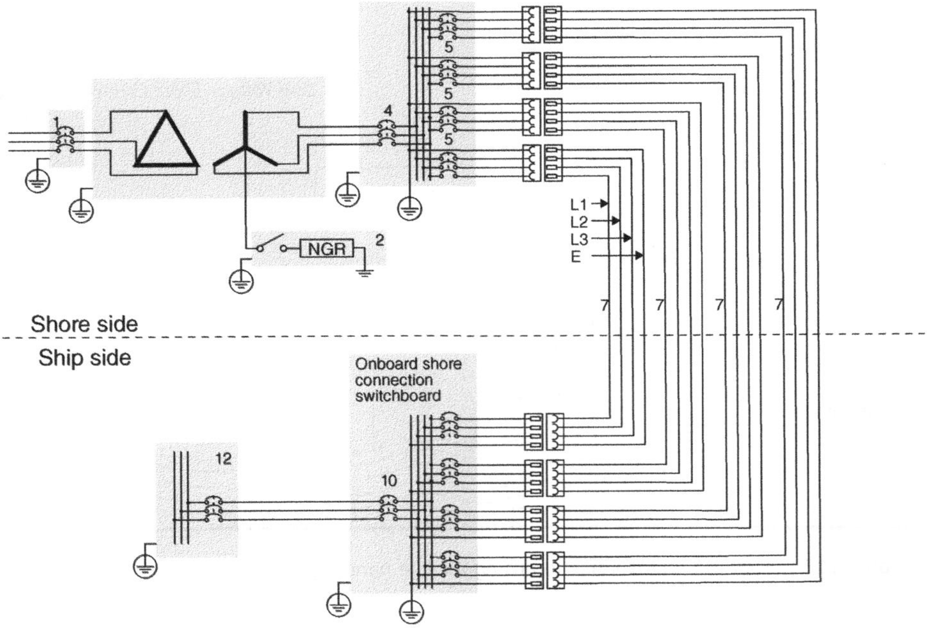

1. Shore supply system circuit breaker
2. Neutral grounding resistor and switch
4. Shore-side circuit breaker
5. Shore-side feeder circuit breaker
7. Shore power cable/cable management system
10. Onboard shore power circuit breaker
12. Onboard shore power receiving circuit breaker from the ship service switchboard

Figure 11.4 IEC/IEEE DIS 80005-3 example LVSC ship-to-shore connection.

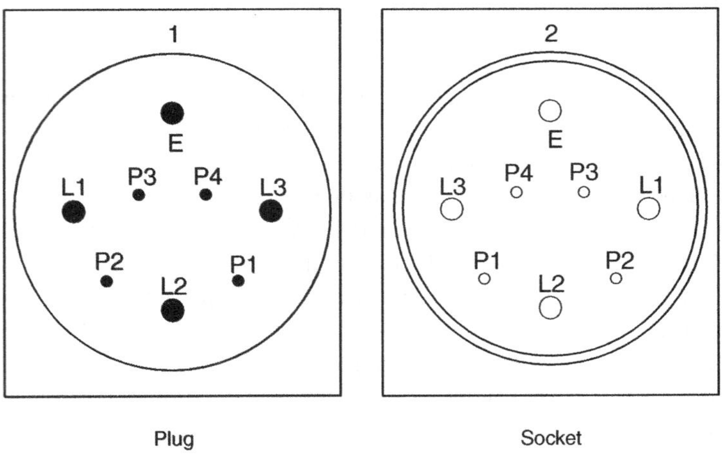

1	Plug/ship inlet face	2	Socket outlet/ship connector face
E	Earth	P1	Pilot 1
L1	Phase 1- A - R	P2	Pilot 2
L2	Phase 2- B - S	P3	Pilot 3
L3	Phase 3- C - T	P4	Pilot 4

Figure 11.5 IEC/IEEE DIS 80005-3 example LVSC ship-to-shore connection plug and socket details.

facility and the person in charge of the ship must agree to operate with the shore supply NGR disconnected. If agreement is not possible, then the shore supply NGR should be left connected and the ship should provide an isolation transformer between the shore connection and the ship power system.

- The minimum rating of the NGR is 16 A for five seconds and 5 A continuous.
- A shoreside circuit breaker supplies feeder circuit breakers for up to five paralleled cables; if one of the feeder circuit breakers trips, the shoreside circuit breaker also trips.
- As shown in Figure 11.5, the cable connection includes the three power contacts, an earth (ground) contact, and four pilot contacts to implement control. The earth contact is a protective earth contact to ensure the hull of the ship and the ground electrode on shoreside are at the same potential.
- The connector is designed so that the order of contact during a connection is: earth contact, power contact, and then pilot contacts. The order during disconnection is: pilot contacts, power contacts, and earth contact.
- The shoreside circuit breaker cannot be closed if a ground fault is detected on the ship distribution system. When the shoreside circuit breaker is closed, it will automatically trip upon detection of a ground fault.
- The pilot contacts ensure the cables are properly connected and that the control systems on both ship and shore do not detect any faults that would not allow for safe operation.

For the US Navy, most ships are provided with 450–480 V AC three-phase ungrounded power from a substation transformer located on the pier. Multiple shore power cables are connected in parallel to provide the amount of current required by the ship; some ships may require 20 or more shore power cables. As detailed in UFC-4-150-02 (2020), each shore power cable connection on the shore station is provided with a circuit breaker with a maximum overcurrent setting of 480 A; the design ampacity for each cable connection is 400 A. The circuit breakers have a shunt trip such that if one circuit breaker trips, they all trip.

Since the shore power connection is ungrounded, the grounding system onboard the ship, if one exists, determines how the system is grounded. Figure 11.6 depicts a typical naval shore power switchboard. Like the shore station, each shore power cable has a dedicated circuit breaker in the shipboard shore power switchboard; if one circuit breaker trips, they all trip. Dedicated circuit breakers are used because the number of shore power cables used will vary depending on the projected maximum load while the ship is in port. Figure 11.7 depicts a sailor connecting a shore power cable onboard a naval ship. The cable connector is detailed in MIL-DTL-24368/1 (2021). The shipboard receptacle is detailed in MIL-DTL-24368/2 (2021). These connectors provide contacts for three power conductors; they do not include contacts for a neutral, ground, or control.

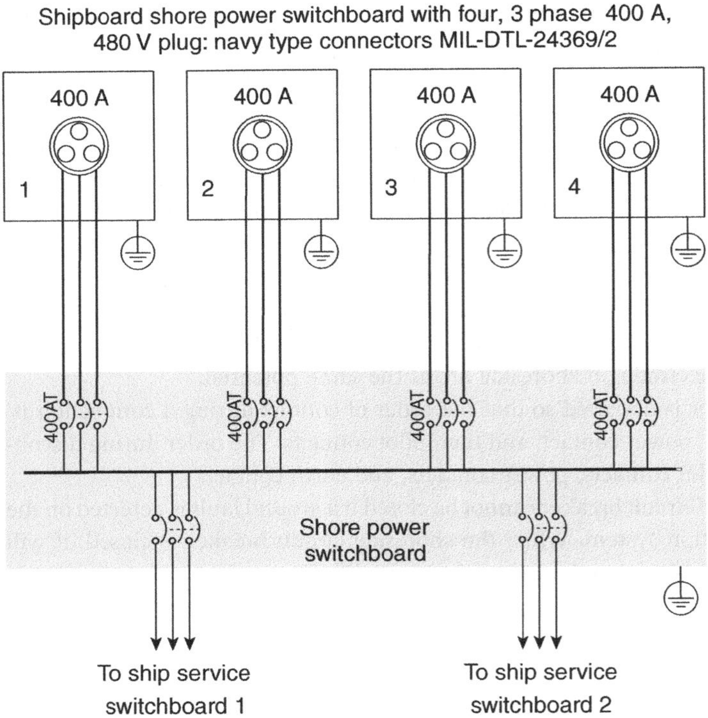

Figure 11.6 Shipboard shore power switchboard for naval applications.

Figure 11.7 Sailor connecting shore power to naval warship. *Source:* MC2 Michael Sanchez/United States Navy/Public domain.

11.3 High-Voltage Shore Connections

IEC/IEEE 80005-1 (2019) is an international standard for high-voltage shore connection (HVSC) systems. Characteristics of this connection include

- The HVSC is intended to provide shore power to ships with an AC primary power bus that requires 1 MVA or more power through the HVSC. The HVSC normally provides three-phase AC power with a nominal system voltage of 6.6 and/or 11 kV AC.
- Details of the HVSC differ for different types of ships; specific requirements are defined for roll-on/roll-off cargo and passenger ships, cruise ships, container ships, liquefied natural gas carriers (LNGC), and tankers. See Table 11.1 and Figures 11.8 through 11.13.
- The frequency provided by the HVSC system and the ship's nominal frequency must match; a frequency converter may be required.
- The shoreside supply is normally via a delta-wye transformer.
- Where an isolated power electronic converter is employed (such as when a change in frequency is required), the shoreside supply may be via a delta-delta transformer.
- If an NGR is required, it is normally connected to the neutral of the delta-wye transformer secondary or to a grounding transformer if an isolated power electronic converter is used.

Table 11.1 IEC/IEC 80005-1 (2019) ship-to-shore connection details by ship type.

	Allowable voltage					
Ship type	**6.6 kV**	**11 kV**	**Number of pilot contacts**	**Transformer NGR resistance (Ω)**	**Number of power cables**	**Control cable**
Roll-on/ roll-off ships	Note 1	Yes	7	335 for 11 kV 200 for 6.6 kV	1	Fiber optic
Cruise ships	Yes	Yes	1	540	4 + 1 neutral cable	24 pin connector
Container ships	Yes	No	3	200	2	Fiber optic
LNGC	Yes	No	7	None	3	Fiber optic
Tankers	Yes	No	3	Not specified	3	Yes

Note 1: May be used in regional waterborne transportation services.

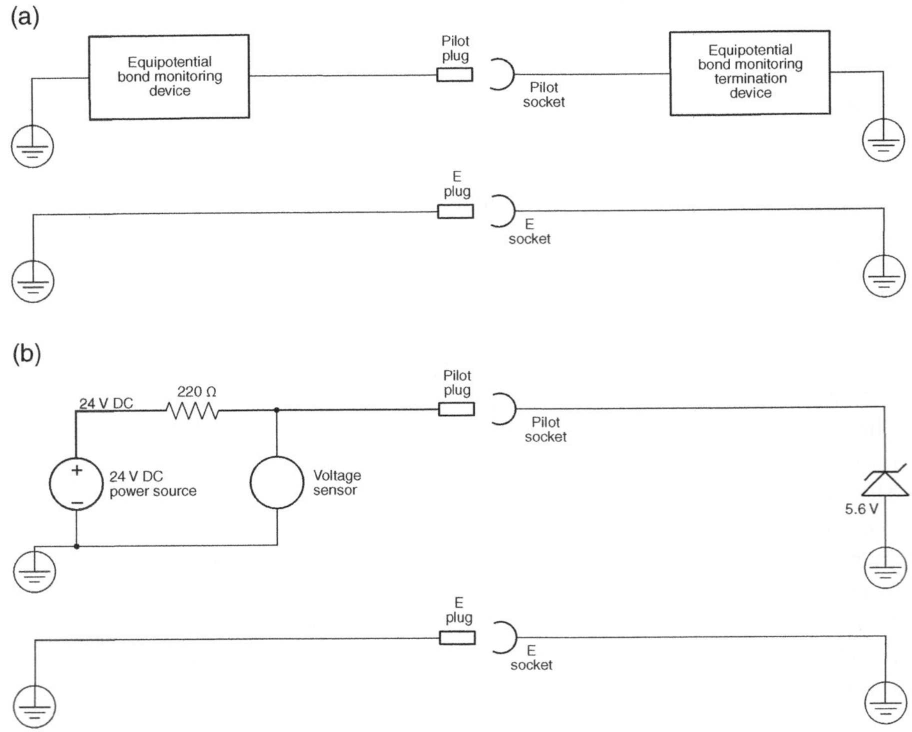

Figure 11.8 Equipotential bond monitoring between shore and ship using pilot contacts (a) general circuit; (b) example implementation.

- An NGR, if provided, should have a rating of not less than 1.25 times the anticipated system charging current; the minimum rating should be 25 A for five seconds.
- If a loss of continuity of the NGR (if provided) is detected, the shoreside circuit breaker should trip.
- The cable connection includes the three 500 A power contacts, an earth (protective earth) contact, and multiple pilot contacts to implement control. The number of pilot contacts depends on the ship type. The earth contact ensures the potential of the ship's hull and the ground conductor electrode at the HVSC switchboard are the same.
- The connector is designed so that the order of contact during a connection is: earth contact, power contact, and then pilot contacts. The order during disconnection is: pilot contacts, power contacts, and earth contact.

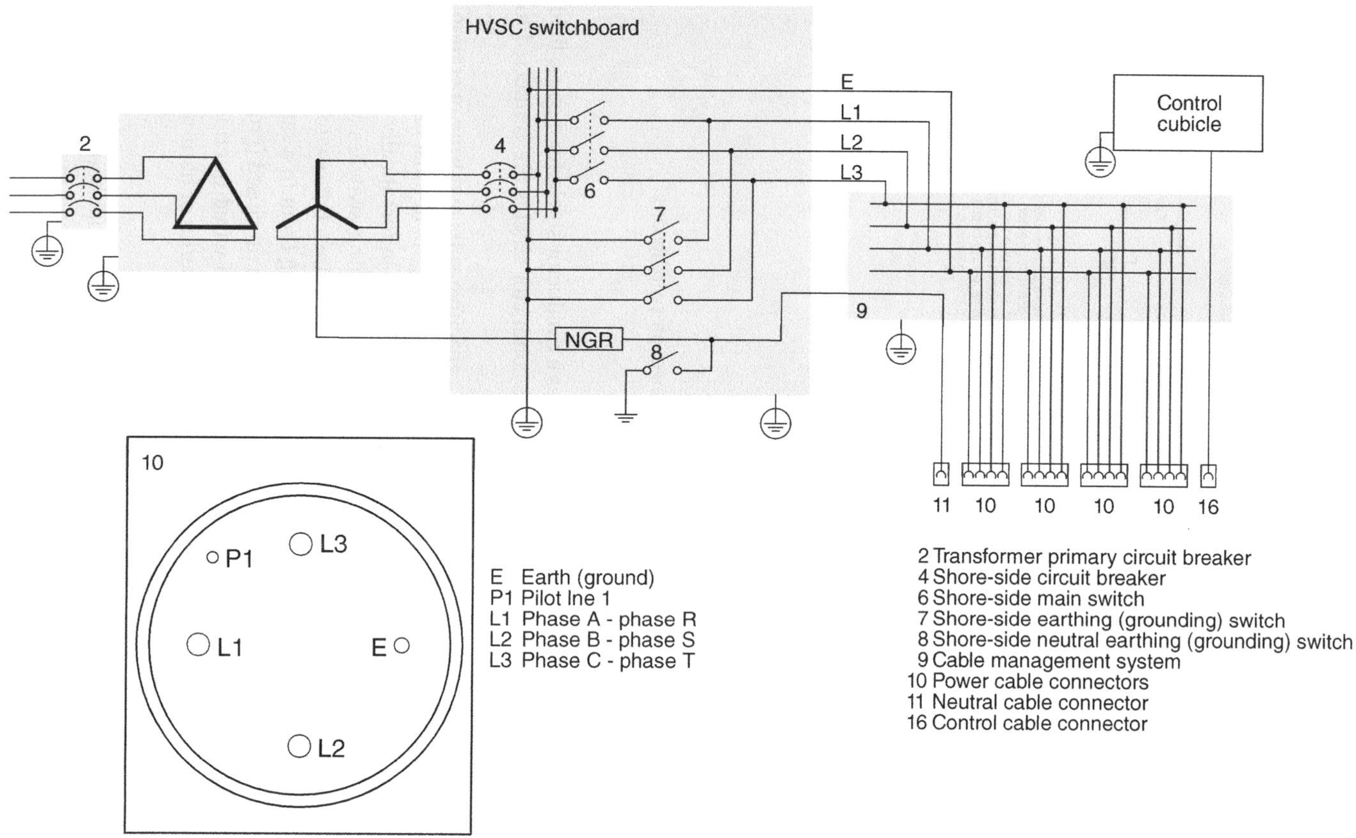

Figure 11.9 Example shore station with transformer for cruise ships.

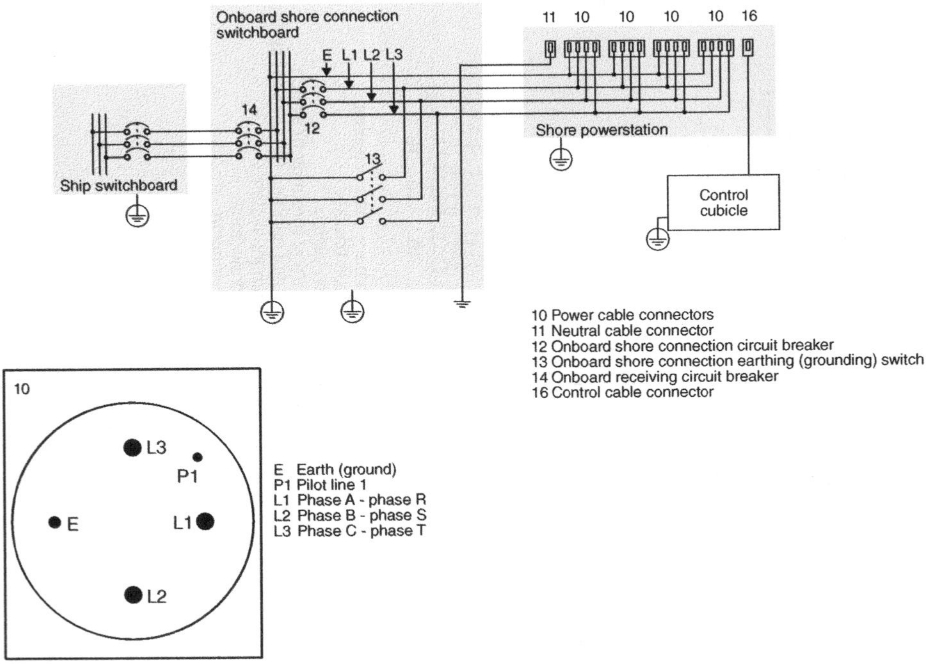

Figure 11.10 Example cruise ship onboard shore connection switchboard.

- Each power conductor within the cables has a conductor shield for electric field protection of the conductor insulation; the conductor shields are connected to ground on the shoreside.
- Since the cables can build up a dangerous capacitive voltage during operation, provisions to discharge the cables when disconnecting them must be provided. Discharging the cables can be accomplished either by automatically grounding them when the cable is de-energized through earthing switches within the HVSC switchboard and within the onboard shore connection switchboard, or provisions should be provided for manually grounding them in a safe manner. If earthing switches are provided and closed, they remain closed (grounding all of the cable power conductors) until a series of hard-wired interlocks enable opening them. Once open, they remain open until a series of hard-wired interlocks enable closing them.

One of the pilot contacts is typically used for equipotential bond monitoring; the circuit determines that the pilot contact is connected, and that the earthing

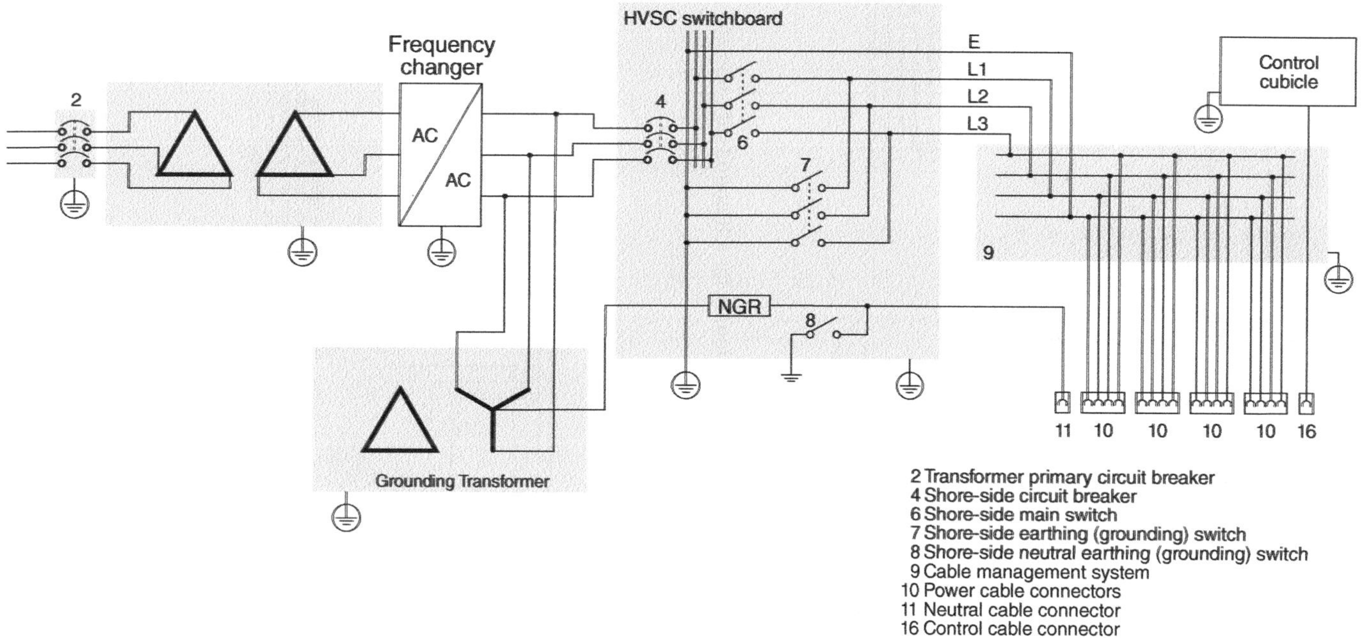

Figure 11.11 Example shore station with frequency changer for cruise ships.

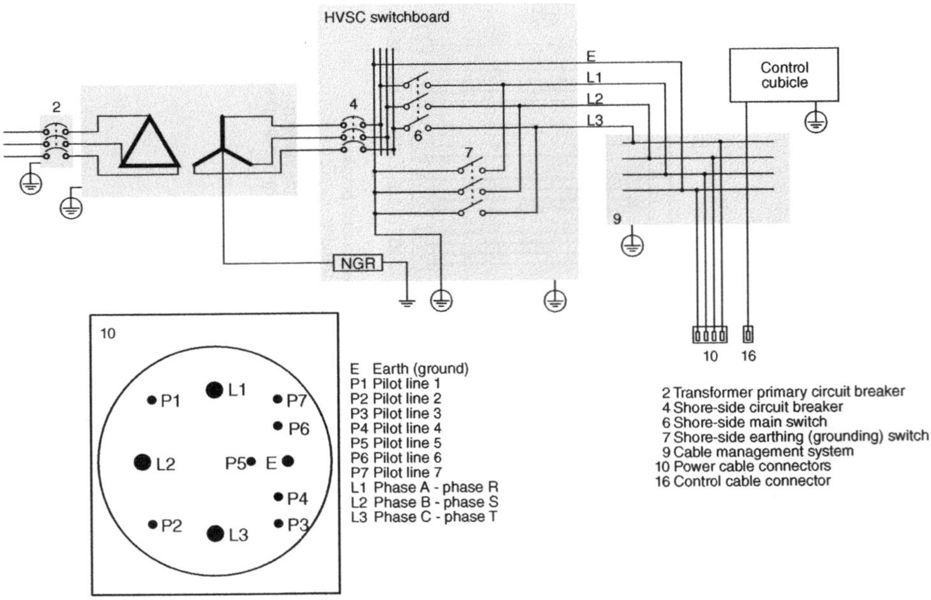

Figure 11.12 Example shore station with transformer for roll-on/roll-off cargo and roll-on/roll-off passenger ships.

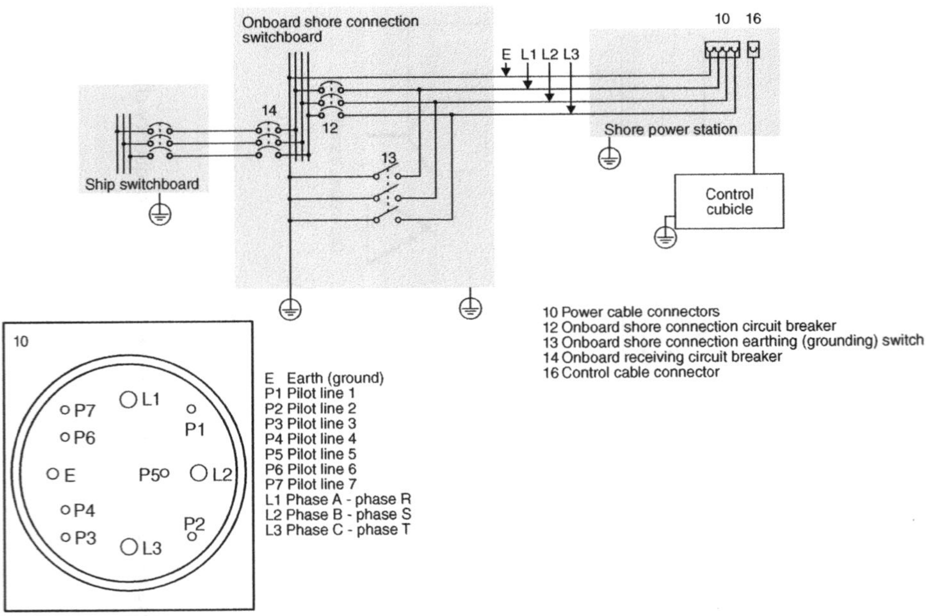

Figure 11.13 Example roll-on/roll-off cargo or roll-on/roll-off passenger ship onboard shore connection switchboard.

(ground) electrode ashore is at nearly the same potential as the hull of the ship. Figure 11.8a depicts a typical requirement for equipotential bond monitoring. Figure 11.8b depicts a possible implementation; if the voltage is significantly below 5.6 V, then the pilot contact conductor is shorted to ground and if the voltage is significantly above 5.6 V, then either the pilot contact conductor is an open circuit, or the earthing connection between the ship and shore is not adequate. A voltage near 5.6 V represents a good connection.

The circuit in Figure 11.8b does not measure the continuity of a particular earth conductor in any one of several cables. If any of the cables successfully creates an equipotential bonding between the hull of the ship and the terrestrial ground, then the equipotential bond monitoring device will indicate a good bonding. The equipotential bond monitoring device will also indicate a good bonding if the bonding is achieved with a connection not even associated with the shore connection system.

Another potential issue is that the equipotential bond monitoring device, owned by the port authority, must be compatible with the equipotential bond monitoring termination device, owned by the ship. Many contemporary ships have not implemented the equipotential bond monitoring termination device. Consequently, equipotential bond monitoring is often not accomplished.

Figure 11.9 depicts a single-voltage shore station to supply cruise ships. The cable management system provides four three-phase 500 A power cables, one neutral single-conductor cable rated at 250 A, and one or more control cables. The power cables include a pilot contact conductor and pilot contact that is used by the control system to ensure the cable is properly connected to the ship receptacle. Figure 11.10 depicts the ship's onboard shore connection switchboard.

Prior to connecting to a ship, the shoreside circuit breaker (4) and shoreside main switch (6) are open. The shoreside earthing switch (7) and shoreside neutral earthing switch are closed (8). On the ship, the onboard shore connection circuit breaker (12) and onboard receiving circuit breaker (14) are open; the onboard shore connection earthing switch (13) is closed.

The normal sequence for establishing a shore connection is

- Cables (10, 11, and 16) from the cable management system are connected to the ship's terminals.
- Diagnostics are performed, and alarm circuits are activated.
- Control data links are established.
- Once it has been determined that the shore connection is safe, the ship activates permission to connect.

- The shoreside earthing switch (7), the shoreside neutral earthing switch (8), and the onboard shore connection earthing switch (13) are opened.
- The shoreside main switch (6) is closed.
- The shoreside circuit breaker (4) and the onboard shore connection circuit breaker (12) are closed.
- If paralleling to the shipboard generators, the onboard receiving circuit breaker (14) is closed when the voltages, frequency, and phase of the shore connection and the shipboard system match. If not paralleling to the shipboard generators, the onboard receiving circuit breaker is closed when it is detected that there is no voltage present on the ship power system (i.e. the main generator breaker is opened).
- If paralleling, load to the shipboard generator is reduced to zero.
- Shipboard generator sets are turned off.

If, during operation, an overcurrent, overvoltage, undervoltage, reverse power, earth fault, or unbalanced cable protection is detected, the shoreside circuit breaker trips.

The normal sequence for disestablishing a shore connection is

- Shipboard generator sets are turned on.
- If paralleling, shipboard generator set is paralleled to the shore connection and load is transferred to shipboard generator set.
- The onboard receiving circuit breaker is opened.
- If not paralleling, main generator breaker is closed to energize shipboard bus.
- The shoreside circuit breaker (4) and the onboard shore connection circuit breaker (12) are opened.
- The shoreside main switch (6) is opened.
- The shoreside earthing switch (7), the shoreside neutral earthing switch (8), and the onboard shore connection earthing switch (13) are closed.
- Cables (10, 11, and 16) from the cable management system are disconnected.

To implement the establishing and disestablishing of a shore connection, interlocking features are incorporated for the circuit breakers and switches.

The shoreside earthing switch (7), the shoreside neutral earthing switch (8), and the onboard shore connection earthing switch (13) cannot be opened if:

- Any of the cable pilot contact circuits of the four power cables are open.
- An emergency-stop has been activated.
- The control cables and associated data links are not operational.
- Diagnostics or alarm circuits detect an error that impacts a safe connection.
- Permission from the ship has not been activated.

The shoreside main switch (6) cannot be closed if:

- Either the shoreside earthing switch (7) or the onboard shore connection earthing switch (13) is closed.
- The safety circuit is not established.
- The control cables and associated data links are not operational.
- Equipotential bonding is not established between the ship's hull and the ground conductor electrode at the HVSC switchboard.

The shoreside circuit breaker (4) and the onboard shore connection circuit breaker (12) cannot be closed if one of the following is true:

- Either the shoreside earthing switch (7) or the onboard shore connection earthing switch (13) is closed.
- The safety circuit is not established.
- An emergency stop has been activated.
- Diagnostics or alarm circuits detect an error that impacts a safe connection.
- The control cables and associated data links are not operational.
- Permission from the ship has not been activated.
- Acceptable voltage on the supply side of the shoreside circuit breaker is not present.
- Equipotential bonding is not established between the ship's hull and the ground conductor electrode at the HVSC switchboard.

In some ports, the frequency of the shore supply does not match the frequency required by the ship. In these cases, a frequency changer is employed in addition to the transformer, as depicted in Figure 11.11. A grounding transformer is used to develop a neutral for use with the NGR.

Figures 11.12 and 11.13 depict an example of a shore supply system and an example of an onboard shore connection switchboard for roll-on/roll-off cargo and passenger ships. The shore supply system differs from the shore supply system for cruise ships in the following ways:

- Only one power cable is used; no neutral cable is provided. The fiber optic control cable is integrated with the power cable. The maximum power demand is 6.5 MVA.
- The pilot contacts on the plug and receptacle are solely used for controlling the shore connection; the control cable is used only for monitoring data.
- Normally, only 11 kV AC is provided; for regional waterborne transportation services, 6.6 kV AC may be used instead.
- The neutral earth resistor (HRG) is permanently connected to the ground (earthing) electrode. The neutral earth resistor has a resistance of 335 Ω for 11 kV and 300 Ω for 6.6 kV.

The shore connection for container ships is very similar to that for the roll-on/roll-off cargo and passenger ships. Differences include

- Two power cables are used instead of one; the maximum power demand is 7.5 MVA.
- The pilot contacts on the cable are configured differently.
- Only 6.6 kV AC is provided.
- The neutral earth resistor has a resistance of 200 Ω.

The shore connection for LNGC is also very similar to that for the roll-on/roll-off cargo and passenger ships. Differences include

- Three power cables are used instead of one; the maximum power demand is 10.7 MVA.
- The pilot contacts on the cable are configured differently.
- Only 6.6 kV at 60 Hz is provided.
- The system is normally ungrounded; a neutral earthing resistor is not normally provided.

For tankers, the differences with respect to the roll-on/roll-off cargo and passenger ships include

- Three power cables are used instead of one; the maximum power demand is 10.8 MVA.
- The pilot contacts on the cable are configured differently.
- Only 6.6 kV AC is provided.
- The presence of the neutral earthing resistor is not definitively specified.
- The control cable may not be integrated with the power cables.

For the US Navy, some ships are provided with either 4.16 or 13.8 kV AC three-phase ungrounded power from a substation transformer located on the pier. Multiple shore power cables, fewer than 10, are usually connected in parallel to provide the amount of current required by the ship. Each shore power cable is protected by a dedicated circuit breaker with a maximum overcurrent setting of 400 A; as with the low-voltage system, if one breaker trips, all breakers trip. The grounding system onboard the ship determines how the system is grounded; MIL-STD-1399 Section 300 Part 2 (2018) states that medium-voltage AC power is high-resistance grounded onboard the ship. Figure 11.14 depicts sailors wearing arc flash personal protective equipment working on a medium-voltage shore connection box.

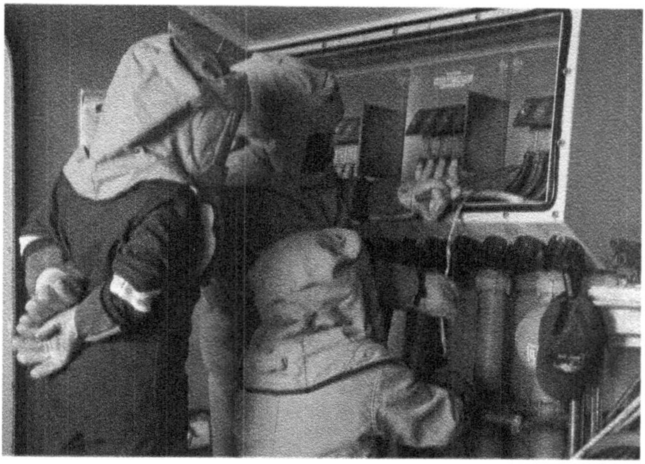

Figure 11.14 Sailors working on a medium-voltage shore connection box. *Source:* MCSN Jarrod A. Schad/United States Navy/Public domain.

References

IEC/IEEE 80005-1-2019 (2019). *Utility Connections in Port – Part 1: High Voltage Shore Connection (HVSC) Systems – General Requirements.* Geneva: International Electrotechnical Commission. Available from http://ieeexplore.ieee.org.

IEC PAS 80005-3:2014 (2014). *Publicly Available Specification Utility Connections in Port – Part 3: Low Voltage Shore Connection (LVSC) Systems – General Requirements.* Geneva: International Electrotechnical Commission. Available from http://webstore.iec.ch.

MIL-DTL-24368/1 (2021). *Detail Specification Sheet Connector Assemblies; Plug Transfer, Shore to Ship and Ship to Ship, 500 Volts, 500 Amperes, 60 Hertz.* Washington, DC: Department of Defense. Available from http://quicksearch.dla.mil.

MIL-DTL-24368/2 (2021). *Detail Specification Sheet Connector Assemblies; Receptacle and Receptacle-Cabled, Power Transfer, Shore to Ship and Ship to Ship, 500 Volts, 500 Amperes, 60 Hertz, Symbol Number 1161.* Washington, DC: Department of Defense. Available from http://quicksearch.dla.mil.

MIL-STD-1399 Section 300 Part 2 (2018). *Medium Voltage Electric Power, Alternating Current.* Washington, DC: Department of Defense. Available from http://quicksearch.dla.mil.

UFC-4-150-02 (2020). *Unified Facilities Criteria (UFC) Dockside Utilities for Ship Service,* Washington, DC: Department of Defense. Available from www.wbdg.org.

12

Vehicle Connections Grounding

12.1 Introduction

Vehicles, such as boats, automobiles, autonomous underwater vehicles (AUVs), unmanned aerial vehicles (UAVs), helicopters, fixed-wing aircraft, amphibious vehicles, and landing craft, may all be found onboard ships. Each of these vehicles has its own dedicated electrical power system; they serve as a separately derived system when not connected to the ship's electrical distribution system. When onboard the ship, many of the vehicles connect to the shipboard power system to recharge energy storage, to perform maintenance and testing, or to start the engine. To avoid damage and injury to personnel, the system ground and protective ground of the vehicle should be integrated with the ship. IEEE Std 45.1-2023 (2023) provides additional recommendations.

Air vehicles often build up a considerable amount of static charge when flying. To avoid personnel injury, this static charge should be safely discharged when the air vehicle comes into electrical contact with the ship.

12.2 Design Considerations

12.2.1 Aircraft Static Electricity

As described in MIL-HDBK-274 (2011), sources of static electricity charge buildup on aircraft include triboelectric charging and induced static charging. Triboelectric charging occurs when specific materials come into contact and then are separated. Examples of triboelectric charging include rubbing glass with fur or using a plastic comb in hair. For aircraft, triboelectric charging can occur when aircraft fly through water droplets, dust, or snow; or alternately, when the aircraft is onboard the ship

Design of Shipboard Power System Grounding/Earthing, First Edition. Norbert Doerry, Mohammed M. Islam, and John Prousalidis.
© 2025 The Institute of Electrical and Electronics Engineers, Inc.
Published 2025 by John Wiley & Sons, Inc.

and strong winds blow dust or snow across the aircraft. The fast rotation of helicopter blades can result in significant charge buildup. The flow of fuel through a refueling hose can also induce charge.

Induced static charge on aircraft typically occurs when an electrical storm results in heavily charged clouds creating an electrostatic field over a large area. This electrostatic field can slowly cause a charge to accumulate on an aircraft parked on a flight deck. Should the charge in the clouds dissipate, such as during a lightning strike, the electrostatic field will disappear, but the charge on the aircraft will only slowly dissipate through the tires if the aircraft does not include a protective ground. In aviation literature, a protective ground is often called a "static" ground.

Aircraft metallic components are normally bonded together and connected to the aircraft's structure. A protective earth connection is required when refueling the aircraft and is implemented via a grounding cable. This grounding cable typically has a clamp on one end to connect to the ship's structure, and on the other end, either a clamp to affix to a part of the aircraft or a grounding plug that is inserted into a grounding receptacle on the aircraft. Although not required by law, it is good practice to keep the grounding cable in place while the aircraft is parked onboard the ship. To enable protective earthing of aircraft, the ship design should include grounding points at appropriate places where aircraft may be parked.

At times, a helicopter may hover over the ship for the purpose of transferring personnel, transferring cargo, or in-flight refueling of the helicopter. Since the helicopter may have accumulated considerable static charge during flight, a helicopter static discharge wand, as depicted in Figures 12.1 and 12.2, is used by the ship's crew to bring the ship's hull and helicopter structure to the same voltage. To facilitate using the static discharge wand, the ship design should provide grounding points at

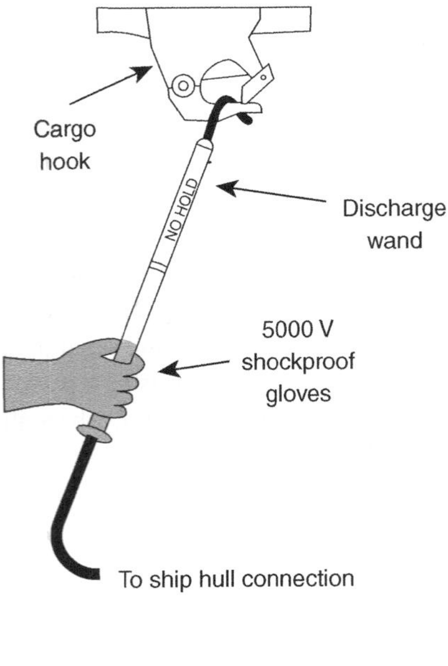

Figure 12.1 Helicopter static discharge wand.

Figure 12.2 A marine holds a static discharge wand against the cargo hook of a CH-53E Super Stallion helicopter. *Source:* Staff Sgt. Artur Shvartsberg 160930-M-VO695-0225A/ U.S. MARINE CORPS/Public domain.

shipboard locations where helicopter operations are anticipated. If the ship's crew is not able to employ the static discharge wand, the helicopter may lower a hoist static discharge cable to make contact with the ship and equalize the voltage.

12.2.2 Vehicle Electrical Systems

The electrical systems used within vehicles employ a wide range of voltages and frequencies; often, the required power quality standards differ from those provided onboard the ship. The vehicle electrical system may or may not be solidly grounded to its own chassis; the vehicle electrical system may be ungrounded. If a three-phase four-wire AC system is used in a grounded vehicle power system, the neutral would normally be solidly grounded. Grounded DC vehicle power systems usually ground the negative conductor to the chassis.

Standard aircraft external power interfaces include

- 115/200 V, 400 Hz power (three-phase four-wire).
- 28 V DC power.
- 270 V DC power.

The external power connectors usually have additional pins for control circuits. Normally, these pins are shorter than the power connectors; these pins connect last when plugged in and disconnect first when unplugged. Typically, the aircraft senses when the control pins are connected to control a master contactor on the aircraft for the main power. The contactor is only closed and power applied to the aircraft loads when the control pins are engaged. Figure 12.3 demonstrates how this is implemented in a 400 Hz three-phase four-wire connection. The voltage

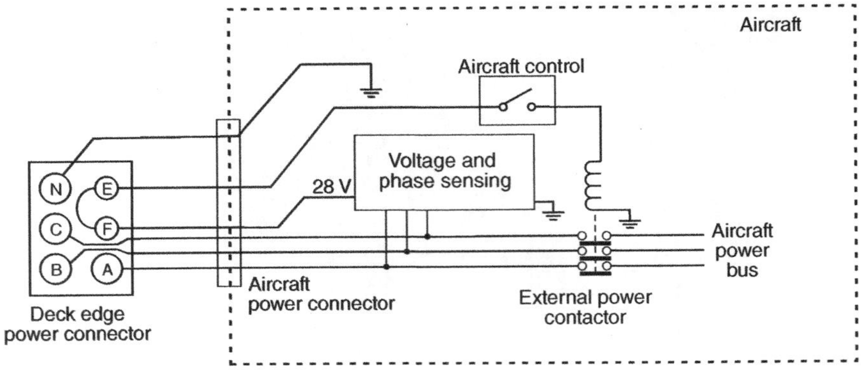

Figure 12.3 Use of control pins on the power cable connector to control the external power contactor.

and phasing sensing circuit ensures the power provided at the connector is within acceptable limits before producing a 28 V DC output signal that flows through the connector and to the external power contactor.

Aircraft power connectors generally do not include a protective earth pin. As discussed in Section 12.2.1, a separate grounding cable fulfills this function.

The interfaces for other types of vehicles, such as unmanned vehicles and boats, may be unique and require careful consideration for establishing a safe grounding system. In general, for power interfaces with a nominal system voltage greater than 40 V, consideration should be given to installing GFCIs into the distribution system.

ABS (2022) recommends that for electric vehicles (automobiles) onboard the ship, cables for powering vehicles should be powered from shipboard sockets that are in a location where the cable can be safely disconnected should the vehicle catch fire. All cables for powering vehicles should be provided by the ship. These recommendations should also be considered for powering other vehicle types as well.

References

ABS (2022). Best Practices for the Transport of Electric Vehicles on Board Vessels. Available from ww2.eagle.org.

IEEE Std 45.1-2023 (2023). *IEEE Recommended Practice for Electrical Installations on Shipboard – Design*. New York: Institute of Electrical and Electronics Engineers. Available from ieeexplore.ieee.org.

MIL-HDBK-274 (2011). *Handbook Electrical Grounding for Aircraft Safety*. Available from quicksearch.dla.mil.

13

Common Mode Grounding: Impact of Common Mode Currents and Voltages on Grounding Systems

13.1 Common Mode Fundamentals

Having a good understanding of common mode (CM) voltages, currents, and impedances is extremely important when designing the grounding system of a ship. In many modern power electronics-intensive power systems, the impact of CM voltages and currents from normal operation can conflate the system response to a ground fault. The ground fault itself is manifest as a CM voltage with respect to ground. Consequently, the grounding methods described in Chapters 2 through 8 impact both the system response to the ground fault and the system response to CM voltage sources that are part of normal operation.

The CM current is the sum of the conductor currents of a set of conductors flowing in the same direction. The return path of the CM current is outside the set of conductors. Another name for CM current is circulating current. Often, the return path is through the ground connection, but not always. CM current is unintended; the intended DM currents of a set of conductors add to zero when measured in the same direction. CM currents are not desired because they do no useful work, can cause corrosion damage if they flow through the ship's hull, can cause localized (and damaging) heating in cables, equipment, and structure, can impact current measurements of protective devices resulting in these devices malfunctioning, and can be a significant source of electromagnetic interference (EMI).

As stated in Chapter 1, the neutral voltage of a set of conductors with respect to a reference voltage potential (often ground) at any instant of time is the average value of the instantaneous conductor voltages with respect to the reference voltage. The neutral voltage is also called a CM voltage. The neutral voltage is a calculated quantity from the voltage measurements of one or more conductors and may not correspond to the voltage of any one conductor.

Design of Shipboard Power System Grounding/Earthing, First Edition. Norbert Doerry, Mohammed M. Islam, and John Prousalidis.
© 2025 The Institute of Electrical and Electronics Engineers, Inc.
Published 2025 by John Wiley & Sons, Inc.

While a conductor may be called a neutral, the voltage on that conductor, at best, will correspond to the neutral voltage of the set of conductors under ideal conditions. This is the case for a three-phase four-wire AC system. Single-phase two-wire systems with one conductor grounded are an extreme example where the neutral conductor is never at the neutral voltage; the conductor that is grounded is called a neutral, while the neutral voltage is the average of the voltages on each of the two-wires with respect to the reference voltage potential.

Power electronic converters are often a source of CM voltage because the inherent topology and switching algorithms result in the neutral voltage on the input being different from the neutral voltage on the output. Figure 13.1 is a simple six-pulse rectifier circuit. The DC output conductor voltages and the CM voltage of the DC output conductors with respect to the AC neutral voltage are depicted in Figure 13.2. The DC output conductor voltages and CM voltage of the AC input conductors with respect to the DC neutral are depicted in Figure 13.3. The middle

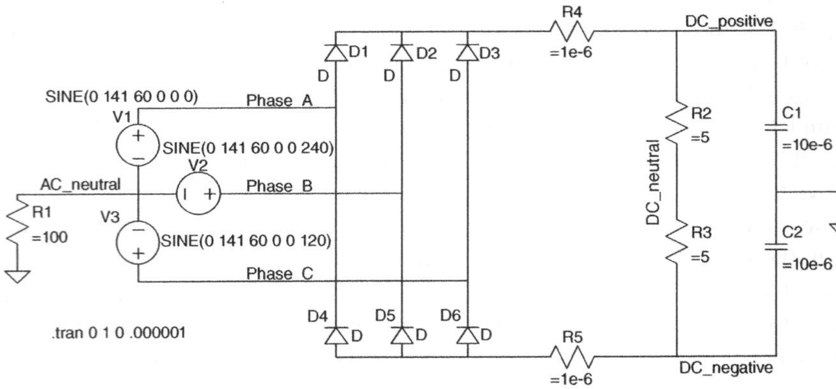

Figure 13.1 Schematic of six-pulse rectifier.

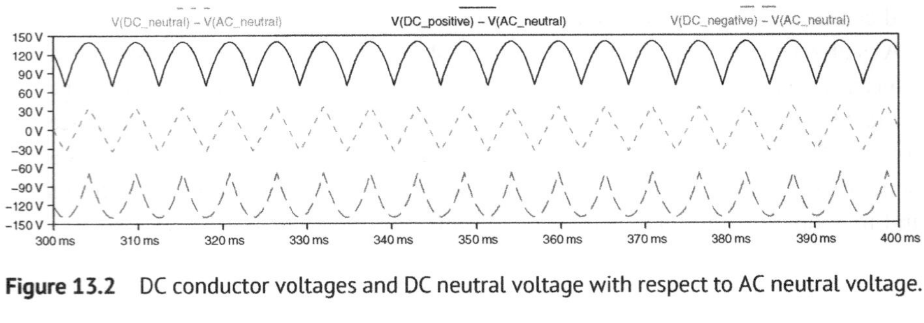

Figure 13.2 DC conductor voltages and DC neutral voltage with respect to AC neutral voltage.

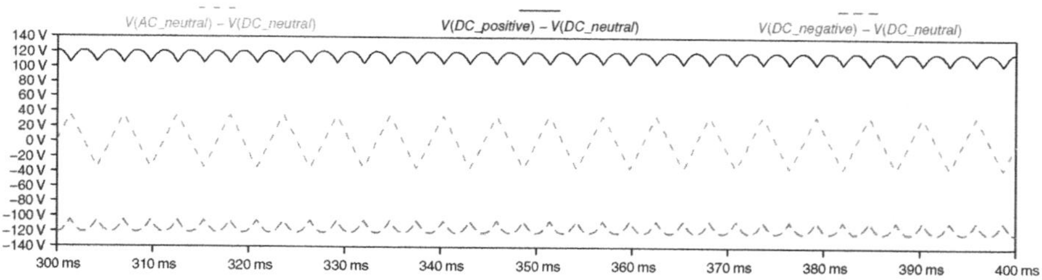

Figure 13.3 DC conductor voltages and AC neutral voltage with respect to DC neutral voltage.

waveform in Figures 13.2 and 13.3 represents the inherent CM voltage source between the neutrals of the input and output of the passive rectifier. The choice of reference voltage has a big impact on the shape of the output voltage waveforms.

Figure 13.4 depicts the DM voltage and current on the DC side of the rectifier. As expected, the voltage waveform depicts six "pulses" in one period of the fundamental voltage waveform (16.7 ms for 60 Hz). The impedance of R2 and R3 is significantly less than the impedance of C1 and C2; hence, the current waveform appears mostly resistive. Figure 13.5 depicts the CM voltage and CM current for the

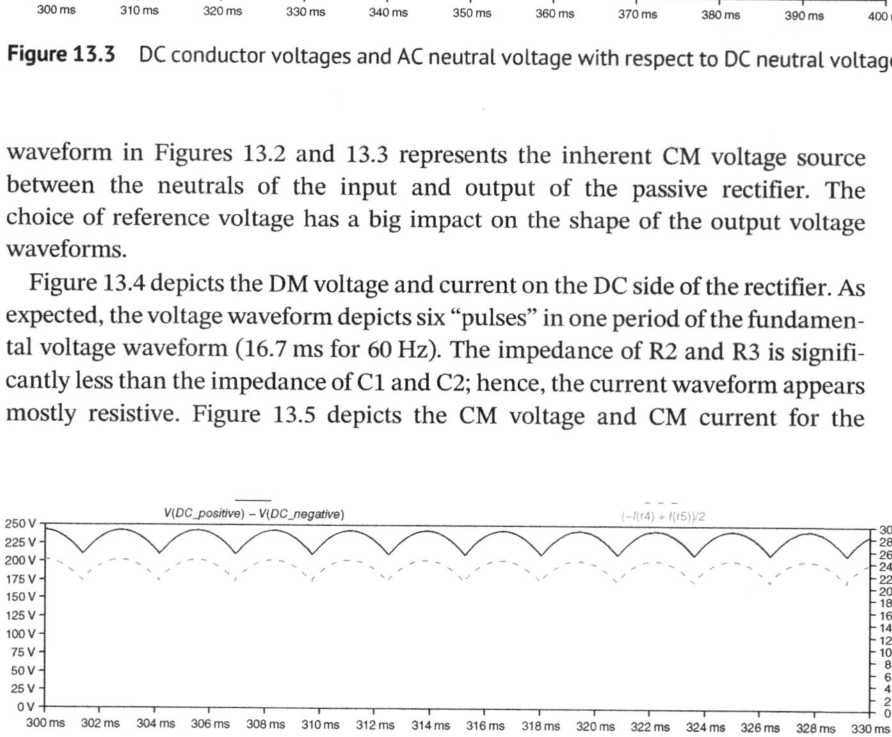

Figure 13.4 DM voltage and current for six-pulse rectifier.

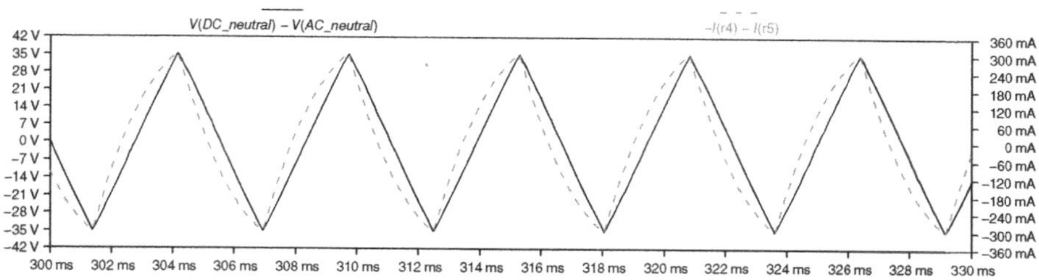

Figure 13.5 CM voltage and current for six-pulse rectifier.

rectifier. In this circuit, the return path for the CM current is through the ground connection.

In typical systems, the DM and CM voltages and currents can be orders of magnitude apart. Because the DM voltages and currents are typically much greater than the CM voltages and currents, the DM characteristics can mask the CM characteristics. For systems where all the impedances are symmetric, the modeling of CM currents and voltages can be simplified and made independent (or nearly independent) of the DM characteristics. Impedances are symmetric when all of the in-line impedances of the conductors are the same, all of the line-to-line impedances of the conductors are the same, and all of the line-to-ground impedances of the conductors are the same. In this simplified CM model, passive components that are in-line or line-to-ground are combined as if they were in parallel. Passive components that are line-to-line (or connected in wye, with the wye not connected to anything else) are ignored. Balanced three-phase sources are eliminated; if connected in wye, the wye point is connected to the CM voltage of the three phases. CM voltage sources are represented as a voltage source. Figure 13.6 provides the simplified CM model, where the CM voltage is

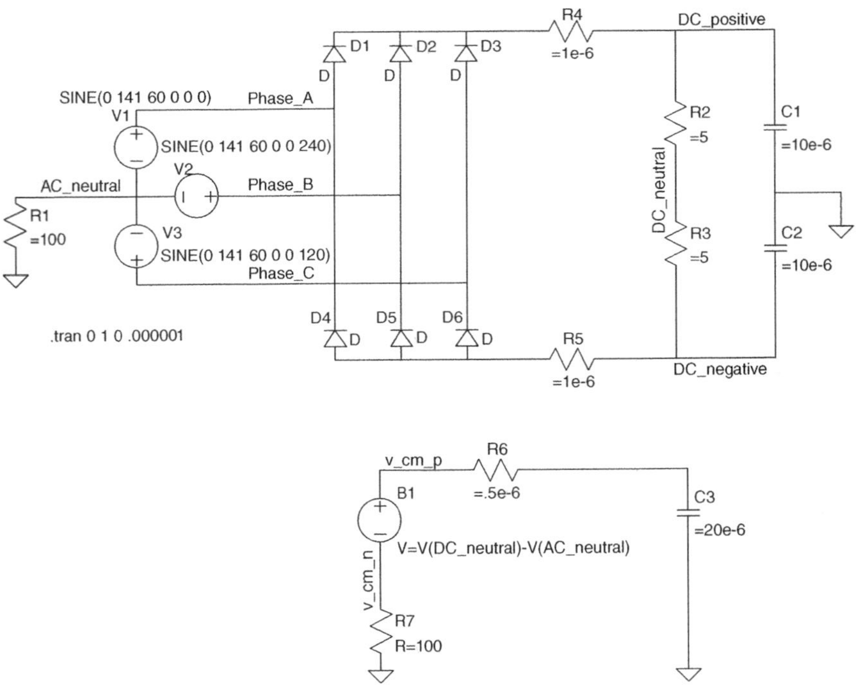

Figure 13.6 Six-pulse rectifier detailed and simplified CM model.

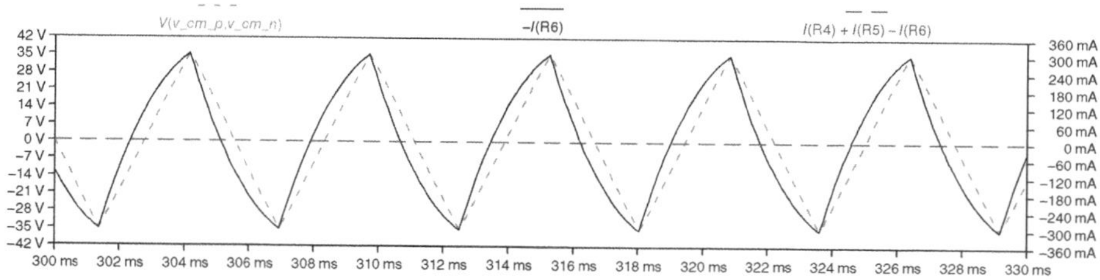

Figure 13.7 CM voltage and current for six-pulse rectifier as derived from the simplified CM model.

obtained from the more detailed model. Figure 13.7 demonstrates that the simplified CM model provides the same values for CM voltages and currents as for the detailed model (Figure 13.5).

If all the impedances and sources are symmetrical, this same technique may be employed to determine the CM currents during a ground fault. Figure 13.8 depicts a simple three-phase system that experiences a ground fault between 100 and 300 ms. The CM currents through the grounding resistor (R1), the three line-to-ground capacitors, and the three load resistors are depicted in Figure 13.9. Before and after the ground fault, the CM currents are all zero because a CM voltage does not exist. During the ground fault, however, the neutral-to-ground

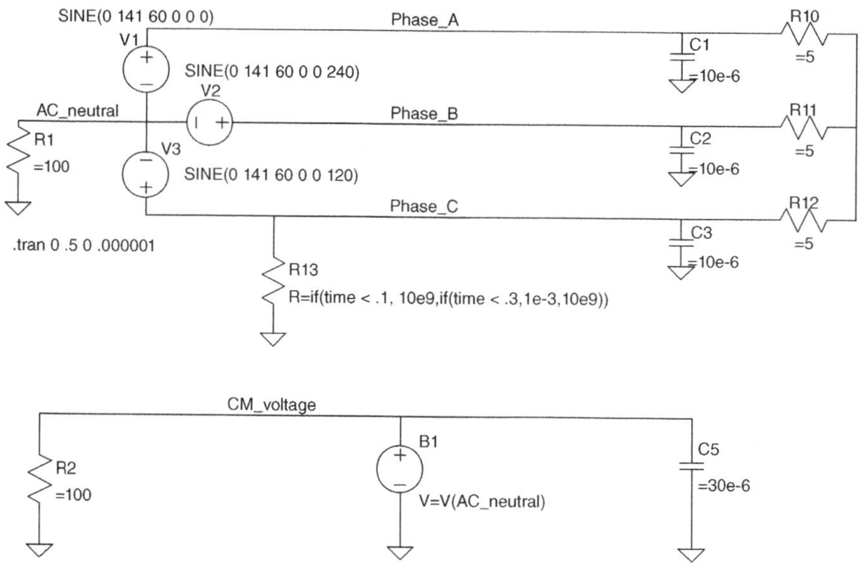

Figure 13.8 CM modeling of a ground fault.

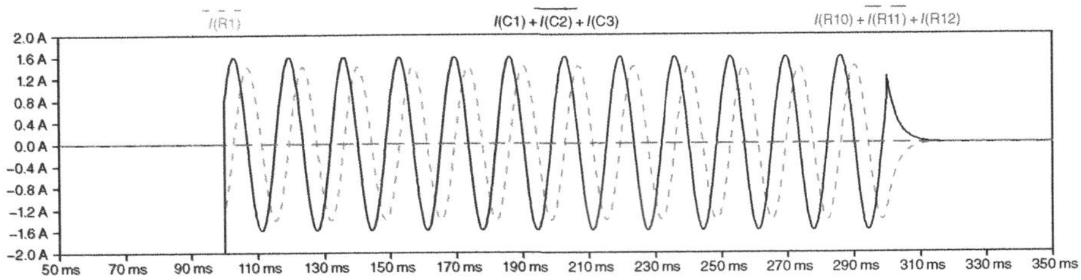

Figure 13.9 CM currents from a three-phase model.

voltage (CM voltage) is fixed to the phase *c* line-to-neutral voltage. This CM voltage results in CM currents through R1 and the line-to-ground capacitances. The load resistances do not experience a CM current because the voltage at their wye point is the same as the neutral voltage of the three-phase voltages. The equivalent simplified CM model is also depicted in Figure 13.8. The grounding resistor, R2, in the CM model has the same value as R1, the grounding resistor in the three-phase model. The neutral-to-ground capacitance (C5) is the parallel combination of the line-to-ground capacitances (C1, C2, and C3). The load resistances are not modeled because they are symmetrical and only connected to the line voltages. As shown in Figure 13.10, the CM currents calculated from the simplified CM model are identical to those from the three-phase model (Figure 13.9).

The lack of symmetrical impedances (the presence of asymmetry) can lead to CM voltages that are linked to the DM circuit. In Figure 13.11, a 10 Ω resistor is placed in one conductor of a two-wire circuit, but not the other. The simplified CM model is also shown. Figure 13.12 shows that this asymmetry results in a CM voltage and current to the right of the asymmetrical resistor. Note that the calculated current for C2 in the simplified CM model is the same as the calculated current for C1 in the detailed model. If one splits the 10 Ω resistor into two equal 5 Ω

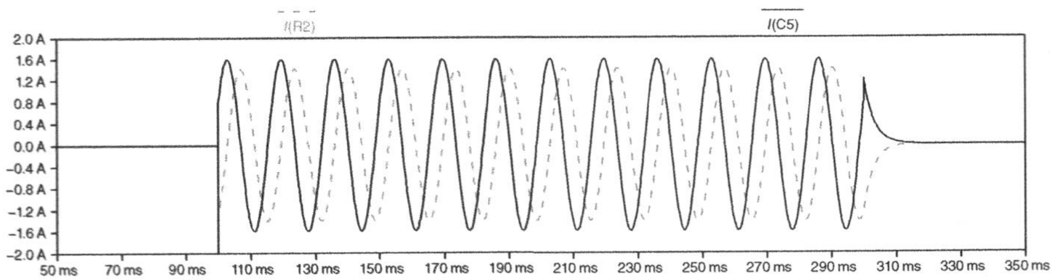

Figure 13.10 CM currents from the simplified CM model.

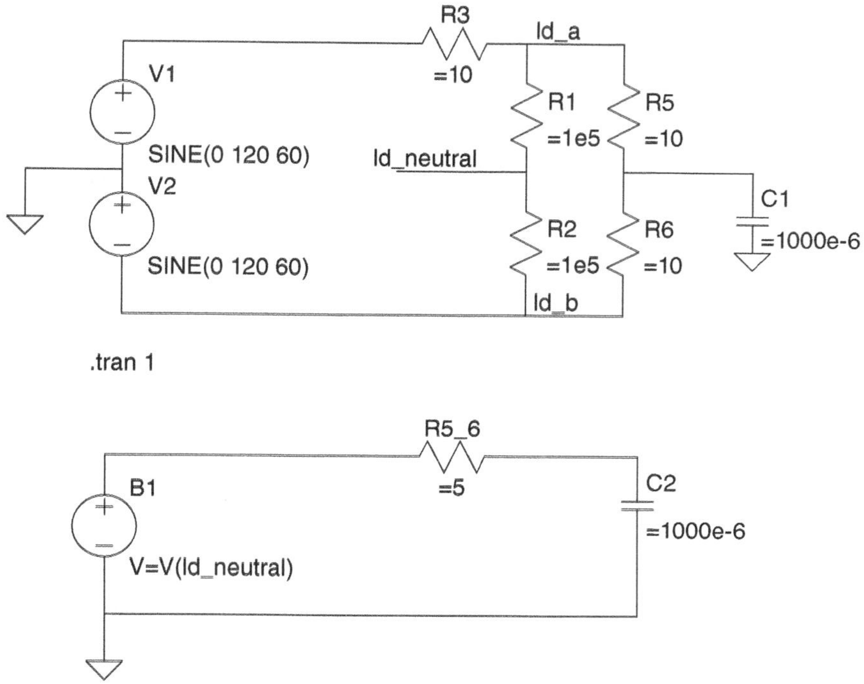

Figure 13.11 Asymmetrical circuit.

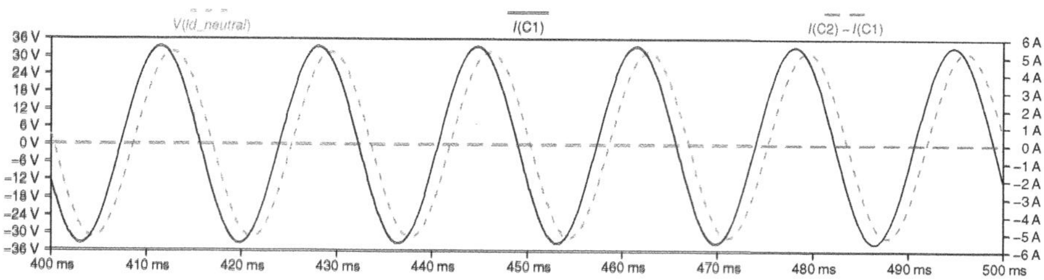

Figure 13.12 Asymmetrical CM voltage and currents.

resistors and places each in one of the two conductors, as shown in Figure 13.13, the system is balanced and a CM voltage does not exist. As shown in Figure 13.14, without a CM voltage, there is no CM current.

Lack of symmetry also applies to how a system is grounded. Figure 13.15 depicts a DC source with a 1 kHz ripple that has its negative conductor grounded. The simplified CM model is also depicted. Figure 13.16 displays the CM current

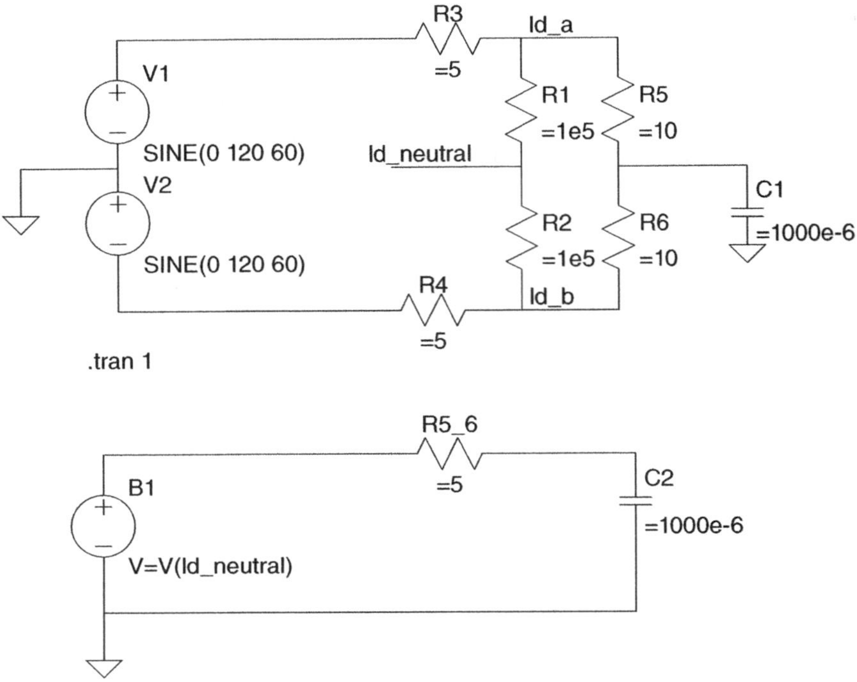

Figure 13.13 Symmetrical circuit.

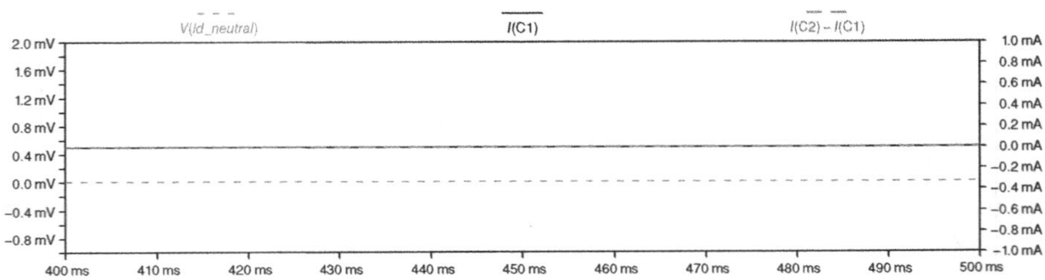

Figure 13.14 Symmetrical CM voltage and currents.

through R4 and R5 and demonstrates that the simplified CM model calculates the same current through R4_5. Figures 13.17 and 13.18 demonstrate that eliminating the asymmetry by splitting the DC supply and grounding at the neutral eliminates the CM voltage source and the resulting CM current.

CM currents need not always flow through the ground connection. Figure 13.19 depicts a circuit where there are two paths to the load (R5 and R6). One path is via

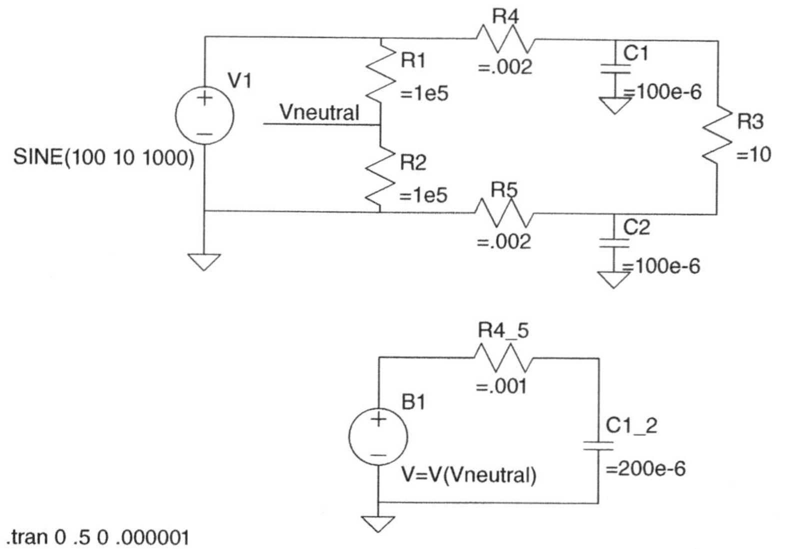

Figure 13.15 DC source with 1 kHz ripple and negative ground schematic.

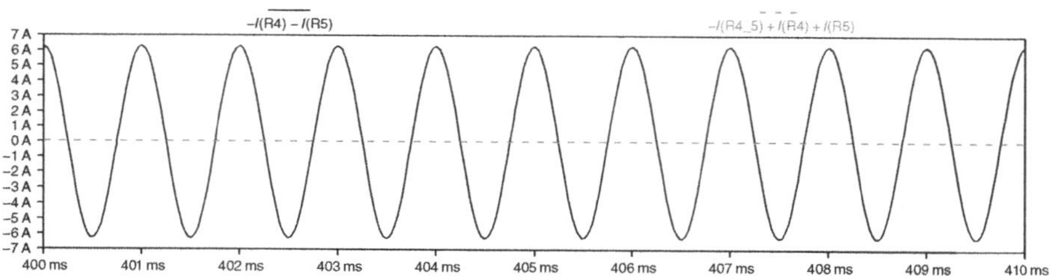

Figure 13.16 DC source with 1 kHz ripple and negative ground CM current.

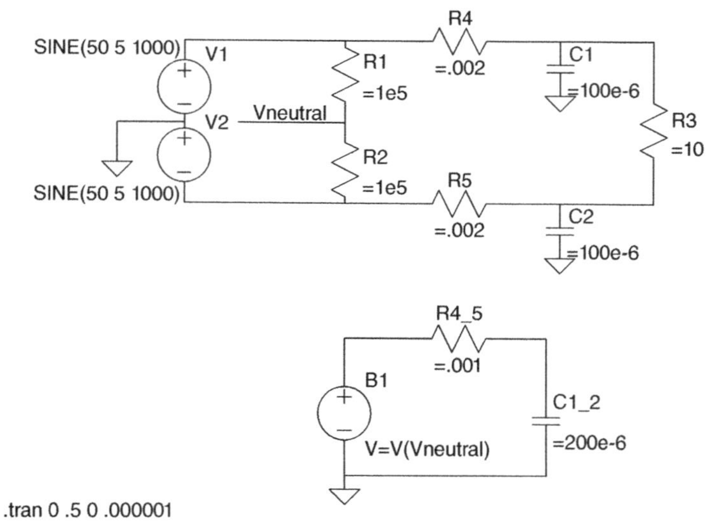

Figure 13.17 Split DC source with 1 kHz ripple.

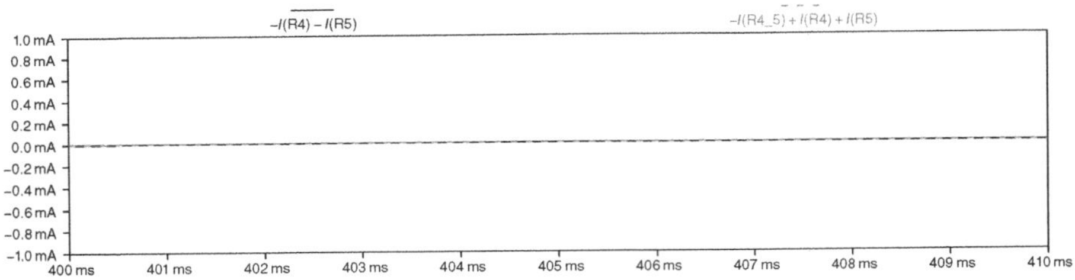

Figure 13.18 Split DC source with 1 kHz ripple CM current.

R1 and R2, and the other is via R3 and R4. This is representative of a simplified ring bus. R9 represents an asymmetry in the resistance of the path via R3 and R4 (perhaps due to a loose connection). This asymmetry can result in a CM current circulating between the two paths. Figure 13.19 also depicts the simplified CM model. Figure 13.20 depicts the CM currents in the two paths and demonstrates that the simplified CM model accurately predicts the CM current. Note that the simplified model impedances do not depend on the load impedance or the source. The impact

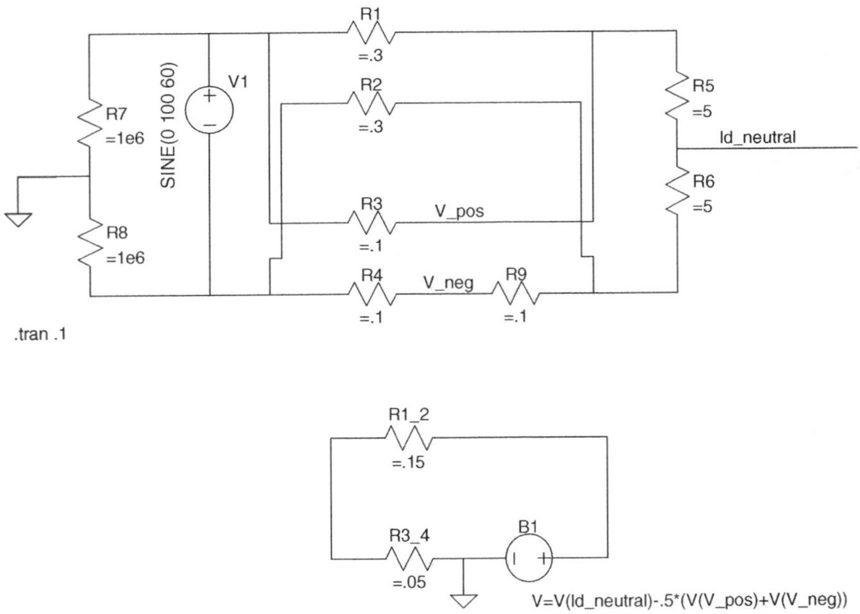

Figure 13.19 Ring bus schematic.

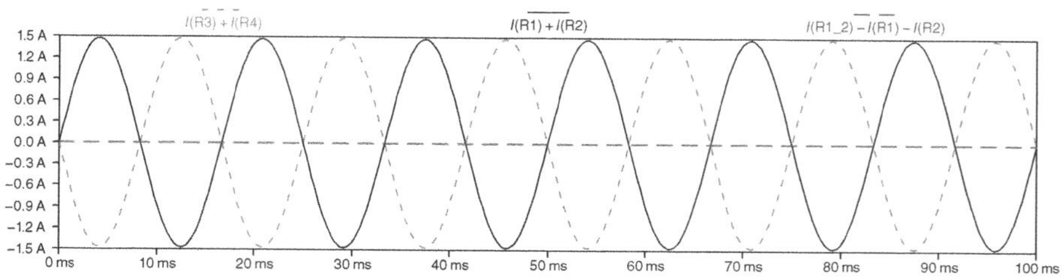

Figure 13.20 Ring bus CM current.

of the source, load, and imbalance on the CM circuit is captured in the magnitude of the CM voltage source. The presence of a circulating current could be an indicator of a loose connection or an arcing in-line fault in a ring bus.

Modeling a cable is not as simple as one might suspect. While cable manufacturers will list an inductance per unit length for cable, this inductance assumes balanced three-phase operation with no CM current. In reality, the inductance of a cable is comprised of a self-inductance L_s and a mutual inductance M with the other conductors. The published inductance per unit length of a multiconductor cable is the self-inductance minus the mutual inductance. If one ignores the impact of mutual inductance with shields, the CM inductance (for a cable with $n_{conductors}$ conductors and phases) is given by

$$L_{CM} = \frac{L_s}{n_{conductors}} + \frac{(n_{conductors} - 1)M}{n_{conductors}}$$

Figure 13.21 depicts a source with a 10 V, 360 Hz CM voltage, a simple cable model with mutual inductance, parasitic capacitance to ground, and a resistive load. It also shows the simplified CM model. Figure 13.22 depicts the CM current through R1, R2, and R3 and demonstrates the simplified model produces the same CM current waveform.

In general, a CM current will flow where there is a CM closed circuit that includes a CM voltage. CM voltages usually are present where there is circuit asymmetry or power electronic switching. While the lack of a CM voltage and a CM closed circuit eliminates the possibility of a CM current, most physical circuits have some asymmetry (which manifests as a CM voltage) as well as parasitic line-to-ground capacitance. The CM voltage is typically a function of time and can be estimated through either simulation or physical measurement of hardware. Figure 13.23 depicts measured values of the CM voltage on the DC side (input) and on the AC side (output) of an inverter. The CM voltage magnitude and the CM impedances derived from the simplified CM model can be used to estimate the CM current magnitude. In this manner, one can identify potential CM current problem areas.

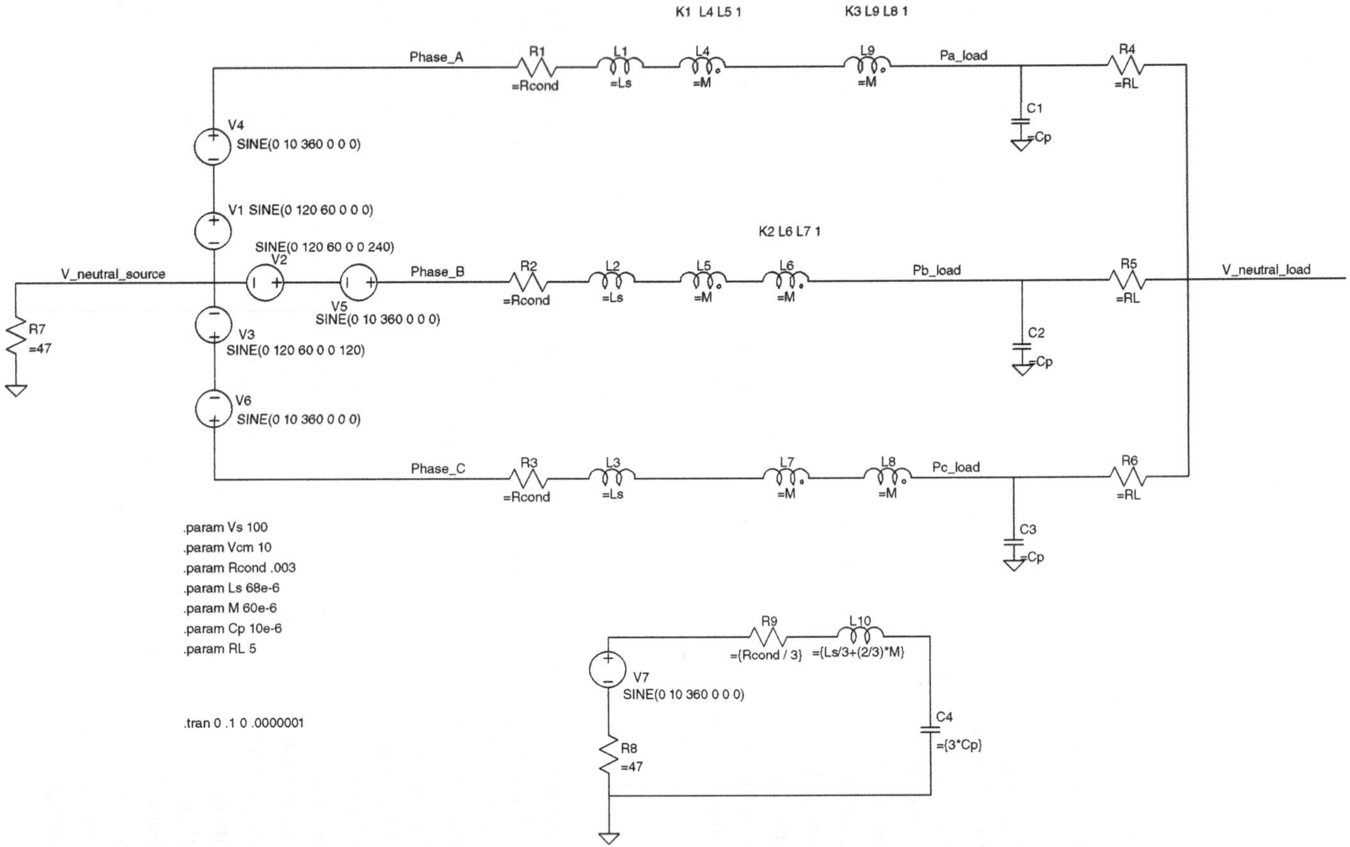

Figure 13.21 Modeling mutual inductance.

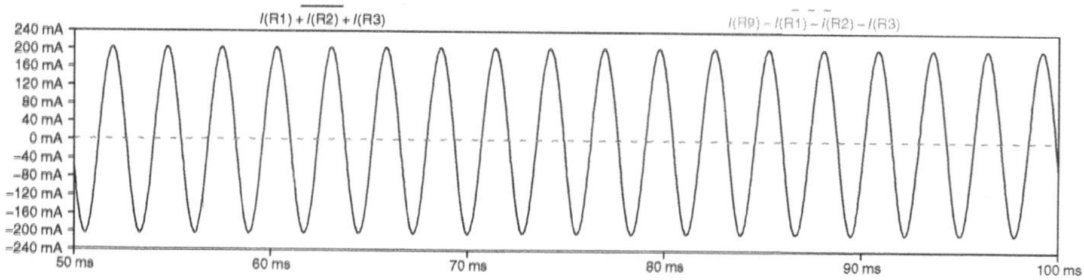

Figure 13.22 CM current with mutual inductance of cable modeled.

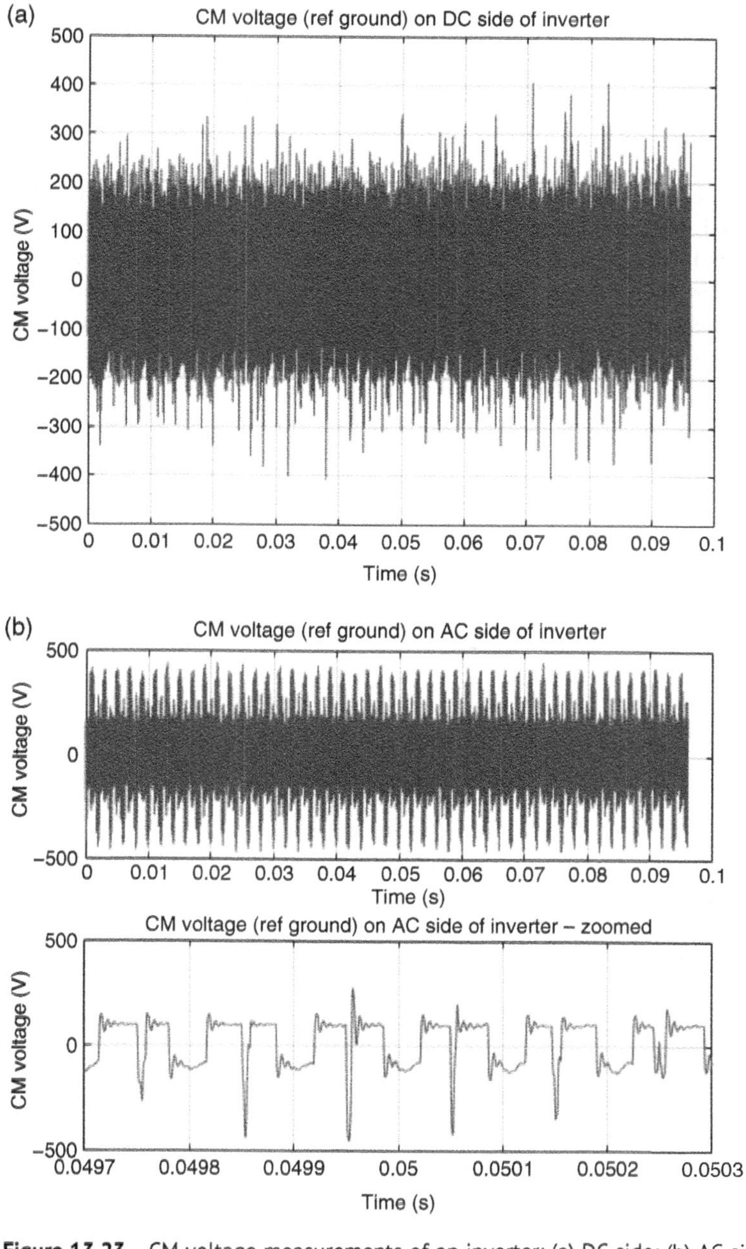

Figure 13.23 CM voltage measurements of an inverter: (a) DC side; (b) AC side.
Source: Courtesy of Steve Pekarek.

13.2 Relationship of CM to EMI and EMC

Electromagnetic Compatibility (EMC) is the ability of equipment and their associated systems to work as intended in the electromagnetic environment, while not creating an electromagnetic environment that causes other equipment and systems to become damaged or malfunction in the form of EMI. EMC is generally achieved by limiting the electromagnetic emissions of equipment, and by ensuring equipment and systems are not susceptible to a specified level of emissions applied at their electrical interfaces. Electromagnetic emissions can be either conducted through the conductors of an interface (Conducted Emissions or CE), or transmitted via electromagnetic radiation (Radiated Emissions or RE). The susceptibility of equipment to CE is called Conducted Susceptibility (CS), while the analogous term for RE is called Radiated Susceptibility (RS).

CM currents in a cable can be a significant source of RE. This is because the radiated fields from each conductor in a cable are additive for CM currents, while they nearly cancel each other out for DM currents. This can be explained by examining the relationship of the magnetic field B produced by each conductor in the cable. For a long conductor, the magnetic field B measured in teslas (T) at a radial distance r (meters) from the conductor center caused by a current I (amps) through the conductor is given by

$$B = \frac{\mu_0 I}{2\pi r}$$

where μ_0 is the permeability of free space: $\mu_0 = 4\pi \times 10^{-7}\ \frac{N}{A^2}$ or $\frac{H}{m}$.

The magnetic field at a point in space from multiple conductors ($n_{conductors}$) in a cable is the sum of the contributions of each of the conductors.

$$B_{cable} = \sum_{i=1}^{n_{conductors}} \frac{\mu_0 I_i}{2\pi(R + r_i)}$$

where

R is the radial distance of the point in space to the center of the cable

$(R + r_i)$ is the radial distance of the point in space to the center of conductor i.

If we assume $R \gg r_i$, then:

$$B_{cable} = \sum_{i=1}^{n_{conductors}} \frac{\mu_0 I_i}{2\pi R\left(1 + \frac{r_i}{R}\right)}$$

$$B_{cable} = \sum_{i=1}^{n_{conductors}} \frac{\mu_0 I_i}{2\pi R} \sum_{j=0}^{\infty} \left(-\frac{r_i}{R}\right)^j$$

$$B_{cable} = \sum_{i=1}^{n_{conductors}} \frac{\mu_0 I_i}{2\pi R} \left(1 - \frac{r_i}{R} + \sum_{j=2}^{\infty} \left(-\frac{r_i}{R}\right)^j\right)$$

$$B_{cable} = \sum_{i=1}^{n_{conductors}} \frac{\mu_0 I_i}{2\pi R} + \sum_{i=1}^{n_{conductors}} \frac{\mu_0 I_i}{2\pi R}\left(-\frac{r_i}{R} + \sum_{j=2}^{\infty}\left(-\frac{r_i}{R}\right)^j\right)$$

Since $R \gg r_i$, the contributions of the higher orders of j can be considered negligible and eliminated. Since the summation of the conductor currents is the CM current (I_{CM}), then:

$$B_{cable} \approx \frac{\mu_0}{2\pi R}\left(I_{CM} - \sum_{i=1}^{n_{conductors}}\left(\frac{r_i}{R}\right)I_i\right)$$

As R increases, the contribution of the summation term decreases by a factor of R^{-2}, while the contribution of the CM current term only decreases by R^{-1}. The DM current only contributes to the summation term, while the CM current contributes to both terms. Hence, CM currents have a much greater impact on radiated emissions than DM currents. It is for this reason that minimizing CM currents, especially at higher frequencies, is critically important to controlling RE and thereby helping to ensure EMC.

CM currents also contribute to CE, however, the limits for CE are typically applied to a combination of CM and DM currents and voltages at frequencies other than the fundamental. Standards, classification society rules, and local regulations are typically the source of EMI limits. Examples include

- MIL-STD-461 (2015).
- CFR Title 47 chapter I subchapter A part 15 subpart B clause 15.107.
- ABS Rules for Building and Classing Marine Vessels (2023).
- IEC 61800-3:2022 (2022).
- IEC 60945:2002 (2002).

13.3 Control of CM Currents and Voltages

Controlling CM voltages is important for minimizing the electric field stress on insulation systems. CM voltages on rotating machinery shafts with respect to ground can lead to CM currents flowing through the bearings, resulting in arcing and pitting damage. Creepage and clearance distances for exposed current-carrying conductors should also be increased to account for the CM voltages. In most systems, particularly those with large amounts of power electronics, CM voltages and CM current paths will exist. Mitigating the negative effects of CM voltages and currents is one of the main reasons for applying measures to control them.

Mitigation efforts consist of eliminating CM voltage sources and controlling CM impedances so that CM voltages and currents are low in specific areas of the power system network. One strategy is to keep CM voltages low at the electrical interfaces and use CM impedances to limit the maximum possible CM current.

13.3.1 Equipment Design

Equipment design has a significant impact on the magnitude of CM voltages and impedances. Eliminating asymmetries in the impedances of the conductors is a straightforward way of avoiding an unnecessary CM voltage source. Circuit topology, coupled with appropriate power electronic gating algorithms, can also reduce or eliminate CM voltages between the input and output of a converter.

Unfortunately, equipment manufacturers typically do not disclose in their datasheets the CM voltages and impedances associated with their products. Direct communication with the equipment manufacturers, or independent measurements of equipment, may be needed to properly characterize the CM performance of equipment.

13.3.2 Cable Shields, Drain Wires, and Ground Conductors

Cable shields are one tool for controlling electric and magnetic fields of power cables. Shields are usually made of either a metallic tape or a metallic mesh and surround one or more conductors. Shields are typically incorporated in power cables with rated voltages greater than 1 kV, and in power cables that are expected to have large CM currents. Figure 13.24 depicts a four-conductor cable where each conductor has a copper foil shield, and the overall cable has a mesh shield.

The conductor shield is normally grounded on one end and is intended to provide a near-uniform voltage potential around the conductor insulation. This uniform voltage potential minimizes the electric field gradient on the insulation; too large an electric field gradient can result in insulation failure. Which end to ground is situationally dependent. Normally, one would choose the side closest to the grounding resistor connection to the ship's hull. If a conductor-to-shield fault should occur, the ground fault current path length through the ship's hull back to the grounding resistor is minimized. For cables that are part of a bus-tie, either end could be closest to the ship's hull grounding point based on the configuration of the electric plant. For these cases, the choice can be arbitrary. The ampacity of the conductor shield should be sufficient to carry the ground fault current. A line-to-conductor shield fault on one phase conductor coupled with a line-to-ground fault (including but not limited to line-to-shield faults) on another phase conductor will result in a line-to-line short circuit. Unless the fault is cleared very rapidly, the fault current will likely exceed the capability of the conductor shield, resulting in a cable failure. Cable selection and installation practices should minimize the risk of a conductor-to-shield fault.

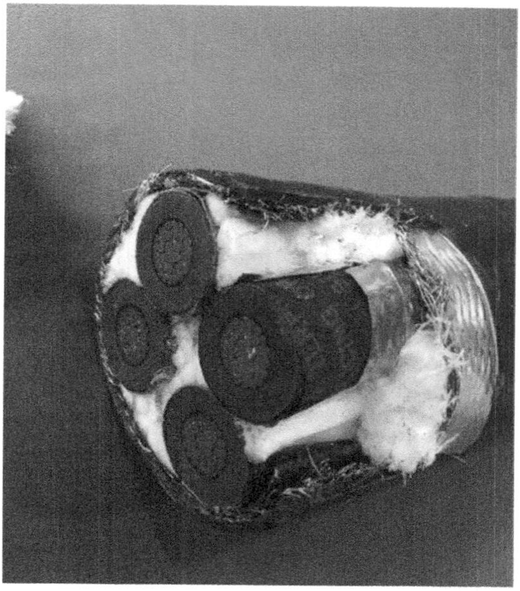

Figure 13.24 Four-conductor cable with conductor shield and cable shield. *Source:* Norbert Doerry.

The cable shield surrounds all the current-carrying conductors. It is normally grounded on both ends and provides an alternate path for CM currents to flow. Figure 13.25 depicts models of a cable with a cable shield (top) and a cable without a cable shield (bottom). This cable is powered with a 60 Hz DM voltage source and a 6 kHz CM voltage source. The sources are solidly grounded. Line-to-ground capacitances are included at the resistive load. Figure 13.26 depicts the CM currents for these two cases. I(R11) is the CM current through the hull for the unshielded cable. I(R4) is the CM current through the hull for the shielded cable. I(R3) is the current through the cable shield. In this example, the hull current is reduced by roughly one-third by including the shield. This also implies that the RE from the shielded cable at 6 kHz will also be reduced by roughly one-third at a sufficiently large distance from the cable.

If one only considered resistances in determining the relationship between CM current flowing through the shield and the hull, one would incorrectly conclude that the vast majority of the CM current would flow through the lower resistance hull path. Instead, the mutual inductances between the conductors and the cable shield are key factors in the effectiveness of the cable shield in diverting CM current from the ship's hull. Because the current through the cable shield is in the opposite direction from the CM current through the conductors, the mutual inductance subtracts from the conductor self-inductance, thereby reducing the overall impedance of this path. The path through the ship's hull is not influenced by

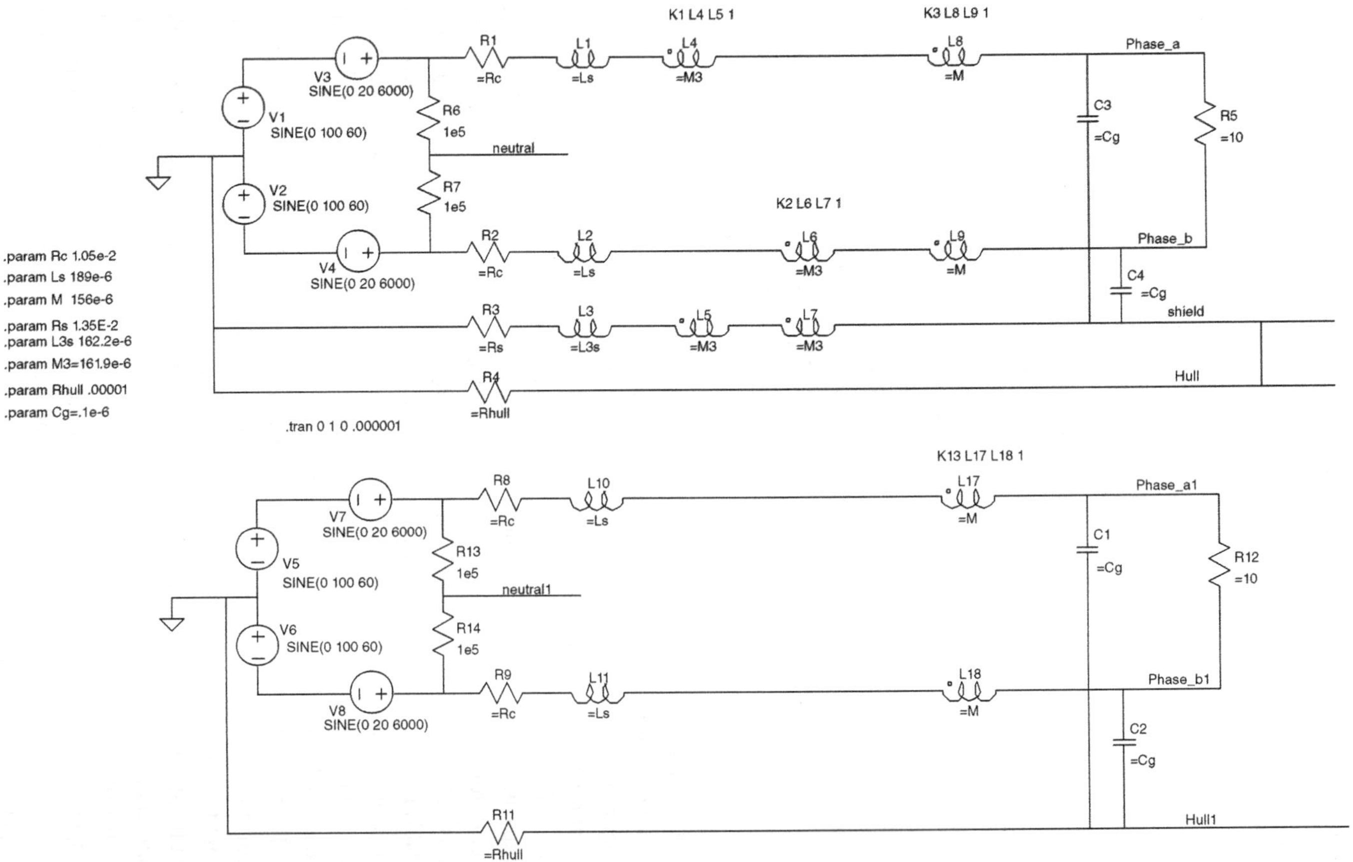

Figure 13.25 Models of a two-conductor cable with and without cable shields.

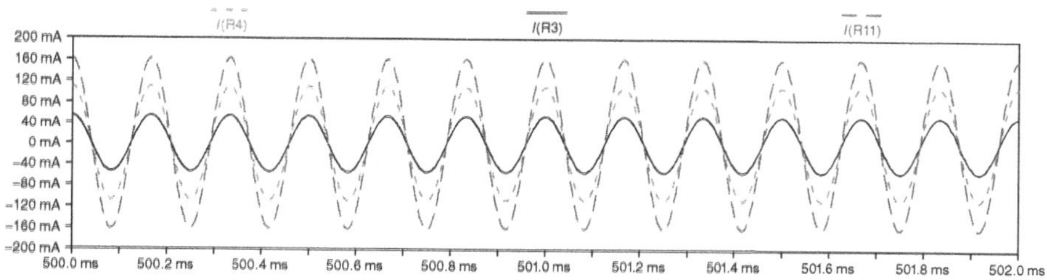

Figure 13.26 CM currents through the ship's hull and cable shield.

mutual inductance, so the full value of the self-inductance contributes to its impedance. While the resistance of the hull is much less than that of the shield, the mutual inductance of the shield results in the impedance of the path through the shield being much closer in magnitude to the impedance of the path through the hull.

Cables designed for use with VFDs will often use drain wires in addition to the cable shield. These conductors are physically adjacent to the phase conductors. Compared to the cable shield, they generally have a much higher current-carrying capacity and a higher mutual inductance with the adjacent phase conductors. See Section 10.6 and IEEE Std. 1580-2021 (2021) for additional details on VFD cable design.

13.3.3 Inductors and CM Chokes

Inductors can be used to increase the impedance for CM currents, but they also increase the impedance for DM currents. Since the impedance of an inductor is proportional to frequency, CM currents at frequencies much higher than the fundamental frequency are attenuated to a much higher degree than the fundamental. Inductors in the form of line reactors are often used on the power system side of a VFD to reduce DM current harmonics and their impact on DM harmonic voltages.

Figure 13.27 depicts a set of balanced inductors for a two-wire system. The constitutive equations for these inductors are

$$v_A = L\frac{di_A}{dt}$$

$$v_B = L\frac{di_B}{dt}$$

These constitutive equations can be expressed in terms of CM voltages and currents.

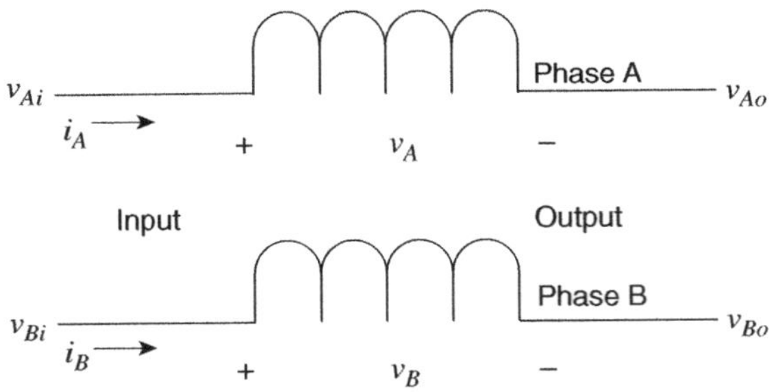

Figure 13.27 Balanced inductors.

$$v_{CMi} = \frac{1}{2}(v_{Ai} + v_{Bi})$$

$$v_{CMo} = \frac{1}{2}(v_{Ao} + v_{Bo})$$

$$v_A = v_{Ai} - v_{Ao}$$

$$v_B = v_{Bi} - v_{Bo}$$

$$v_{CMi} - v_{CMo} = \frac{1}{2}(v_{Ai} + v_{Bi}) - \frac{1}{2}(v_{Ao} + v_{Bo})$$

$$v_{CMi} - v_{CMo} = \frac{1}{2}(v_A + v_B)$$

$$v_{CMi} - v_{CMo} = \frac{1}{2}\left(L\frac{di_A}{dt} + L\frac{di_B}{dt}\right)$$

$$v_{CMi} - v_{CMo} = \frac{1}{2}L\frac{di_{CM}}{dt}$$

The CM impedance at frequency f (Hz) is thus given by

$$Z_{CM} = \frac{1}{2}L(2\pi f)\hat{j}$$

The constitutive equations can also be expressed in terms of the DM voltages.

$$v_{DMi} = v_{Ai} - v_{Bi}$$

$$v_{DMo} = v_{Ao} - v_{Bo}$$

$$v_{DMi} - v_{DMo} = v_{Ai} - v_{Bi} - v_{Ao} + v_{Bo}$$

$$v_{DMi} - v_{DMo} = v_A - v_B$$

$$v_{DMi} - v_{DMo} = L\frac{di_A}{dt} - L\frac{di_B}{dt}$$

$$i_B = -i_A$$

$$v_{DMi} - v_{DMo} = L\frac{di_A}{dt} + L\frac{di_A}{dt}$$

$$v_{DMi} - v_{DMo} = 2L\frac{di_A}{dt}$$

The DM impedance at frequency f (Hz) is, therefore:

$$Z_{DM} = 2L(2\pi f)\hat{j}$$

A CM choke is an improvement over using individual phase inductors for controlling CM currents. A CM choke presents a high impedance to CM currents but a low impedance to DM currents at any frequency. Figure 13.28 depicts a CM choke for a two-wire system. The constitutive equations for this choke are

$$v_A = L\frac{di_A}{dt} + M\frac{di_B}{dt}$$

$$v_B = M\frac{di_A}{dt} + L\frac{di_B}{dt}$$

These constitutive equations can be expressed in terms of CM voltages and currents.

$$v_{CMi} - v_{CMo} = \frac{1}{2}(v_A + v_B)$$

$$v_{CMi} - v_{CMo} = \frac{1}{2}\left(L\frac{di_A}{dt} + M\frac{di_B}{dt} + M\frac{di_A}{dt} + L\frac{di_B}{dt}\right)$$

$$v_{CMi} - v_{CMo} = \frac{1}{2}(L + M)\frac{d(i_A + i_B)}{dt}$$

$$v_{CMi} - v_{CMo} = \frac{1}{2}(L + M)\frac{di_{CM}}{dt}$$

The CM impedance at frequency f (Hz) is thus given by

$$Z_{CM} = \frac{1}{2}(L + M)(2\pi f)\hat{j}$$

Figure 13.28 CM choke.

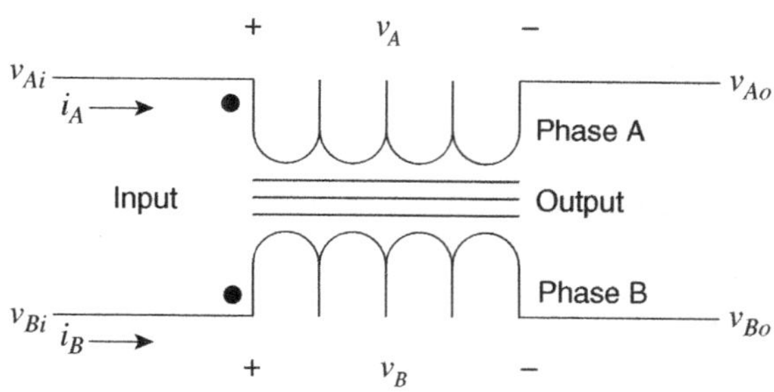

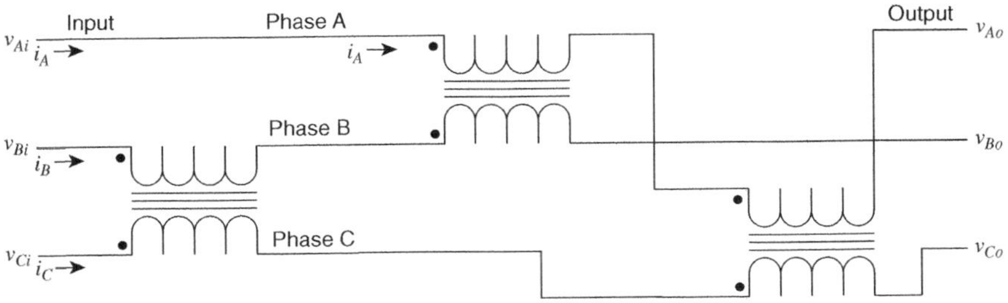

Figure 13.29 Three-phase CM choke.

The constitutive equations can also be expressed in terms of the DM voltages.

$$v_{DMi} - v_{DMo} = v_A - v_B$$

$$v_{DMi} - v_{DMo} = L\frac{di_A}{dt} + M\frac{di_B}{dt} - M\frac{di_A}{dt} - L\frac{di_B}{dt}$$

$$i_B = -i_A$$

$$v_{DMi} - v_{DMo} = L\frac{di_A}{dt} - M\frac{di_A}{dt} - M\frac{di_A}{dt} + L\frac{di_A}{dt}$$

$$v_{DMi} - v_{DMo} = 2(L - M)\frac{di_A}{dt}$$

The DM impedance at frequency f (Hz) is, therefore:

$$Z_{DM} = 2(L - M)(2\pi f)\hat{j}$$

For a CM choke design, the goal is to make the self-inductance L and mutual inductance M as large as possible and as equal as possible. This ensures that their sum is large for a very high CM impedance, and their difference is very small for a very low DM impedance. As depicted in Figure 13.29, CM chokes can also apply to three-phase systems. To a certain degree, the mutual inductance of multiconductor cable results in the cables themselves acting as CM chokes.

In some cases, it may be desirable to combine the functionality of line reactors for controlling DM current harmonics and CM chokes into a single device; the device would be very similar to a choke but would be designed to have a DM impedance needed to control the DM current harmonics.

13.3.4 Capacitors and CM Shunts

A CM shunt presents a low impedance to CM currents but a high impedance to DM currents. Shunts used in combination with CM chokes can be used to limit the CM voltage at an interface.

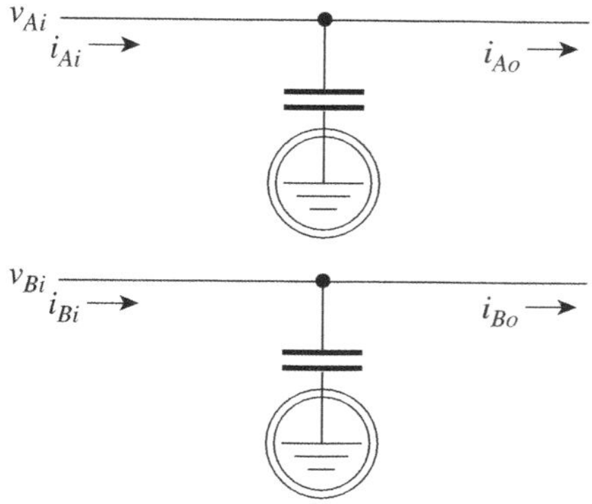

Figure 13.30 Capacitor shunts.

Capacitor shunts can be used to shunt high-frequency CM currents. They do, however, also provide a low impedance to ground for high-frequency DM currents. The capacitors must be rated to handle the AC currents from both the CM currents and the DM currents.

The constitutive equations for a two-wire capacitor shunt, as depicted in Figure 13.30, are given by

$$i_{Ai} - i_{Ao} = C\frac{dv_{Ai}}{dt}$$

$$i_{Bi} - i_{Bo} = C\frac{dv_{Bi}}{dt}$$

$$i_{CMi} - i_{CMo} = C\frac{dv_{Ai}}{dt} + C\frac{dv_{Bi}}{dt} = 2C\frac{dv_{CMi}}{dt}$$

The CM impedance at frequency f (Hz) is

$$Z_{CM} = -\frac{1}{2(2\pi f)C}\hat{j}$$

The DM impedance at frequency f (Hz) can be derived:

$$i_{Ai} - i_{Ao} - i_{Bi} + i_{Bo} = C\frac{dv_{Ai}}{dt} - C\frac{dv_{Bi}}{dt}$$

$$i_{Ai} - i_{Ao} = \frac{C}{2}\frac{dv_{DM}}{dt}$$

$$Z_{DM} = -\frac{2}{(2\pi f)C}\hat{j}$$

An inductor-based CM shunt is depicted in Figure 13.31. The constitutive equations for this shunt are

$$v_{Ai} = L\frac{d(i_{Ai} - i_{Ao})}{dt} - M\frac{d(i_{Bi} - i_{Bo})}{dt}$$

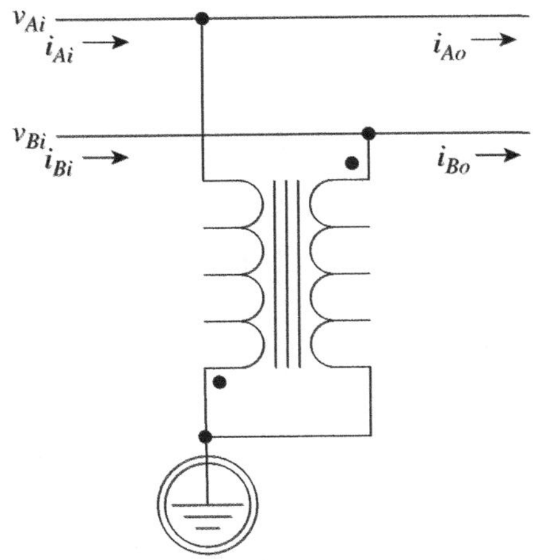

Figure 13.31 Inductor-based shunt.

$$v_{Bi} = L\frac{d(i_{Bi} - i_{Bo})}{dt} - M\frac{d(i_{Ai} - i_{Ao})}{dt}$$

These equations can be expressed in terms of CM voltages and currents.

$$v_{Ai} = \frac{v_{Ai} + v_{Bi}}{2} = \frac{1}{2}L\frac{d(i_{Ai} - i_{Ao} + i_{Bi} - i_{Bo})}{dt} - \frac{1}{2}M\frac{d(i_{Ai} - i_{Ao} + i_{Bi} - i_{Bo})}{dt}$$

$$v_{CM} = \frac{v_{Ai} + v_{Bi}}{2} = \frac{1}{2}(L - M)\frac{d(i_{CMi} - i_{CMo})}{dt}$$

The CM impedance at frequency f (Hz) is

$$Z_{CM} = \frac{1}{2}(L - M)(2\pi f)\,\hat{j}$$

Expressed in terms of DM voltages and currents results in

$$v_{ADM} = v_{Ai} - v_{Bi} = L\frac{d(i_{Ai} - i_{Ao} - i_{Bi} + i_{Bo})}{dt} + M\frac{d(i_{Ai} - i_{Ao} - i_{Bi} + i_{Bo})}{dt}$$

$$i_{Ai} = -i_{Bi}$$

$$i_{A0} = -i_{Bo}$$

$$v_{ADM} = 2(L + M)\frac{d(i_{Ai} - i_{Ao})}{dt}$$

The DM impedance at frequency f (Hz) is

$$Z_{DM} = 2(L + M)(2\pi f)\,\hat{j}$$

As is the case for a CM choke design, the goal for the inductor-based shunt design is to make the self-inductance L and mutual inductance M as large as possible and as equal as possible. This ensures that their sum is large for a very high

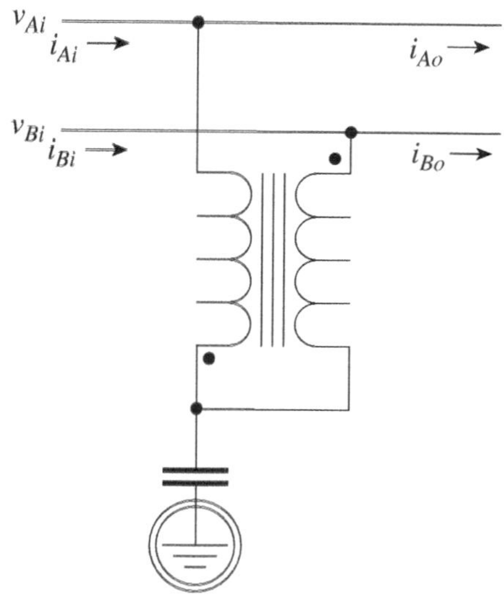

Figure 13.32 Combination inductor-based shunt and capacitor.

DM impedance and that their difference is very small for a very low CM impedance.

If the DM inductance is too low at low frequencies, a capacitor can be inserted in the ground circuit as shown in Figure 13.32. In this combination shunt, the capacitor ensures high DM impedance at low frequencies, and the inductor ensures high DM impedance at high frequencies. The DM circuit will have a resonance when

$$-\frac{2}{(2\pi f)C}\hat{j} + 2(L+M)(2\pi f)\hat{j} = 0$$

$$f = \frac{1}{(2\pi)}\sqrt{\frac{1}{C(L+M)}}$$

To preclude large DM currents, it is important to ensure that the frequency component of the DM voltage at or near this resonant frequency is very low, ideally at zero.

Each of the two-wire shunts can be extended to three or more phases. For DC systems, capacitors are normally employed, while for AC systems, any one of the shunt approaches may be employed.

13.3.5 CM Control Philosophies

13.3.5.1 Cable Only Solution

If the CM currents are projected to be modest, one approach for minimizing the impact of the CM currents on RE is to use cable (such as VFD cable) with a cable shield, drain wires, or ground wires that are grounded at both ends. These cables are used on both sides of a CM voltage source until that point in the system where the CM currents are negligible. The grounded shields and conductors will divert some of the ground current from the ship's hull, as well as reduce the magnetic fields due to the CM current from the main conductors. In some cases, the VFD cable will need to be employed all the way from the original source of power (transformer or generator) for the separately derived system to the final load equipment.

13.3.5.2 Cable and CM Choke Solution

If the CM currents are larger than desired, but there are only a few CM voltage sources, CM chokes in combination with VFD (or similar) cable may be sufficient (Figure 13.33). The CM chokes reduce the magnitude of the CM current to a reasonable level, and the grounded shield/conductor diverts some of the remaining ground current from the ship's hull. A challenge with this approach is determining an appropriate power rating for the CM choke, since the power it must handle is a function of the CM characteristics of all the components in the system.

13.3.5.3 Controlling CM at the Interfaces

In a power system with many CM voltage sources, controlling the CM at equipment interfaces may prove to be an effective solution. Under normal steady-state operation, equipment should be designed such that, at their power interfaces, a CM short to ground will result in a relatively small CM current through the power interface, and a large CM impedance to ground will result in a CM voltage that does not stress the insulation of the cable or equipment, or cause arcing in rotating machine bearings. With appropriate interface controls, the use of VFD cable may not be required. Figure 13.34 depicts Option A, where CM voltage control at Interface 1 is required, but not at Interface 2. If one of the interfaces is DC, it is generally easier to implement a CM shunt or choke on the DC interface. The CM choke may be on the side of the equipment with the CM shunt, as depicted in Figure 13.34, or it could be on the opposite side of the equipment from the CM shunt, as depicted in Figure 13.35. If the CM voltage must be minimized on both interfaces, Option C, as depicted in Figure 13.36, may be employed. If the CM characteristics of the rest of

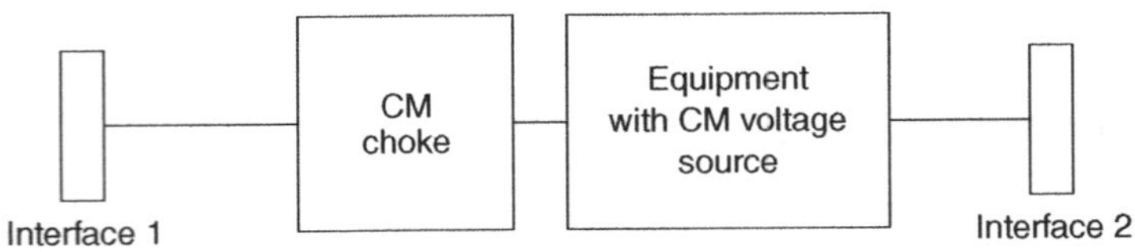

Figure 13.33 Use of CM choke to control CM currents.

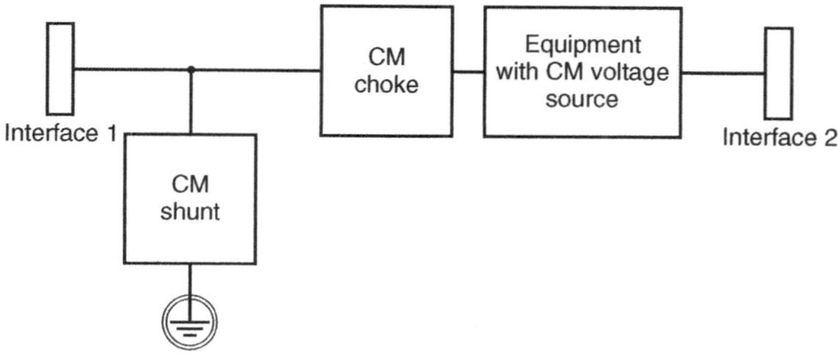

Figure 13.34 Use of CM choke and CM shunt to control CM at interfaces: Option A.

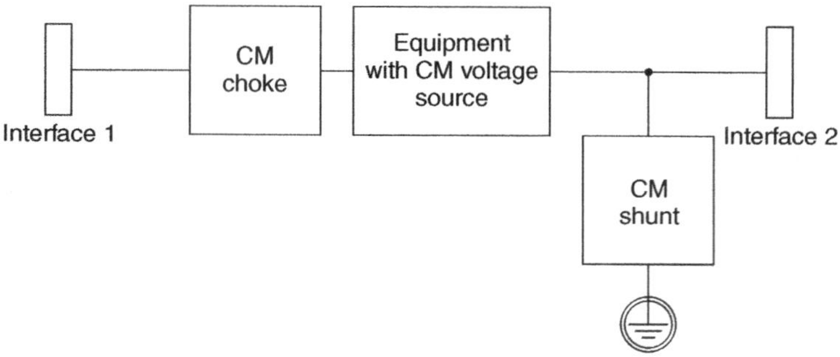

Figure 13.35 Use of CM choke and CM shunt to control CM at interfaces: Option B.

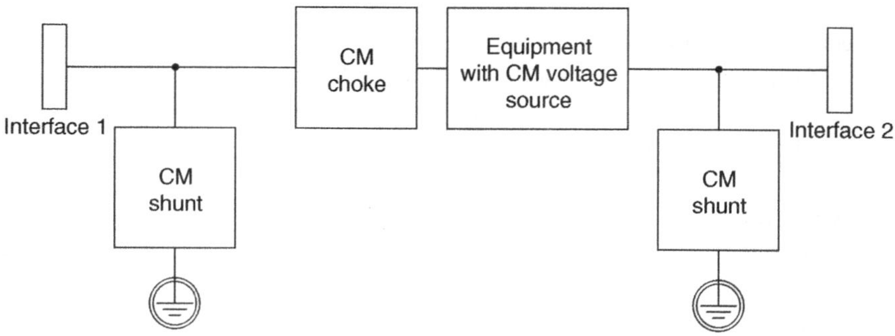

Figure 13.36 Use of CM choke and CM shunt to control CM at interfaces: Option C.

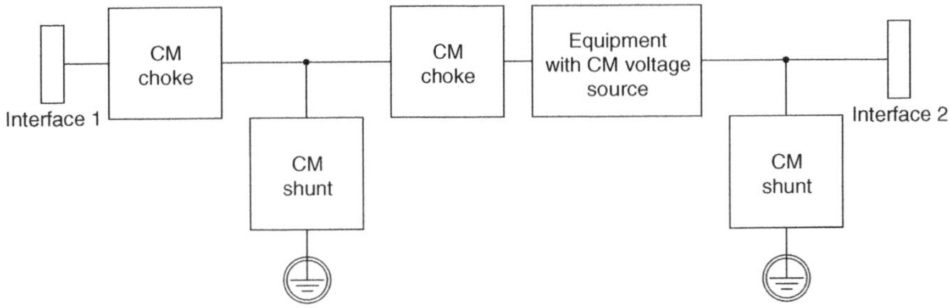

Figure 13.37 Use of CM choke and CM shunt to control CM at interfaces: Option D.

the system are not known, then adding an additional CM choke, as depicted in Figure 13.37, may prove beneficial.

If any of the options depicted in Figures 13.34 through 13.37 (or similar) are employed, a considerable number of analyses and simulations should be performed to verify the design will work as intended and that all of the components are rated appropriately. In particular, the designs should be analyzed for performance under a sustained ground fault anywhere in the separately derived system. The impact on ground fault detection and localization systems should also be assessed.

13.3.5.4 Controlling CM at the HRG

It is also possible to modify the CM behavior of a separately derived system by modifying the impedance of a high-resistance grounding resistor. For example, a capacitor in parallel with the grounding resistor will provide a low impedance for high-frequency CM currents, while the grounding resistor provides a minimum impedance at low frequencies. The system grounding design should be analyzed for performance under a sustained ground fault anywhere in the separately derived system. The impact on ground fault detection and localization systems should also be assessed.

13.3.5.5 Controlling CM Within Motors and Generators

Rotating electrical machines such as motors and generators can couple CM currents and voltages onto the rotating shaft. Figure 13.38 depicts the CM features of a motor that incorporates CM mitigation. The mitigation features include insulating the bearings from the frame, using a shaft brush to provide a low-resistance connection to ground, and an insulated shaft coupling. These features are not usually included in motor specifications; if desired, they should be added to the specifications. In some cases, instead of insulating the bearing housing, ceramic bearings may be employed to provide insulation. However, manufacturers may

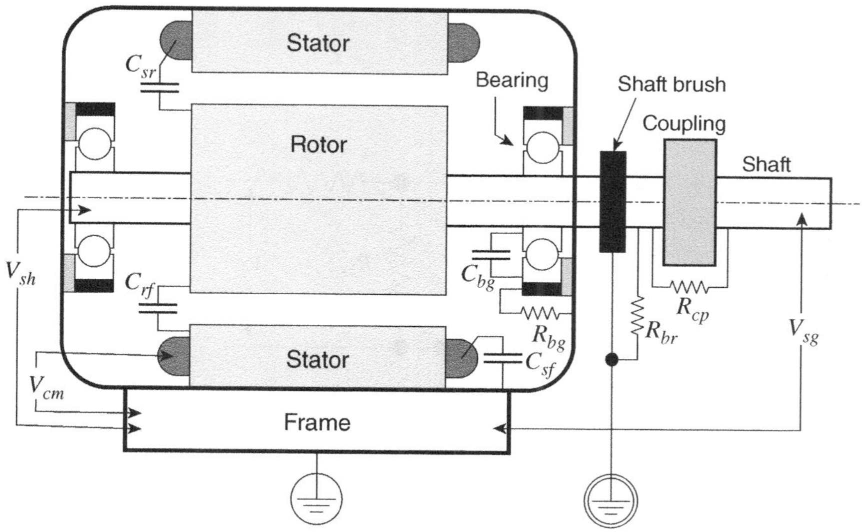

Figure 13.38 CM features of a motor.

not be willing to provide insulated bearings. In this case, the manufacturer should recommend the type of grease to use in the bearings that minimizes the chances of bearing pitting damage and bearing corrosion due to CM voltages and currents. The insulated coupling is recommended if the mechanical load on the motor is sensitive to CM currents; reduction gears in particular may be damaged by CM currents. Figure 13.39 depicts a schematic of the CM model of the motor. The parameters for this model are not typically provided by a manufacturer; they usually must be either estimated or measured once the motor is acquired. IEEE 1349-2021 (2021) provides guidance for estimating these parameter values; IEEE 1349-2021 (2021) also provides a link to a freely available spreadsheet for their estimation. Section 15.4.4 provides guidance for determining a safe bearing current to prevent pitting damage.

13.3.6 CM AC Limits

In designing a system to keep CM voltages and currents low at interfaces, defining what "low" means is imperative. Unfortunately, this definition does not explicitly exist. Instead, we must infer the limits from existing regulations and guidance. Examples include

- The IMO Rules and Regulations for the Classification of Ships part 6, chapter 2, section 6.4 (2022) require that the current through any intentional connection to

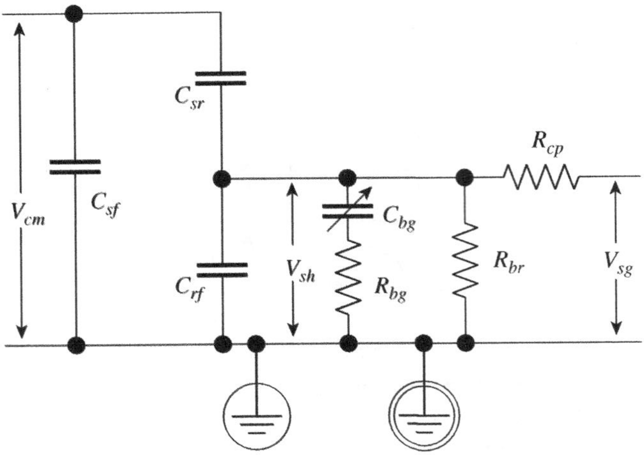

Figure 13.39 CM model of a motor.

earth be continuously monitored. If the current through any earth connection exceeds 5 A, then protective action must be initiated (sounding an alarm and either limiting the current through the connection to earth to below 5 A or isolating the source of the CM current).

- IEEE 3003.1-2019 (2019) recommends ground fault currents in high-resistance grounded systems should not exceed 10 A, although systems with nominal system voltages in the 15 kV class may require higher values. The 10 A limit is intended to minimize damage to equipment.
- EMI requirements often combine the impacts of nonfundamental frequency components of CM and DM currents. MIL-STD-461 (2015) CE-101 requirements provide limits for the frequency content of the feeder cable current (both CM and DM) over the frequency band of 30 Hz to 10 kHz. For frequencies below 2.6 kHz, the maximum current magnitude in one conductor for any one frequency component is 56 mA for load currents no greater than 3 A and 3.16 A for load currents no less than 185 A. If all of the frequency content were due to CM currents, then the overall CM current would be the limit for one conductor multiplied by the number of power conductors in the cable.
- Appendix C describes the impact of electrical currents on the human body. These impacts are used as the basis for tripping GFCIs (4–6 mA), RCDs and RDBOs (10 mA), and ALCIs (30 mA). These tripping current levels are generally limited to AC low-voltage secondary distribution systems.

None of these requirements provide an explicit limit on the CM current in a cable supplied by a source or a feeder cable to a load under normal conditions. Instead, each design should devise an approach that will facilitate meeting the various requirements through a systems integration process. Separate approaches apply to interfaces to the power system and interfaces to dedicated loads (such as motors driven from a VFD).

The interfaces to the dedicated loads should depend on the susceptibility of the load to CM voltages and currents. This can only be determined through discussions with the manufacturer or via testing.

One approach to CM limits for interfaces to the power system establishes the maximum value for the CM current a source or load may supply into a CM short circuit to ground on the feeder cable. This CM short circuit to ground is part of the test apparatus and does not include any equipment installed onboard the ship. This CM short-circuit current can be analyzed by simulation in the design phases and verified experimentally through the use of CM shunts to ground. For this approach, reasonable limits for the rms ground current through the CM shunt under steady-state un-faulted operation would be of the form:

- Not exceed 500 mA for loads rated greater than 185 A.
- Not exceed 0.27% of the rated load current for loads with a rated current no greater than 185 A but greater than 3 A.
- Not exceed 8.1 mA for loads with a rated current no greater than 3 A.

Under normal operation, the contribution of the equipment to CM current on the bus should be less than the CM current into the CM short circuit. However, the design of the CM shunts must also account for CM current contributions from the bus in addition to the CM currents from the equipment. Based on the IMO requirement, the CM current through the shunt should be limited to 5 A. Hence, the CM choke, if needed, should be designed to limit the CM current from equipment that is shunted to a value less than 5 A to allow the CM choke to also shunt CM current from the bus. Simulation should be used to verify that elements of the CM mitigation circuit are sufficiently rated.

13.3.7 CM DC Limits

DC currents in DC distribution cables result in a DC magnetic field around the distribution cable, approximated by the same equation used for AC fields from Section 13.2:

$$B_{cable} \approx \frac{\mu_0}{2\pi R}\left(I_{CM} - \sum_{i=1}^{n_{conductors}} \left(\frac{r_i}{R}\right) I_i\right)$$

The field from the cable should be limited to address:

- Not interfering with degaussing systems.
- Not impacting the operation of equipment.

The degaussing system is intended to reduce or eliminate a remnant magnetic field in the ship's hull; this remnant field is usually a result of exposure to the earth's magnetic field. The earth's magnetic field has a magnetic flux density of about 50 μT; restricting B_{cable} to 10% of the earth's magnetic flux density, or 5 μ T, should prevent interference with the degaussing system:

$$\frac{2\pi B_{cable}}{\mu_0} > \frac{I_{CM}}{R}$$

$$\frac{2\pi 5 \times 10^{-6}}{4\pi \times 10^{-7}} > \frac{I_{CM}}{R}$$

$$25 > \frac{I_{CM}}{R}$$

If we assume the cable is kept at least 20 cm from the ship's structure, then an upper limit for DC CM currents under normal conditions to avoid interfering with the degaussing system for I_{CM} is 5 A DC.

Some shipboard equipment may be susceptible to DC magnetic fields. Naval equipment is required by MIL-STD-1399-407 (2022) to operate satisfactorily when exposed to a magnetic field strength (H) of 1600 A/m. Since multiple cables can contribute to the magnetic field strength, the limit for each should be much less; limiting the field strength of a single cable to 160 A/m should minimize the probability of equipment malfunction due to the magnetic field. The relationship between magnetic flux density (B) and magnetic field strength (H) is given by

$$B = \mu H$$

In air, the permeability (μ) is assumed to be μ_0.

The limit then becomes

$$\frac{2\pi \mu_0 H_{cable}}{\mu_0} = 2\pi H_{cable} > \frac{I_{CM}}{R}$$

or

$$2\pi\, 160 \approx 1005 > \frac{I_{CM}}{R}$$

If we assume the cable is kept at least 20 cm from any susceptible equipment, then the upper limit for DC CM current under normal conditions to avoid interfering with susceptible equipment is 201 A DC; limiting CM DC currents based on not interfering with degaussing will likely be sufficient for ensuring equipment can operate properly.

Additionally, DC CM current should be limited to minimize the probability that hull currents will result in galvanic corrosion and to minimize the probability of electrical shock.

If we assume all of the CM current is a corrosion current for steel, then from Chapter 15, the amount of mass lost (M_{iron}) in a given amount of time (t) would be

$$M_{iron} = kI_{CM}t$$

where $k = 0.000289\ \text{g}/\text{A} \cdot \text{s}$.

The thickness of the iron lost T (m) in time t is given by

$$T = \frac{kI_{CM}t}{\rho A}$$

where

ρ, density of iron ($7.874e^6$ g/m^3)
A, surface area of the iron (m^2).

The thickness of material lost depends strongly on the surface area. If we assume the corrosion thickness losses are localized to $A = 1\ \text{m}^2$ and $t = 1$ year ($31.56e^6$ seconds), then:

$$\frac{T}{I_{CM}} = \frac{0.000289 \times 31.56}{7.874 \times 1} = .001597$$

MIL-STD-889 indicates that corrosion of less than 0.009 mil/ear ($2.286e^{-7}$ m) is acceptable. This rate is achieved when I_{CM} equals 0.20 mA. This value is very conservative since it is very unlikely the CM current would fully participate in galvanic corrosion. However, if the DC CM current can be kept to less than 0.20 mA under normal conditions, then the probability that the DC CM current will contribute to corrosion is minimal.

Appendix C describes the impact of electrical currents on the human body. In general, the human body can tolerate higher DC currents for the same level of response as compared to AC currents. Based on the data presented in Appendix C, a CM limit of 40 mA DC in a cable is likely safe. This level of current should be evaluated during a ground fault condition. Using an HRG system as described in Chapter 7 should result in a CM limit on the order of 10 mA during a ground fault; hence, a safe CM current level during a ground fault should not be an issue if the recommendation for grounding resistance is followed.

13.4 Advanced CM Modeling

The CM performance of a system may be estimated by combining frequency-based Thévenin CM models of individual components. This type of analysis can provide best-case and worst-case estimates of CM voltages and currents as a function of frequency. IEEE Std 45.1-2023 (2023) provides an introduction to this type of analysis as well as additional references. Consideration should be given to using Thévenin CM models for separately derived systems with many CM voltage sources or high levels of CM current.

13.5 Design Considerations

Because the CM performance of power electronic equipment and rotating machines typically is not known during the early stages of design, establishing space and weight reservations for CM chokes and shunts is recommended. Once the CM performance of equipment has been assessed and verified, the appropriate CM chokes and shunts can be incorporated into the design, or the space and weight reservations can be given up.

References

ABS (2023). *Rules for Building and Classing Marine Vessels (MVR)*. Spring, TX: American Bureau of Shipping. Available from http://eagle.org.

CFR Title 47 Chapter I subchapter A Part 15 Subpart B clause 15.107. *Conducted Limits*. Code of Federal Regulations. Available from https://www.ecfr.gov/current/title-47/chapter-I/subchapter-A/part-15/subpart-B/section-15.107.

IEC 60945:2002 (2002). *Maritime Navigation and Radiocommunication Equipment and Systems – General Requirements – Methods of Testing and Required Test Results*. Geneva: International Electrotechnical Commission. Available from http://webstore.iec.ch.

IEC 61800-3:2022 (2022). *Adjustable Speed Electrical Power Drive Systems – Part 3: EMC Requirements and Specific Test Method for PDS and Machine Tools*. Geneva: International Electrotechnical Commission. Available from http://webstore.iec.ch.

IEEE Std 45.1-2023 (2023). *IEEE Recommended Practice for Electrical Installations on Shipboard – Design*. New York: Institute of Electrical and Electronics Engineers. Available from http://ieeexplore.ieee.org.

IEEE Std. 1349-2021 (2021). *Guide for the Application of Electric Machines in Zone 2 and Class I, Division 2 Hazardous (Classified) Locations*. New York: Institute of Electrical and Electronics Engineers. Available from http://ieeexplore.ieee.org.

IEEE Std 1580-2021 (2021). *IEEE Recommended Practice for Marine Cable for Use on Shipboard and Fixed or Floating Facilities*. New York: Institute of Electrical and Electronics Engineers. Available from http://ieeexplore.ieee.org.

IEEE Std 3003.1-2019 (2019). *IEEE Recommended Practice for System Grounding of Industrial and Commercial Power Systems*. New York: Institute of Electrical and Electronics Engineers. Available from http://ieeexplore.ieee.org.

IMO (2022). *Rules and Regulations for the Classification of Ships*. London: International Maritime Organization.

MIL-STD-461 (2015). *Interface Standard Requirements for the Control of Electromagnetic Interference Characteristics of Subsystems and Equipment*. Washington, DC: Department of Defense. Available from http://quicksearch.dla.mil.

MIL-STD-1399 Section 407 (2022). *Interface Standard Section 407 DC Magnetic Field Environment*. Washington, DC: Department of Defense. Available from http://quicksearch.dla.mil.

14

Protective Earthing: Bonding

14.1 Introduction

As stated in Chapter 1, bonding is the act of deliberately connecting exposed metal parts that are not designed to carry electrical currents under normal operation to the hull of the ship via a low-impedance path. Bonding is a safety measure to prevent electrical shock to personnel caused by capacitively coupled voltages to the exposed metal parts, inductively coupled currents in exposed metal parts, or insulation failures. Dangerous static charges can result from the flow of fluid in piping; piping and tanks typically require bonding to the hull when the static charge could pose a hazard. While bonding will most likely conduct CM currents from equipment to ground, this feature is not the intended use of bonding. Where possible, separate CM circuits should be established to minimize CM currents during normal operation.

The terms "equipment earthing" and "equipment grounding" are equivalent to bonding; they are used by classification societies and in some standards. The term "equipotential" means that two conductors have equal voltages with respect to a reference voltage. Two conductors that are electrically connected with a low impedance are equipotential. "Equipotential bonding" and "equipotential grounding" are equivalent to "bonding"; the conductors are at the same voltage as the hull of the ship. The key is that the impedance must be low enough such that the voltage difference between the conductors is less than a safety limit (typically on the order of 40 V) when the current between the conductors is at a maximum value: typically, the fault current.

As described in Chapter 11, a special case for bonding may exist in shore power connections. For this special case, the hull of the ship and the terrestrial ground may be equipotentially bonded via a conductor in the shore power cables. This equipotential bond may be continuously monitored via a conductor and pilot

Design of Shipboard Power System Grounding/Earthing, First Edition. Norbert Doerry, Mohammed M. Islam, and John Prousalidis.
© 2025 The Institute of Electrical and Electronics Engineers, Inc.
Published 2025 by John Wiley & Sons, Inc.

contact connections in the shore power cable, an equipotential bond monitoring termination device, and an equipotential bond monitoring device.

14.2 Design Considerations

ABS Rules for Building and Classing Marine Vessels provides direction for equipment earthing (ABS 2023). Electrical equipment is required to be bonded unless the supply voltage is not more than 50 V (DC or AC_{rms}), the supply voltage does not exceed 250 V AC_{rms} and is provided by an isolation transformer supplying only one consuming device, or employs double insulation. The minimum cross-sectional area of the earthing conductor is also specified for various conditions.

ABS also requires the use of bonding to control static electric charges from the flow of fluid in piping or from the flow of air in certain ventilation fans.

IEC 60092-401:1980 (1980) provides detailed requirements for equipment earthing on ships. In particular, it requires "handles, handrails, etc., of metal on the bridge or upper decks" to be bonded to the hull or superstructure to "minimize shock from high-frequency voltage induced by the radio transmitter." Figure 14.1 is an example of a bonding strap on a superstructure door.

Figure 14.2 depicts a bonding cable for a gasketed equipment cover to the equipment box. Similarly, Figure 14.3 depicts a bonding cable for bonding an enclosure to the foundation. Figure 14.4 depicts bonding multiple protective earth cables to a single protective earth bus within an electrical apparatus enclosure. Figure 14.5

Figure 14.1 Bonding cable on superstructure door. *Source:* Norbert Doerry.

Figure 14.2 Bonding single-conductor cable for gasketed equipment cover to the equipment box. *Source:* Mohammed M. Islam.

Figure 14.3 Bonding equipment frame to the foundation. *Source:* Mohammed M. Islam.

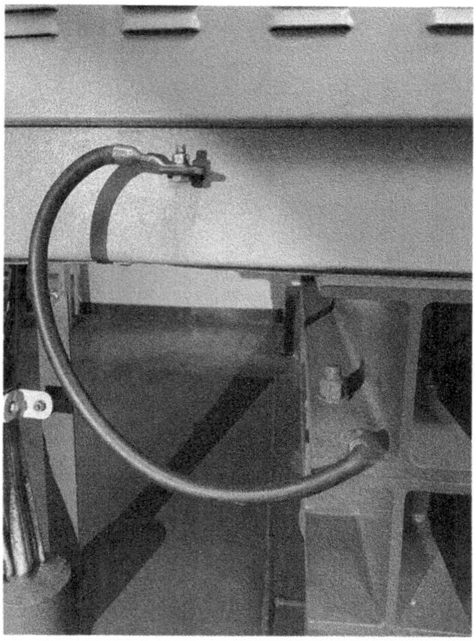

Figure 14.4 Bonding to a common protective earth bus. *Source:* Mohammed M. Islam.

Figure 14.5 Use of a bonding strap to bond the equipment cover to the equipment enclosure. *Source:* Mohammed M. Islam.

depicts the use of a bonding strap to bond an equipment cover to the equipment enclosure.

IEEE Std 45.1-2023 (2023) and IEEE Std 3003.2-2014 (2014) are additional sources of recommendations for equipment bonding. The National Electric Code, NFPA 70 (2023), article 250 should also be consulted for understanding equipment grounding and bonding.

MIL-STD-1310 (2009) provides guidance for bonding on naval ships. It defines three classes of bonding methods and specifies the conditions each should be used:

Class A Metallic surfaces bonded by welding or brazing.

Class B Metallic surfaces (bright metal) bonded by bolting or clamping.

Class C Metallic surfaces (bright metal) bonded by bridging them with a metallic (conductive) bond strap.

MIL-STD-2003-2 (2020) provides additional requirements for implementing bonding. Details for employing the different classes of bonding are depicted in Figures 14.6 through 14.9. Figure 14.10 depicts bonding of equipment box covers.

MIL-DTL-24749 (2019) is a specification for grounding straps for electrical bonding; Figures 14.11 through 14.13 depict types of bond straps detailed in the specification.

The resistance of the connection between the equipment and the ship's hull should be small enough such that fault current through the connection will be sufficient to properly coordinate protective devices. Unfortunately, consensus does not exist as to the minimum cross-sectional area of the bonding connection/bond strap to ensure proper protective device coordination. The applicable specifications, classification society rules, and regulations should be used.

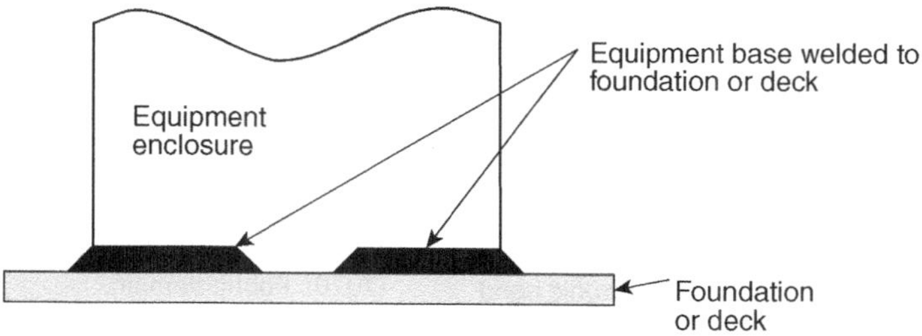

Figure 14.6 MIL-STD-1310 Class A bond. *Source:* Adapted from MIL-STD-2003-2. Public domain.

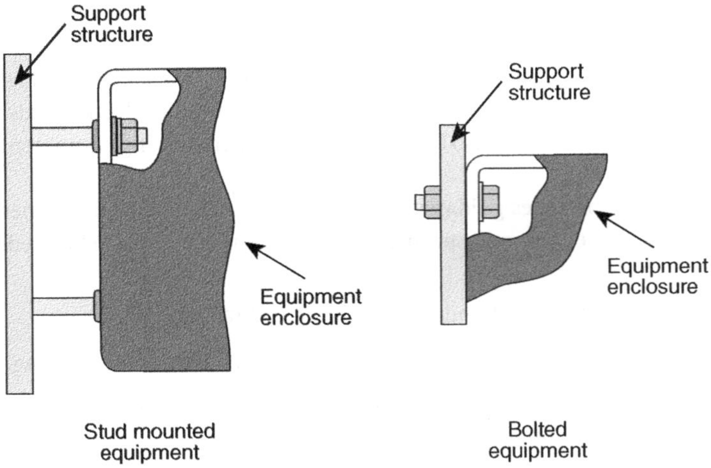

Figure 14.7 MIL-STD-1310 Class B bond. *Source:* Adapted from MIL-STD-2003-2. Public domain.

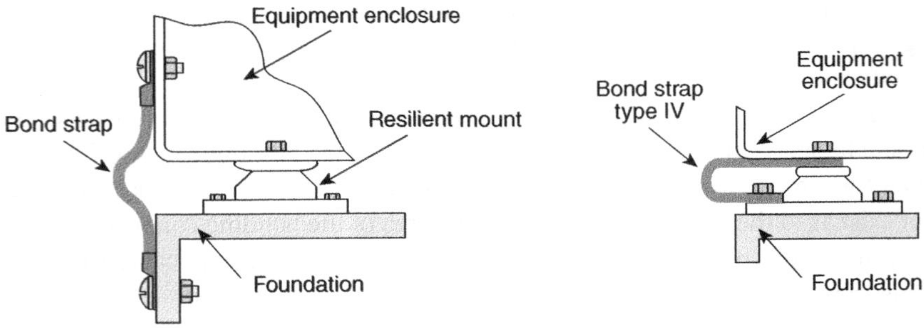

Figure 14.8 MIL-STD-1310 (2009) Class C bond for isolated electrical equipment. *Source:* Adapted from MIL-STD-2003-2 (2020). Public domain.

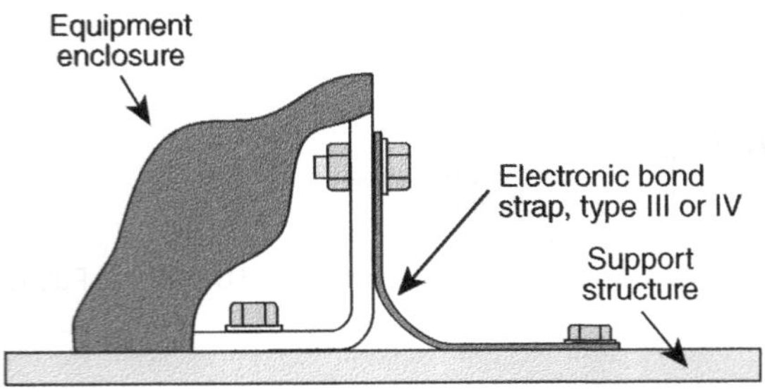

Figure 14.9 MIL-STD-1310 (2009) Class C bond for bolted electrical equipment. *Source:* Adapted from MIL-STD-2003-2 (2020). Public domain.

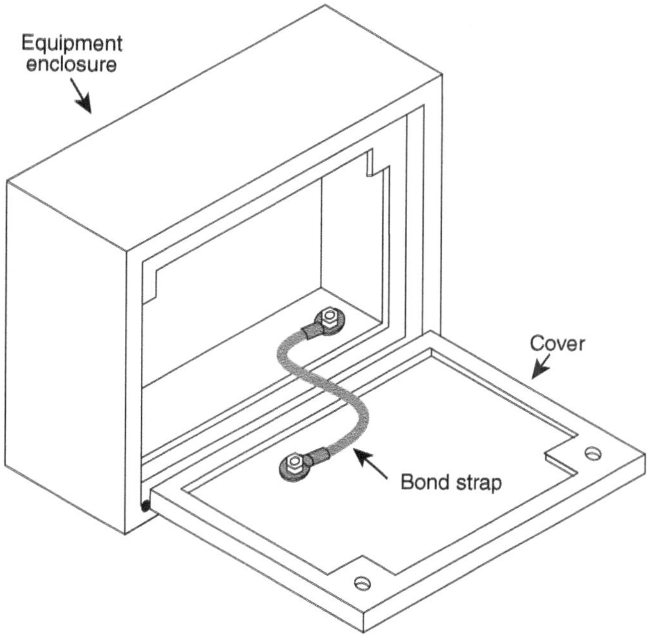

Figure 14.10 Bonding of electrical equipment box covers. *Source:* Adapted from MIL-STD-2003-2 (2020). Public domain.

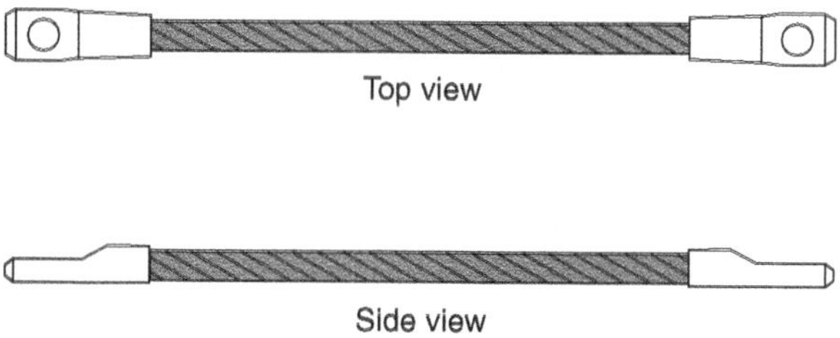

Figure 14.11 Wire cable bond strap, type I. *Source:* Adapted from MIL-DTL-24749 (2019). Public domain.

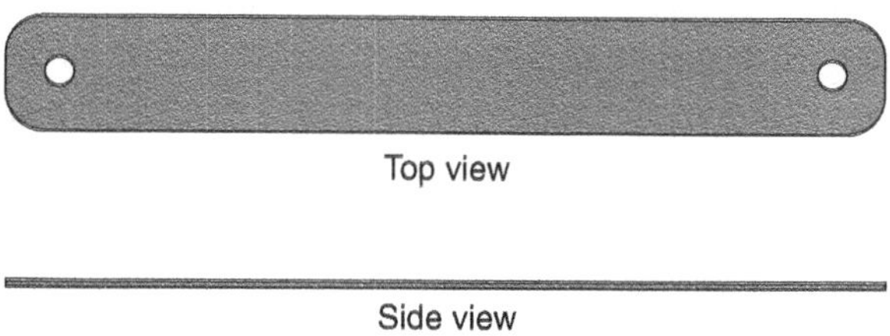

Figure 14.12 Metal strip bond strap, type II and type III. *Source:* Adapted from MIL-DTL-24749 (2019). Public domain.

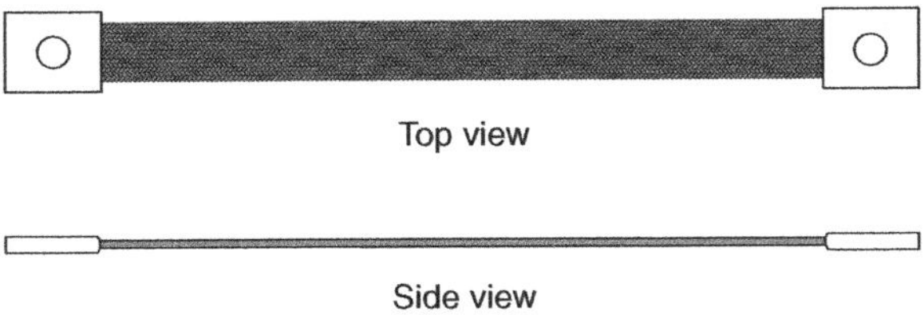

Figure 14.13 Metal braid bond strap, type IV. *Source:* Adapted from MIL-DTL-24749 (2019). Public domain.

14.3 Testing

MIL-STD-1399 Section 300 Part 1 (2018) details a "Simulated human body impedance ground current test." This test is intended for low-voltage AC equipment. It measures the CM current that would flow through a simulated human body should the bonding connection to ground be disconnected. Two limits are specified: 5 mA for frequencies up to 700 Hz, and 70 mA for frequencies above 700 Hz and less than 100 kHz.

Ground continuity tests employ a low-voltage (typically DC) source that is current limited to a low value (typically less than 10 mA); the voltage source is connected between the exposed conductors and the hull of the ship. If the current flows at the current limit, then ground continuity is confirmed; if little to no current flows, then the bonding is insufficient. Often, the voltage is measured, and Ohm's law is employed to determine the resistance of the connection to the hull.

Equipotential bonding verification tests are similar to ground continuity tests except that the current levels are much higher (10–100 A). A bonding strap that is nearly corroded through may pass a ground continuity test but may fail with the higher currents of the equipotential bonding verification tests. The voltage is normally measured, and the resistance calculated to ensure compliance with applicable standards.

References

ABS (2023). *Rules for Building and Classing Marine Vessels (MVR)*. Spring, TX: American Bureau of Shipping. Available from eagle.org.

IEC 60092-401:1980 (1980). *Electrical Installations in Ships – Part 401: Installation and Test of Completed Installation*. Geneva, Switzerland: International Electrotechnical Commission. Available from webstore.iec.ch.

IEEE Std 45.1-2023 (2023). *IEEE Recommended Practice for Electrical Installations on Shipboard – Design*. New York: Institute of Electrical and Electronics Engineers. Available from ieeexplore.ieee.org.

IEEE Std 3003.2-2014 (2014). *Recommended Practice for Equipment Grounding and Bonding in Industrial and Commercial Power Systems*. New York: Institute of Electrical and Electronics Engineers. Available from ieeexplore.ieee.org

MIL-DTL-24749 (2019). *Detail Specification Grounding Straps and Studs, Electrical Bonding*. Washington, DC: Department of Defense. Available from quicksearch.dla.mil.

MIL-STD-1310 (2009). *Standard Practice Shipboard Bonding, Grounding, and Other Techniques for Electromagnetic Compatibility, Electromagnetic Pulse (EMP) Mitigation, and Safety*. Washington, DC: Department of Defense. Available from quicksearch.dla.mil.

MIL-STD-1399 Section 300, Part 1 (2018). *Low Voltage Electric Power, Alternating Current*. Washington, DC: Department of Defense. Available from quicksearch.dla.mil.

MIL-STD-2003-2 (2020). *Standard Practice Electric Plant Installation Standard Methods for Surface Ships and Submarines (Equipment)*. Washington, DC: Department of Defense. Available from quicksearch.dla.mil.

NFPA 70 2023 edition (2023). *National Electrical Code*. Quincy, MA: National Fire Protection Association. Available from www.nfpa.org.

15

Current-Related Corrosion

15.1 Introduction

Preventing, managing, and removing corrosion from metal components, structures, and hull plating is a continuous process onboard a ship. In the United States, corrosion is defined in law (Title 10 U.S.C. § 2228) as "the deterioration of a material or its properties due to a reaction of that material within its chemical environment." While corrosion can result from many processes, including mechanical stress, erosion, and leaching, this chapter concentrates on two mechanisms that are strongly influenced by electric current: galvanic corrosion and pitting due to lubricating oil breakdown.

This chapter provides an introduction to current-induced corrosion on ships. It is intended to provide the reader with sufficient understanding to enable an intelligent conversation with a corrosion engineering expert, as well as to provide the reader with motivation to minimize hull currents.

15.2 Galvanic Corrosion Theory

Galvanic corrosion occurs when dissimilar metals in electrical contact are also in contact with a fluid electrolyte. This combination of dissimilar metals and electrolyte is called a galvanic cell; one of the metals serves as an anode, and the other as a cathode. At the anode, metal dissolves into the electrolyte to form positively charged metal ions and electrons. The electrons flow through the electrical connection to the cathode, where they enter the electrolyte. Within the electrolyte, the electrons participate in a cathode reaction. For ships, the electrolyte is usually seawater; the cathode reaction in seawater is typically either the reduction of dissolved H^+ to H_2 or O_2 and H_2O to OH^-.

Design of Shipboard Power System Grounding/Earthing, First Edition. Norbert Doerry, Mohammed M. Islam, and John Prousalidis.
© 2025 The Institute of Electrical and Electronics Engineers, Inc.
Published 2025 by John Wiley & Sons, Inc.

Because ships carry many different types of fluids, the electrolyte in a particular situation may have a very different composition than seawater; a different cathode reaction may result. This chapter focuses on galvanic corrosion in a seawater environment.

Table 15.1 lists common metals found onboard a ship. For a pair of metals in contact, the metal listed first in the table will serve as the anode, and the metal listed second will serve as the cathode. Alloys of materials can significantly change their position within the table as compared to the base material. Aluminum alloys, for example, can be more anodic than zinc or more cathodic than tin, depending on their chemical composition.

Table 15.1 does not, however, provide any guidance as to the rate of corrosion of the anode when in seawater. MIL-STD-889 (2021) states that dissimilar metals are compatible if the anode corrosion rate is no greater than 0.009 mil/year for a specific test. Compatible metals do not require corrosion protection. Unfortunately, the common metals found onboard a ship (aluminum, steel, brass, bronze, copper, and stainless steel) are not compatible; corrosion protection is required.

Table 15.1 Galvanic series.

Active (anodic) ***Source:*** **MIL-STD-889 (2021)**	**Half-cell potential in salt water (V)** ***Source:*** **Flinn and Trojan (1981)**
Magnesium	−1.73
Zinc	−1.10
Cadmium	−0.82
Aluminum	—
Steel	−0.58
Tin	−0.49
Brass	—
Bronze	—
Copper	−0.20
Stainless steel	−0.09
Monel®	—
Passivated stainless steel	—
Silver	−0.08
Graphite	—
Platinum	—
Less active (cathodic)	—

MIL-STD-889 (2021) provides the following guidance for joining conductive metals that are not compatible:

a) Compatible conductive materials shall be selected such that the corrosion rate when coupled does not exceed 0.009 mil/year in immersed conditions, and coupled conductive materials shall be designed to ensure that the area of the cathode is smaller than the area of the anode. For example, bolts or screws of stainless steel may be used for fastening aluminum sheet, but not the reverse, since stainless steel is cathodic to the aluminum.
b) For couples that serve as an electrical connection, a compatible conductive gasket or washer shall be interposed between the dissimilar faying surface materials prior to fastening, or the cathodic member shall be plated with a material compatible with the anode.
c) For couples that are not serving as an electrical conductor, a non-absorbing, inert gasket material or washer shall be interposed between the dissimilar conductive materials prior to connecting them. For immersion applications, full electrical isolation with insulating sleeves and washers on the bolts is required, along with an insulating spacer. In addition, an applicable organic coating shall be applied to the faying surfaces prior to assembly. In situations where large faying surfaces are involved, insert a barrier layer such as sprayable or pre-cured urethane sealants, which retain flexibility and are resistant to shrinking, aging, and cracking.
d) All faying edges shall be sealed to preclude the entrance of liquids.
e) For joints that are not required to be electrically conductive, applicable corrosion-inhibiting compounds shall be applied under the heads of bare, previously plated, and treated fasteners that are inserted into dissimilar conductive material surfaces.
f) Where practicable, or where it will not interfere with the proposed use of the assembly, the external joint shall be adequately protected using protection schemes (e.g., sealants, primers, coatings, etc.) sufficient for the level of exposure they will experience, or sufficient for the environment in which they are required to operate and approved by the CEA[1] or appropriate design authority.
g) Welded or brazed dissimilar conductive material assemblies shall be coated with a paint system or other suitable protective coating that covers the weld and all heat-affected zones.

The rate of corrosion depends on a number of factors, including:

- Properties of the anode and cathode materials (captured in polarization curves)
- Properties of the electrolyte (captured in polarization curves)

1 CEA = Cognizant Engineering Authority.

- Geometry and surface areas of the anode and cathode
- Electrical resistance between anode and cathode
- Aeration and motion of the electrolyte
- Temperature
- Surface roughness, including pitting
- Crevices
- Manufacturing processes, including treatments, coatings, and joining methods
- Presence of organic matter
- Stray currents

As shown in Eqs. (15.1) and (15.2), the amount of anode material corroded is proportional to the anode current and the amount of time the current flows.

$$M = kI_{anode}t \tag{15.1}$$

$$k = \frac{(Atomic\ weight\ of\ metal)}{(Number\ of\ electrons\ transferred) \times 96{,}500\ \text{A-sec}} \tag{15.2}$$

where

M, mass of metal dissolved (g)
I_{anode}, anode current (A)
t, time (s).

The number of electrons transferred depends on the specific anodic reaction. For iron, the number is typically 2.

If the two metals are not in contact, no current flows between them. At the surface of each metal, though, both anode and cathode reactions take place; the currents associated with the anode and cathode reactions of each metal are equal (I_{corr}) and opposite. Both metals will corrode. The voltage between the two metals will equal the difference between their half-cell potentials. The half-cell potential depends on the metal and the electrolyte and is usually measured with respect to a hydrogen electrode within the electrolyte. Table 15.1 lists the half-cell potentials for a number of metals in salt water.

If a half-cell consisting of a metal electrode immersed in the electrolyte and a hydrogen electrode has a voltage source applied between the two electrodes that differs from the half-cell potential, a current will flow. The plot of the voltage versus the current is called a polarization curve. For many materials, the polarization curve adheres to Eq. (15.3), the Butler-Volmer equation, in the steady-state. The Tafel constants are properties of the metal and the electrolyte. If the applied voltage varies, dynamic effects can result in the current varying significantly from the Butler-Volmer equation.

$$I = I_{corr}\left(e^{\frac{2.303(E-E_{corr})}{\beta_a}} - e^{\frac{-2.303(E-E_{corr})}{\beta_c}}\right) \tag{15.3}$$

where

I, current (A) flowing through the voltage source

I_{corr}, corrosion current (A) associated with the anode and cathode reactions when the metal is not electrically connected to another electrode or voltage source

E, magnitude of the applied voltage (V) between the metal electrode and hydrogen electrode

E_{corr}, magnitude of the half-cell potential (V) between the metal electrode and hydrogen electrode when the metal is not electrically connected to another electrode or voltage source

β_a, anodic Tafel constant (V/decade)

β_c, cathodic Tafel constant (V/decade).

An example polarization curve based on the Butler-Volmer equation is depicted in Figure 15.1. For many materials, different electrochemical reactions will take place at different voltages; different electrochemical reactions result in the anodic Tafel constant and the cathodic Tafel constant varying with respect to voltage instead of remaining a constant. Material will corrode if the applied voltage is along the anodic curve and will not corrode if the applied voltage is along the cathodic curve.

If two different metals (a galvanic pair) in electrical contact are also in contact with an electrolyte, a corrosion current will flow through both the electrical contact and the electrolyte. If the surface areas of the two metals are the same, the current density of the corrosion current at both the metal surfaces will also be the same (Figure 15.2). Thus, the corrosion current density will be the intersection of the polarization curves of the two metals. The cathodic curve is used for the metal (cathode) with the higher half-cell potential, and the anodic curve

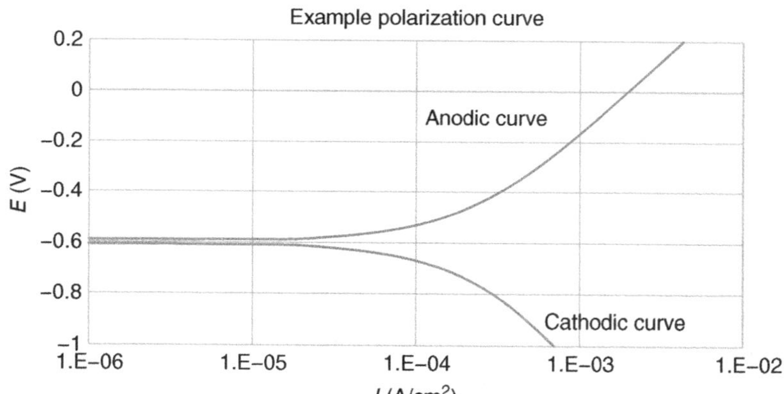

Figure 15.1 Example polarization curve.

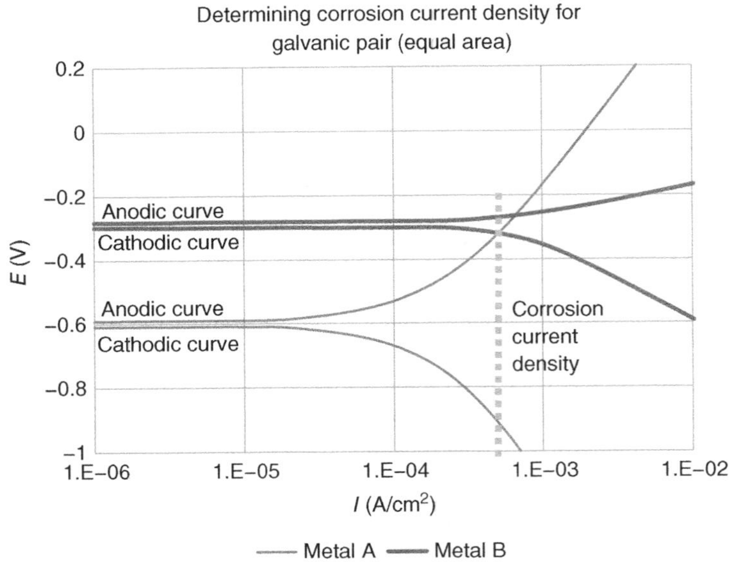

Figure 15.2 Corrosion current density for two metals of equal area.

is used for the metal (anode) with the lower half-cell potential. The corrosion current density is multiplied by the area of the anode to determine the corrosion current. Equation (15.1) is then used to calculate the amount of material lost to corrosion in a given time *t*.

If the surface area of the cathode and anode are not the same, then the current densities at the anode and cathode will be inversely proportional to their surface areas. As depicted in Figure 15.3, if the cathode surface area is one-tenth the anode surface area, then the cathode current density will be 10 times the anode current density. The anode corrosion current density is determined by finding the voltage where the cathode corrosion current density is 10 times larger than the anode current density. As before, the anode current is calculated by multiplying the anode current density by its surface area. If, on the other hand, the cathode surface area is 10 times the anode surface area (Figure 15.4), then the cathode current density will be one-tenth the anode current density. The anode current density corresponds to the voltage where the cathode corrosion current density is one-tenth the anode current density.

For example, when copper and iron form a galvanic pair in seawater, copper serves as the cathode and the iron serves as the anode. If a copper plate with a large surface area is nailed to the bottom of a wooden ship with iron nails, the iron nails will quickly corrode due to the large ratio of the cathode area (copper plate) to the

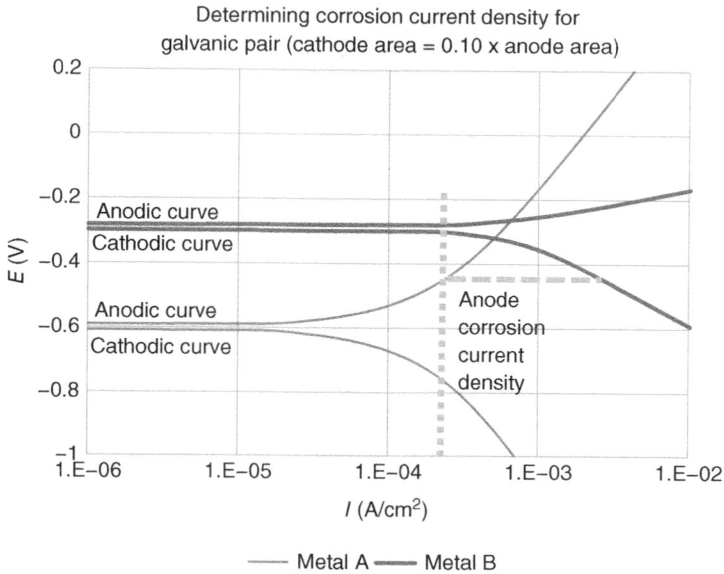

Figure 15.3 Corrosion current density for two metals with the cathode one-tenth the surface area of the anode.

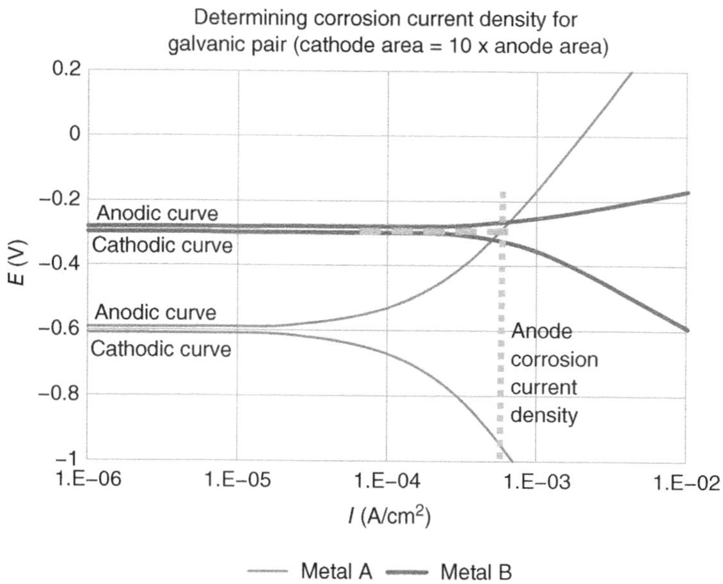

Figure 15.4 Corrosion current density for two metals with the cathode 10 times the surface area of the anode.

anode area (iron nail). On the other hand, if a steel plate with a large surface area is nailed to the bottom of a wooden ship with copper nails, the copper nails will not corrode substantially, and the corrosion of the iron plate will not be severe; the small surface area of the copper nails precludes a large corrosion current to develop. In general, to reduce corrosion of an anode, keep the size of the cathode as small as possible.

15.3 Impact of Current on Galvanic Corrosion

15.3.1 DC Current

If a DC current flows through the hull of a ship, then due to the resistance of the hull material, a voltage difference will exist between different regions of the hull. Regions where the voltage is more positive are anodic regions, and those where the voltage is more negative are cathodic regions. As depicted in Figure 15.5, for cathodic and anodic regions of the same size of a hull of a single metal, the voltage difference between the anodic region and the cathodic region determines the corrosion current density.

The possibility of galvanic corrosion due to DC currents in the hull is a good reason to minimize DC CM currents. Other sources of DC currents are those that enter and exit the hull from the sea. These currents can originate from cathodic protection systems on other ships or infrastructure in a port. Another possibility is a power system that uses the ground as a return path; if the ground path includes the water in the port, then this current can flow through the ship's hull.

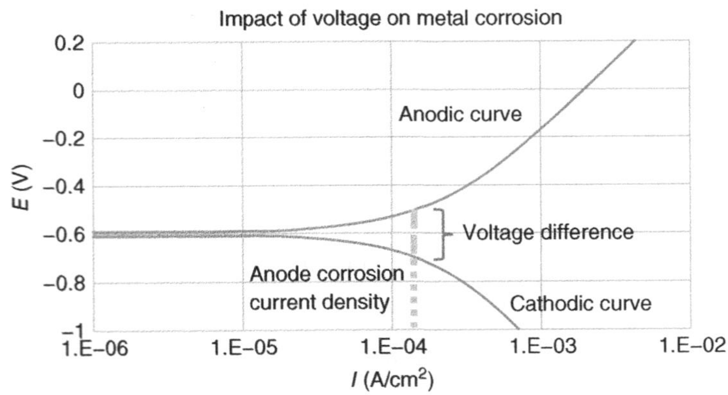

Figure 15.5 Impact of voltage on metal corrosion.

15.3.2 AC Current

The discussion so far has been limited to steady-state behavior with DC currents. The dynamic response to sinusoidal voltages and currents are not incorporated in the polarization curves. The time constants of the various electrochemical reactions can be expected to bracket the time period of the fundamental frequency of the power system (50–60 Hz). Consequently, the polarization curves are not appropriate for use with AC currents.

Corrosion does occur in the presence of AC currents; the corrosion rates are, however, much less for AC currents compared to DC currents. The theory for quantitative analysis for AC currents is not as well developed as it is for DC currents. The possibility of AC current corrosion in the ship's hull is good reason to minimize AC CM currents as well as other sources of AC hull currents.

15.4 Shipboard Corrosion

15.4.1 Hull and Structure Corrosion

A bare steel hull, not in contact with any other metals, but in contact with seawater, will corrode; both anodic and cathodic reactions occur such that the anodic and cathodic currents are equal and opposite. The anodic reactions result in corrosion. A protective coating (epoxy paint, for example) is usually the most cost-effective way of reducing this corrosion; the protective coating minimizes the surface area of the steel that is exposed to the electrolyte. Ideally, all hull plating, structure, sea chests, ballast tanks, and appendages in contact with seawater would have a protective coating. However, bare metal may be exposed to seawater in areas where the protective coating is damaged (scratched) or has failed.

Seawater pipes are typically in contact with the ship's hull and are usually made of copper-nickel, which is more cathodic than steel. Similarly, components made of other metals more cathodic than steel may also be in contact with both the hull and seawater. Consequently, the ship's hull (and anything in electrical contact with it and the seawater) may be subject to galvanic corrosion. A small scratch in the protective coating of the steel can result in a high rate (at least initially) of corrosion because the exposed cathodic surface area may be much larger than the exposed anodic surface area of the scratch.

If possible, metals that can serve as cathodes should have a protective coating to reduce the overall cathode surface area. Reducing the cathode surface area exposed to the electrolyte (seawater) will reduce the corrosion on any exposed anodic metals.

For many aluminum alloys used for ship structures and components, the corrosion products adhere strongly and form a physical barrier to further corrosion in a process called passivation. Passivation can also be artificially induced by exposing the bare surface to another material (such as a weak acid or a chromate conversion coating) that reacts with the bare metal to form a protective barrier. Smaller aluminum components can be anodized to form a thick oxide layer that passivates the bare aluminum.

To prevent galvanic corrosion, most ships with a metallic hull are protected with a cathodic protection system. These systems prevent corrosion of the hull by inserting other materials or an electric current to ensure the ship's hull always serves as a cathode.

15.4.2 Propulsion Shaft Corrosion

Materials used for propellers include alloys of aluminum, stainless steel, and bronze (such as nickel aluminum bronze). Steels, including stainless steel, are typically used for shafting. The dissimilar metals form a galvanic pair, which may result in corrosion. Furthermore, since the lubricating film in the bearings is typically insulating, the shaft is not in direct electrical contact with the hull; cathodic protection of the ship's hull does not automatically apply to the propeller and propeller shaft.

To enable the ship's cathodic protection system to also protect the shaft and propeller and to protect the bearings from bearing corrosion, a shaft grounding system is typically employed, as shown in Figure 15.6. The system consists of the following elements:

- A silver alloy slip ring attached electrically and mechanically to the shaft.
- Brush holders that hold the brushes in the proper position and apply pressure to the brushes for good electrical contact with the slip ring.

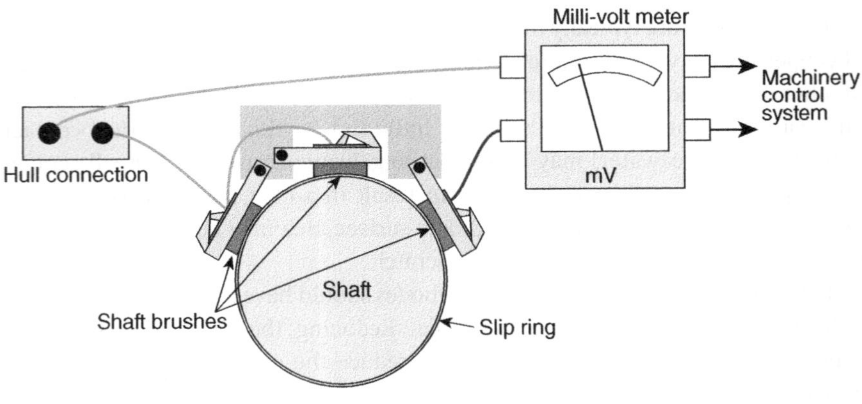

Figure 15.6 Shaft grounding system.

- High-conductivity brushes (e.g., silver graphite): one or more connected to ground; one connected to a millivolt meter.
- Slip ring cleaning device.
- Ground wire.
- Millivolt meter (optional interface to machinery control system for remote monitoring).

The millivolt meter is used to ensure the shaft ground is an effective ground. Typically, a reading of less than 40 mV is considered acceptable. Readings above 80 mV can lead to bearing damage. If the shaft voltage exceeds 40 mV, the system should be cleaned, all screw connections checked to ensure they are tight, and the spring pressure checked to ensure it is in accordance with the manufacturer's recommendations.

15.4.3 Electrical Connection Corrosion

Corrosion of an electrical connection is possible if the connection is exposed to an electrolyte such as condensate or humid air. Shipboard electrical wiring is usually copper, while the lugs used to connect to terminals are often tin-plated. Bus bars are generally copper but may be silver-plated. Nuts and bolts used in bus bar connections may be corrosion-resistant steel (CRES), silicon bronze, or zinc-plated steel. Because the design of equipment will undoubtedly contain multiple metals in electrical contact, preventing an electrolyte from coming into contact with the connection is paramount. Equipment enclosures should be designed with an appropriate ingress protection rating (see IEC 60529:1989 1989). Equipment cooled by a fluid such as chilled water, fresh water, or seawater should have means incorporated to prevent condensation. Condensation can be prevented by controlling the flow of cooling fluid to prevent the interior temperature of the equipment from falling below the dewpoint or by using heaters to maintain the interior temperature above the dewpoint.

Bonding straps, particularly in weather-exposed locations, are subject to galvanic corrosion. Corrosion-resistant fasteners (such as CRES) are typically employed.

15.4.4 Bearing Corrosion

Bearing corrosion is generally not caused by galvanic corrosion but is a product of the currents flowing through the bearing. A bearing is composed of one or more sets of metal surfaces in relative motion with each other and separated by a lubricating fluid such as oil. The metal surfaces and lubricating fluid form a capacitor; this capacitor allows AC currents to flow from the shaft to the bearing housing. If this current is of sufficient magnitude, corrosion in the form of pitting can occur on the metal surfaces.

The magnitude of a "safe" bearing current depends on the design of the bearing and the properties of the lubricating liquid. Based on a number of experimental studies, Muetze et al. (2004) suggest that for ball bearings, a bearing current density of not more than 0.1 A/mm^2 does not influence bearing life, while bearing currents of not less than 0.7 A/mm^2 may significantly reduce bearing life. The bearing current density J_b is defined as

$$J_b = \frac{i_b}{A_H} \tag{15.4}$$

where

i_b, peak current (A) flowing through the bearing
A_H, Hertz'ian contact area.

The Hertz'ian contact area depends on the material properties of the bearing, the bearing dimensions, and the forces exerted on the bearing. The bearing manufacturer should be consulted to determine an appropriate value to use for A_H.

Section 13.3 provides guidance for calculating and controlling the bearing current.

15.5 Cathodic Protection Systems

Cathodic protection systems protect the ship's hull, structure, and system components in contact with the sea by altering the electrochemical environment such that all protected metals are cathodic. Cathodic protection is implemented either through a passive system employing sacrificial anodes or through an active system employing impressed current cathodic protection (ICCP). Often, both methods are used at the same time; the active system provides an overall level of protection, while the passive system provides additional protection in specific areas such as propellers, rudders, bilges, and sea chests.

As discussed in ABS (2017), cathodic protection systems seek to fix the potential of the ship's hull with respect to an Ag/AgCl/seawater reference electrode to between −0.8 and −1.10 V for most hull materials. At the more negative voltages, high-strength steels may be susceptible to hydrogen-induced stress cracking; in this case, the cathodic protection system should fix the potential to between −0.8 and −0.83 V.

ABS (2017) should be consulted for additional information.

15.5.1 Sacrificial Anodes

Sacrificial anodes are metal bars that are anodic compared to the metals they are intended to protect. They are physically and electrically connected to the ship hull, rudders, propellers, etc. to provide cathodic protection. Most sacrificial anodes are

Figure 15.7 Sacrificial anodes on ship's rudder. *Source:* Norbert Doerry.

made of alloys of zinc, magnesium, or aluminum. Magnesium anodes are typically only used in fresh water applications, while aluminum anodes are not normally used in fresh water. Figure 15.7 shows sacrificial anodes attached to a ship's rudder. ABS (2017) should be consulted to determine the number of sacrificial anodes needed and where they should be located on the ship. Sacrificial anodes are usually designed to have a service life at least as long as the period between dry dockings. They should be mounted in such a manner to provide good electrical contact as well as the ability to be easily replaced. Figure 15.8 depicts a MIL-DTL-18001 (2015) type ZBS sacrificial anode.

Figure 15.8 MIL-DTL-18001 type ZBS sacrificial anode.

15.5.2 Impressed Current Cathodic Protection

An ICCP system consists of the following elements:

- Power supply and controller
- Anode electrodes
- Reference electrodes
- Dielectric shield

An ICCP system is used to provide protection within a cathodic protection zone. Small ships may have only one cathodic protection zone and one corresponding ICCP system. Large ships may have two or more cathodic protection zones and corresponding ICCP systems. A typical installation has two cathodic protection zones: one to protect the stern, rudder, propeller, struts, and shafting; and the other to protect the remainder of the ship. Figure 15.9 depicts a ship with a single cathodic protection zone.

The power supply and controller produce a DC current that draws electrons from the anode electrode and inserts them into the ship's hull. The controller measures the potential between the reference electrode and the hull to adjust the anode electrode current to regulate the hull potential to the desired value. For most ships, the hull voltage with respect to the reference electrode is regulated to between −0.8 and −1.10 V.

The anode electrode should resist corrosion and have a life greater than the dry-docking interval. As described in ABS (2017), a typical ICCP anode electrode consists of a titanium, tantalum, or niobium core with a platinum coating. Anode electrodes are normally installed in sets of two; one for each side of the ship at the same distance aft of the forward perpendicular and the same distance above the baseline. One power supply and controller may be used for multiple anode

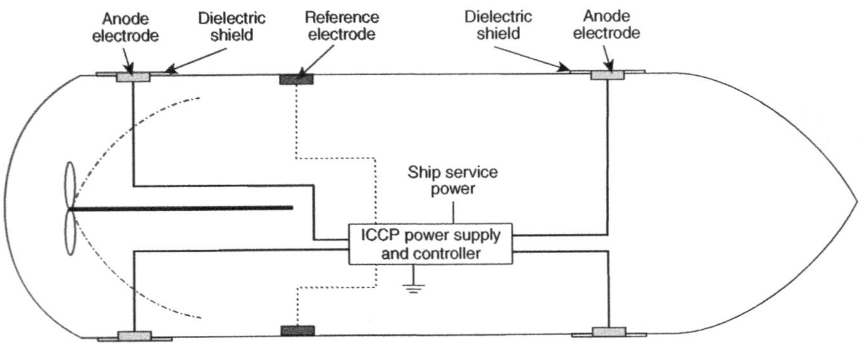

Figure 15.9 Shipboard ICCP system.

electrode sets; two sets are typical. The anode electrodes should be positioned to provide an even current distribution to the hull.

The reference electrodes are used to establish the voltage reference for measuring the hull potential. Reference electrodes are typically composed of either zinc or silver/silver chloride. Normally, a reference electrode is located on each side of the ship. If two anode electrode sets are used by a power supply and controller, on each side of the ship, the reference electrode is usually located midway between the anode electrodes.

A dielectric shield may be used around the anode electrode to prevent high hull potential magnitudes near the electrode and to promote a more even current distribution into the hull. High hull potential magnitudes can cause deterioration of protective coatings.

References

ABS (2017). *Guidance Notes on Cathodic Protection of Ships*, Spring, TX: American Bureau of Shipping. Available from eagle.org.

Flinn, Richard A., Trojan, Paul K. (1981). *Engineering Materials and Their Applications*, Second edition, Boston, MA: Houghton Mifflin Company.

IEC 60529:1989 (1989). *Degrees of Protection Provided by Enclosures (IP Code)*. Geneva, Switzerland: International Electrotechnical Commission. Available from webstore.iec.ch.

MIL-DTL-18001 (2015). *Detail Specification Anodes, Sacrificial Zinc Alloy*. Washington, DC: Naval Sea Systems Command. Available from quicksearch.dla.mil.

MIL-STD-889 (2021). *Standard Practice Galvanic Compatibility of Electrically Conductive Materials*. Washington, DC: Department of Defense. Available from quicksearch.dla.mil.

Muetze, Annette, Binder, Andreas, Vogel, Hans-jorg et al. (2004). Experimental Evaluation of the Endangerment of Ball Bearings due to Inverter-Induced Bearing Currents. *Conference Record of the 2004 IEEE Industry Applications Conference, 2004. 39th IAS Annual Meeting*. Seattle, WA, vol. 3, pp. 1989–1995. https://doi.org/10.1109/IAS.2004.1348741.

16

Lightning Protection Systems

16.1 Introduction

Lightning at sea (Figure 16.1) and lightning strikes on ships at sea or in port are not unusual. If electrically connected with low-resistance bonds, the hull, superstructure, and mast of a metal ship will normally provide a level of protection against the impacts of a lightning. However, induced currents and direct strikes on exposed electrical equipment and antennas can result in significant surges in the electrical system. For this reason, elements of a lightning protection system (LPS) should be incorporated in all ships. While a well-designed and installed LPS will not prevent lightning strikes, it should be able to reduce or eliminate damage from the strike.

A lightning strike is the manifestation of electrical charges in the atmosphere exceeding the dielectric strength for the atmosphere; the lightning strike is a large current discharge (Figure 16.2) lasting only microseconds that results from the dielectric breakdown. The peak current magnitude of the current waveform can be in the order of hundreds of kiloamps. In many cases, a series of strikes occurs in quick succession. The manner in which the dielectric breakdown occurs is complex, and there is no simple explanation that covers all cases. In general, though, lightning usually starts with an initial leader traveling from the clouds to the ground. As the leader head nears the surface, it creates an electric field gradient around every object electrically connected to the ground. Where the local electric field gradient exceeds the dielectric breakdown, a leader is formed from the object; the head of this leader is attracted by the head of the leader from the clouds. When the two connect, a low-impedance ionization path is created, enabling the large current discharge.

A LPS consists of up to four elements:

- Air terminal
- Down conductors

Design of Shipboard Power System Grounding/Earthing, First Edition. Norbert Doerry, Mohammed M. Islam, and John Prousalidis.
© 2025 The Institute of Electrical and Electronics Engineers, Inc.
Published 2025 by John Wiley & Sons, Inc.

Figure 16.1 Lightning at sea. Source: Heidi J. Giacalone/United States Navy/Public Domain.

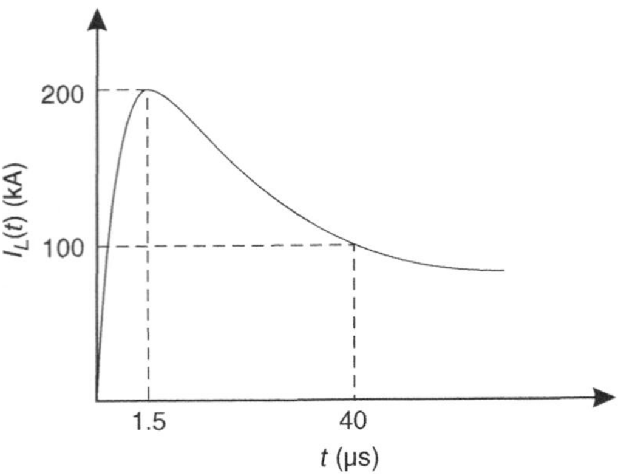

Figure 16.2 Representative lightning current waveform. *Source:* Adapted from MIL-HDBK 274. Public domain.

- Low-impedance ground system
- Surge protection

For many ships, installing only the surge protection element is adequate; for other ships, particularly those with nonmetallic hulls, superstructures, or masts, installing all four elements is warranted and usually is required by regulation or classification society rules. Many class rules and standards only address LPS for ships with nonmetallic masts, superstructures, or hulls; ships with considerable topside electrical equipment, such as radars, radio antennas, and cameras, or that potentially could exhaust flammable gases, should conduct a risk assessment to determine which elements of an LPS to incorporate.

An air terminal is essentially a modern version of Benjamin Franklin's "lightning rod." It is typically a copper spike installed at the top of a mast. The point on the air terminal is intended to cause a high electric field stress so that the dielectric breakdown is likely to occur at the air terminal rather than at the surface of equipment within its zone of protection. Air terminals are usually used to protect antennas and other electrical equipment mounted on the mast at a lower level.

The down conductors electrically connect the air terminal to the low-impedance ground system. For metal hull ships with continuous metal masts, the down conductor is typically the combination of the metal mast, metal superstructure, and metal hull. For other ships, a dedicated copper cable or tape is employed to bridge nonmetal sections.

The low-impedance ground system provides a current path to the water. For ships with metal hulls, the hull itself is the low-impedance ground system. Other ships employ a metal plate on the bottom of the hull that is connected to the down conductors.

Surge protection reduces the probability of damage to electrical equipment from overvoltage due to the lightning strikes. High lightning-induced currents and voltages in distribution systems can result from either direct strikes on connected equipment or inductive currents from strikes nearby or to the ship's hull. Ship power generation and distribution equipment, such as generators, transformers, switchgear, and cables, are protected through the specification of Basic Insulation Levels (BIL). Protecting electrical loads such as fathometers and speed logs that are mounted on the outside of the hull is very important; their destruction due to a high overvoltage could result in physical damage that leads to flooding of the ship. Other sensitive equipment such as radars, radios, and engine controllers should also be protected. Protection of specific equipment is often accomplished with the installation of SPDs.

Each air terminal provides a zone of protection; equipment within this zone is not likely to be damaged by a direct strike from lightning. A direct lightning strike to equipment will not only cause extensive damage to that equipment, but large currents may flow in its electrical connections to other equipment. A radio antenna that is struck by lightning may result in severe damage to the radio transmitter, even if the radio transmitter is within a zone of protection. For this reason, all exterior equipment with an electrical interface should be located within a zone of protection.

The exterior metal structure of a ship's hull, superstructure, and mast also provides zones of protection. Any sharp edge or point (such as those associated with the mast, stanchions, deck edges, and ladders) will serve to establish a high electric field gradient; they will likely not be as effective as an air terminal. Suitable placement of exterior equipment with an electrical interface is often possible without having to install air terminals or dedicated down conductors.

Dehn (2014) provides general guidance for lightning protection.

16.2 Design Considerations

16.2.1 Zone of Protection

In terrestrial applications, only the air terminals are assumed to provide zones of protection. Several methods exist for determining the geometry of the zones of protection for terrestrial applications. One method establishes a grid of air terminals on the roof of a building, with terminals spaced every 25 ft. This method is typically not applied to ships. Another method assumes each air terminal creates a protection zone consisting of a 45° cone with the vertex at the tip of the air terminal. This method, if used, is typically only applied when the structure does not exceed 50 ft above ground level.

A third method relies upon the probability that a lightning leader from the cloud connecting to a leader from the air terminal becomes significant at distances of 150 ft. If the lightning leader from the cloud is 150 ft away from the air terminal, it is much more likely to connect to a leader from the air terminal than to one from equipment or structures located more than 150 ft away; if the lightning leader from the cloud is 150 ft or less from the ground, it is more likely to connect to leaders from the ground than from a leader from equipment or structures located more than 150 ft away. The rolling sphere method uses a sphere with a 150 ft radius that rolls over the ground toward the building until it touches one or more air terminals. The sphere then rolls over the building, always touching the tip of at least one air terminal; the tip of an air terminal is never allowed to be in the interior of the sphere. The zone of protection is the volume that is never in the interior of the sphere.

NFPA 780 requires reducing the radius of the sphere to 100 ft (30 m) for watercraft. This reduction in radius from the 150 ft used in terrestrial applications is likely due to many of the ship features used in lieu of air terminals are not as effective at establishing an electric field gradient as an air terminal.

As part of the risk assessment to determine which elements of a LPS should be incorporated, the rolling sphere method with a sphere radius of 100 ft should be employed, as depicted in Figure 16.3; the sphere should instead make contact with, but not enclose, edges of the hull or superstructure electrically in contact with the sea (mast, deck edges, etc.), nonelectrical equipment and fittings electrically in contact with the sea (stanchions, davits, ladders, etc.), as well as any air terminals. NFPA 780 (2023) provides guidance for determining whether a given fitting should

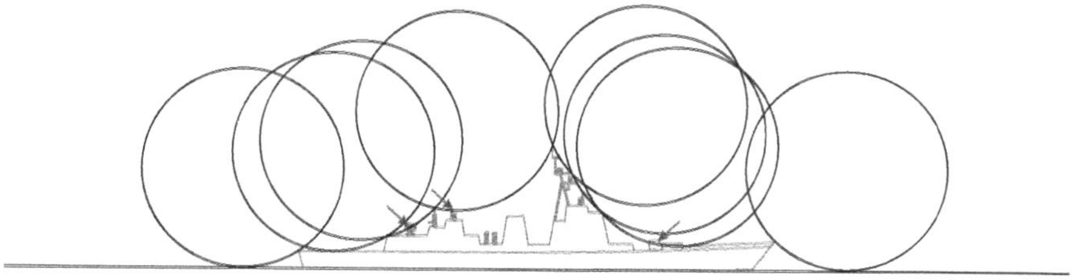

Figure 16.3 Rolling sphere method.

be considered equivalent to an air terminal. Consideration should be given to ensuring all external electrical equipment is within the zone of protection by relocating external electrical equipment outside of the zone of protection, relocating existing air terminals, or adding additional air terminals.

In Figure 16.3, the arrows point to potential vulnerable equipment; they should be evaluated to determine if additional lightning protection is warranted. Additional lightning protection is likely not needed if:

- The equipment provides significant insulation between the power system and the external environment (for example: a radome enclosing a radar antenna).
- The distance from the vulnerable equipment to the boundary of a zone of protection is not large enough such that the resistance of the insulation is much greater than the resistance of the air between the equipment and the boundary of the zone of protection.
- Any exposed conductors are bonded to the ship's hull.

For protecting features such as air terminals, applicable nonelectrical equipment and fittings electrically in contact with the sea, and edges (including deck hedges) with heights less than the sphere radius (100 ft), one can analytically determine if a lower point is within a zone of protection using Eq. (16.1). When using Eq. (16.1), the units must be consistent.

$$d = \sqrt{h_1(2R - h_1)} - \sqrt{h_2(2R - h_2)} \tag{16.1}$$

where

R, sphere radius (normally 100 ft)

h_1, height above waterline of the air terminal or edge (protecting feature)

h_2, height above waterline of the top of the component to be protected

d, horizontal distance of the edge of the zone of protection from the protecting feature.

If the component to be protected has a horizontal distance from the protecting feature less than d, then it is within the zone of protection of the protecting feature.

Large ships may have masthead heights greater than 100 ft and must use the geometric method directly.

16.2.2 Air Terminals

An air terminal is typically a copper spike that projects at least 150 mm (6 in) above the uppermost part of the vessel or mast. The air terminal may project to a higher level if a larger zone of protection is needed. The air terminal connects to a metallic mast or to a down conductor if the mast is not metallic. See Figure 16.4 for a typical air terminal installation on a naval frigate.

IEC 60092-401 (1980) requires the air terminal to be a copper spike not less than 12 mm in diameter and affixed to the down conductor via copper rivets or with copper clamps. IEEE Std 45.1-2023 (2023) recommends the copper spike to be not less than 13 mm in diameter.

The Canadian TP 127E (2018) requires

> "Vessels in which tank vent outlets for flammable gases is located near, or at the top of a non-conductive mast are to be protected by an air terminal at least 2 metres above the vent outlet; on a steel mast the steel mast must extend at least 2 metres above the vent outlet."

Figure 16.4 Arrow pointing to air terminal (lightning rod) on frigate mast. *Source:* Norbert Doerry.

Some ships with two masts connect the two with a metallic cable connected to the LPS. A relatively large zone of protection is created underneath this cable between the two paths. The cable should have a resistance no greater than the equivalent length of down conductor cable.

16.2.3 Down Conductors

Dedicated copper cable or tape is employed to bridge nonmetal sections of the path from air terminals or masts to the low-impedance ground connection.

ABS MVR (2023) requires installing a copper conductor of at least 8 mm^2 (No. 8 AWG) on each nonmetallic mast; the conductor should follow a straight as practicable route to the metal structure of the vessel and avoid metal objects.

NFPA 780 (2023) requires a copper down conductor to have a cross-sectional area of at least 0.013 in^2 (8.3 mm^2).

IEC 60092-401 (1980) requires the down conductor to be continuous copper tape or rope with a cross section of at least 75 mm^2. As with ABS MVR (2023), the conductor should be run as straight as possible; sharp bends should be avoided. IEC 60092-401 (1980) does not allow solder joints in the down conductor.

The Canadian TP 127E (2018) standard requires the down conductor to have a cross-sectional area of at least 25 mm^2 (No. 4 AWG). It also states

> "The resistance of the lightning conductor, measured between the mast head and the position on the ground plate or hull to which the lightning conductor is grounded, shall not exceed 0.02 ohm."

A down conductor should be used to bridge sections of the path from the air terminals or masts to the low-impedance ground connection if the path (without a down conductor) has a resistance per meter that is not significantly less than the required copper down conductor. Sections of the path where the material is a poor conductor, such as wood, plastics, or most composites, should be bypassed with a down conductor. If the cross-sectional area of the metallic path is significantly greater than that indicated in Table 16.1 for the material of the path, then a down conductor does not provide significant benefit for that portion of the path. Most metal masts should have a cross-sectional area greater than shown in Table 16.1; hence a dedicated down conductor is likely not needed to bypass the mast.

Down conductors should be routed away from other electrical cables to avoid inductive coupling of the lightning current to other equipment. Down conductors should also be routed away from combustible materials to avoid fires due to the lightning strike.

Table 16.1 Properties of materials to determine if down conductor should be used.

Material	Resistivity (Ωm at 20 °C)	Resistance per meter for 8 mm² cross section (Ω/m at 20 °C)	Cross-sectional area required for resistance to equal 8 mm² annealed copper at 20 °C (mm²)	Cross-sectional area required for resistance to equal 25 mm² annealed copper at 20 °C (mm²)	Cross-sectional area required for resistance to equal 75 mm² annealed copper at 20 °C (mm²)
Annealed copper	1.72*E*−08	0.0022	8.0	25.0	75.0
Aluminum	2.65*E*−08	0.0033	12.3	38.5	115.6
Carbon steel	1.43*E*−07	0.0179	66.5	207.8	623.5
Titanium	4.20*E*−07	0.0525	195.3	610.5	1831.4
Stainless steel	6.90*E*−07	0.0863	320.9	1002.9	3008.7

16.2.4 Low-Impedance Ground Connection

As stated earlier, metallic hull ships normally use the hull itself as the low-impedance ground connection. Nonmetallic hull ships, however, require a copper ground plate to be installed on the exterior surface of the hull below the light-load waterline and in a manner that ensures it is submerged under all trim and heel conditions. The minimum required size of the copper plate varies based on the standard, regulation, or class rules used; in general, the required size falls between 1 ft^2 (0.093 m^2) and 0.2 m^2.

The ground plate used for lightning protection should not be used for any other grounding purpose.

Figure 16.5 depicts how the ground plate integrates with all the other elements of the LPS for a nonmetallic hull ship. For this example, an air terminal on the main mast, two other metallic masts, and the metal engine exhaust funnel serve as air terminals. The air terminal on the main mast connects to the metallic portion of the main mast via a down conductor. The metallic portion of the main mast, the two other metallic masts, and the metal engine exhaust funnel are connected to the ground plate via down conductors.

16.2.5 Surge Protection

Surge protection should be accomplished through a layered approach. Specifications for generators, cables, switchgear, transformers, and motors include a requirement for BIL, which provides an initial level of protection. SPDs are also

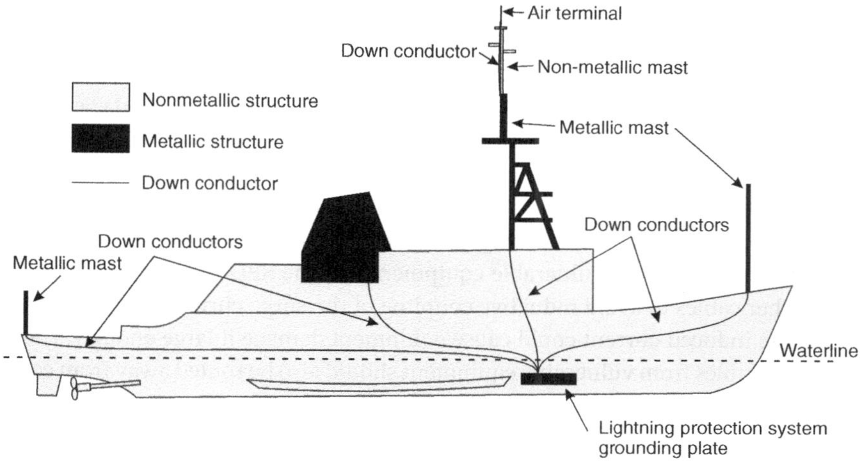

Figure 16.5 Lightning protection system for nonmetallic hull ship.

employed to ensure surges remain below the specified BIL, as well as to protect specific equipment; power electronic equipment cannot typically survive being subjected to BIL requirements. SPDs may also be employed to limit surges due to switching events and ground faults.

As detailed in IEEE Std 45.1-2023 (2023), the following standards provide guidance for the employment of SPDs:

- IEEE Std C62.42.0 (2016) provides an overview of SPD applications.
- IEEE Std C62.41.1 (2002), IEEE Std C62.41.2 (2002), and IEEE Std C62.41.3 (2020) detail the surge environment.
- IEEE Std C62.44 (2016) and IEEE Std C62.72 (2016) provide guidance for AC system below 1 kV.
- IEEE Std C62.11 (2020) and IEEE Std C62.22 (2009) provide metal oxide surge arrestor guidance and requirements.
- IEEE Std C62.37.1 (2012) provides thyristor SPD guidance.
- IEEE Std C62.21 (2003) provides SPD guidance for rotating machines 1 kV and higher.

IEEE 45.1-2023 (2023) also states

> "To provide isolation in the event of a surge protective device s failure and to facilitate insulation testing, surge protective devices should have a mechanism for disconnecting from the power system. In power systems with a nominal system voltage above 400 volts, a means should also be provided to detect and alert the operator of a surge protective device failure."

SPDs should be considered for protecting equipment that may result in serious damage to the equipment, serious damage to other equipment, injury to personnel, or loss of hull integrity when exposed to a lightning induced surge. Examples of equipment for which SPDs should be provided include fathometers and speed logs that are mounted on the outside of the hull, exterior electrical equipment not in a zone of protection, and critical equipment such as engine controllers, radars, and radios. The shore power connection should also incorporate a SPD.

Any electrical cabling, including power cables, signal cables, and radio frequency cables, between vulnerable equipment and the SPD should be segregated from other cables to avoid inductive coupling of the surge current to other equipment; the induced current could cause equipment damage if large enough. These electrical cables from vulnerable equipment should also be routed away from combustible materials to avoid fires due to a lightning strike.

References

ABS (2023). *Rules for Building and Classing Marine Vessels (MVR)*. Spring, TX: American Bureau of Shipping. Available from eagle.org.

Dehn (2014). Lightning Protection Guide. Third updated edition. Publication No. DS702/E/2014. Available from www.dehn-international.com.

IEC 60092-401:1980 (1980). *Electrical Installations in Ships – Part 401: Installation and Test of Completed Installation*. Geneva, Switzerland: International Electrotechnical Commission. Available from webstore.iec.ch.

IEEE Std C62.11 (2020). *IEEE Standard for Metal-Oxide Surge Arresters for AC Power Circuits (>1 kV)*. New York: Institute of Electrical and Electronics Engineers. Available from ieeexplore.ieee.org.

IEEE Std C62.21 (2003). *IEEE Guide for the Application of Surge Voltage Protective Equipment on AC Rotating Machinery 1000 V and Greater*. New York: Institute of Electrical and Electronics Engineers. Available from ieeexplore.ieee.org.

IEEE Std C62.22 (2009). *IEEE Guide for the Application of Metal-Oxide Surge Arresters for Alternating-Current Systems*. New York: Institute of Electrical and Electronics Engineers. Available from ieeexplore.ieee.org.

IEEE Std C62.37.1 (2012). *IEEE Guide for the Application of Thyristor Surge Protective Device Components*. New York: Institute of Electrical and Electronics Engineers. Available from ieeexplore.ieee.org.

IEEE Std C62.41.1 (2002). *IEEE Guide on the Surge Environment in Low-Voltage (1000 V and Less) AC Power Circuits*. New York: Institute of Electrical and Electronics Engineers. Available from ieeexplore.ieee.org.

IEEE Std C62.41.2 (2002). *IEEE Recommended Practice on Characterization of Surges in Low-Voltage (1000 V and Less) AC Power Circuits*. New York: Institute of Electrical and Electronics Engineers. Available from ieeexplore.ieee.org.

IEEE Std C62.41.3 (2020). *IEEE Guide for Interactions Between Power System Disturbances and Surge-Protective Devices*. New York: Institute of Electrical and Electronics Engineers. Available from ieeexplore.ieee.org.

IEEE Std C62.42.0 (2016). *IEEE Guide for the Application of Surge-Protective Components in Surge-Protective Devices and Equipment Ports – Overview*. New York: Institute of Electrical and Electronics Engineers. Available from ieeexplore.ieee.org.

IEEE Std C62.44 (2016). *IEEE Guide for the Application of Low-Voltage (1000 V_{rms} or Less) Surge Protective Devices Used on Secondary Distribution Systems (Between the Transformer Low-Voltage Terminals and the Line Side of the Service Equipment)*. New York: Institute of Electrical and Electronics Engineers. Available from ieeexplore.ieee.org.

IEEE Std C62.72 (2016). *IEEE Guide for the Application of Surge-Protective Devices for Use on the Load Side of Service Equipment in Low-Voltage (1000 V or Less, 50 Hz or 60 Hz) AC Power Circuits*. New York: Institute of Electrical and Electronics Engineers. Available from ieeexplore.ieee.org.

IEEE Std 45.1-2023 (2023). *IEEE Recommended Practice for Electrical Installations on Shipboard – Design*. New York: Institute of Electrical and Electronics Engineers. Available from ieeexplore.ieee.org.

MIL-HDBK-274 (2011). *Handbook Electrical Grounding for Aircraft Safety*. Washington, DC: Department of Defense. Available from quicksearch.dla.mil.

NFPA 780 (2023). *Standard for the Installation of Lightning Protection Systems*. Quincy, MA: National Fire Protection Association. Available from www.nfpa.org.

TP 127E (2018). *Ships Electrical Standards*. Ottawa, Ontario, Canada: Transport Canada. Available from tc.canada.ca.

17

Grounding Systems for Nonmetallic Hull Ships

17.1 Design Considerations

Metallic hull ships use the ship's hull as the means of establishing the ground potential. In nonmetallic hull ships, a ground plate installed on the submerged exterior of the ship is used instead. This ground plate should be made of copper, and its area should be at least 0.2 m^2. This plate should remain submerged under all expected drafts, heel angles, and trim angles. A through bolt is brazed to this ground plate to serve as a system ground and a protective earth connection for bonding. As depicted in Figure 17.1, conductors used to connect systems grounds and CM grounds should be connected together at a separate junction point from a junction point for conductors used for bonding. These two junction points are connected together and connected to the through bolt on the ground plate.

All power cables should have a protective earth conductor dedicated to bonding; all exposed metal conductors of equipment should be connected to the protective earth conductors of cables connected to the equipment.

A dedicated grounding bus that interconnects all switchboards and load centers should be provided to connect to system grounds and for the ground connection for CM and EMI filters.

This ground plate for system grounding and bonding is in addition to the ground plate used as part of a lightning protection system.

Requirements for grounding nonmetallic hull ships are provided in IEEE Std 45.1-2023 (2023).

With the exception of ground plates, external metal appendages with through bolts should not be connected to the protective earthing system via bonding; they are already in contact with the sea. Connecting external metal appendages to the protective earthing system could result in stray currents from the exterior of the ship entering through one appendage, traveling through the protective earthing

Design of Shipboard Power System Grounding/Earthing, First Edition. Norbert Doerry, Mohammed M. Islam, and John Prousalidis.
© 2025 The Institute of Electrical and Electronics Engineers, Inc.
Published 2025 by John Wiley & Sons, Inc.

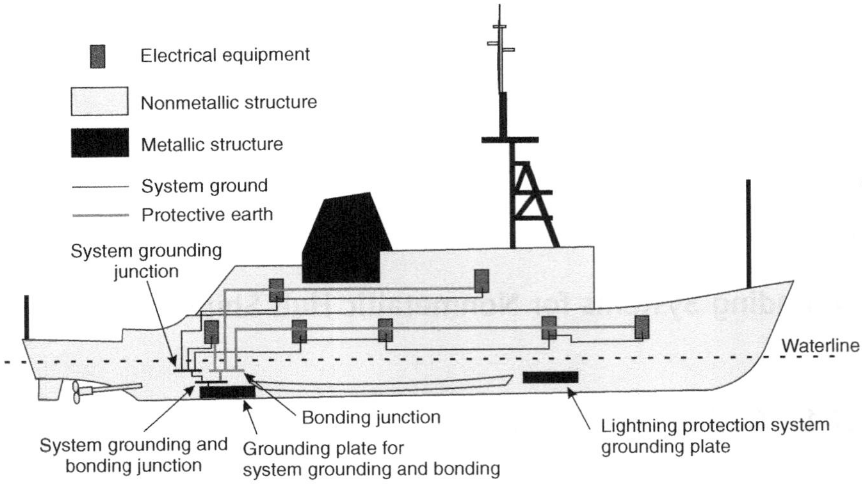

Figure 17.1 System grounding and bonding on a nonmetallic hull ship.

system, then exiting through another appendage; in doing so, galvanic corrosion may be accelerated on one of the appendages.

If the ship employs only ungrounded power distribution systems and does not employ any CM shunts or EMI filters that require a connection to ground, then there is no need for a system ground. Protective earthing via bonding is still required, and a ground plate is still needed, as depicted in Figure 17.2.

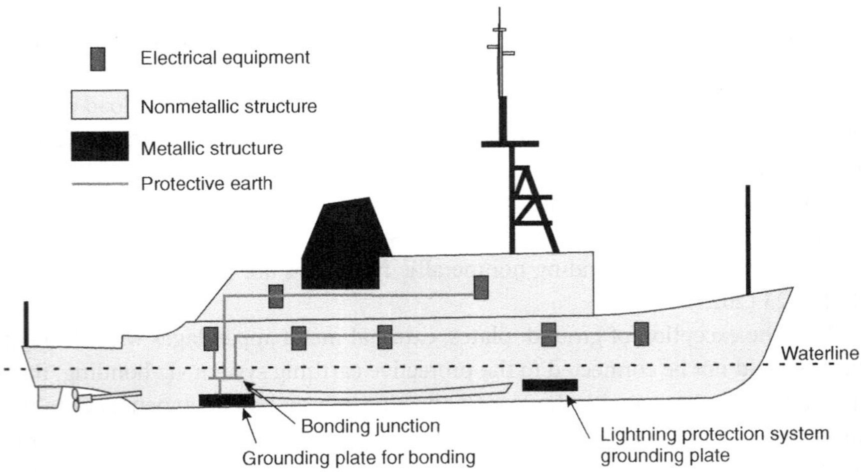

Figure 17.2 Protective earthing on a nonmetallic hull ship with an ungrounded power distribution system.

External metallic items that are intended to serve as part of the lightning protection system (see Chapter 16) should be electrically connected to the lightning protection system ground plate and should not be bonded to the protective earthing system. This practice helps ensure that large lightning currents are not coupled into electrical components via the protective earthing system.

Reference

IEEE Std 45.1-2023 (2023). *IEEE Recommended Practice for Electrical Installations on Shipboard – Design*. New York: Institute of Electrical and Electronics Engineers. Available from ieeexplore.ieee.org.

Appendix A

Glossary

absence of voltage A permanently installed test device that indicates when hazardous voltages are present within a piece of equipment and, when a test is initiated, verifies the absence of voltage within the equipment.

active rectifier A power electronic converter for converting AC power into DC power that incorporates voltage regulation of the DC output and power electronic switching at a frequency much higher than the fundamental AC frequency of the input power.

air terminal A lightning rod; typically, a copper spike installed at the top of a mast. The point on the air terminal is intended to cause a high electric field stress so that dielectric breakdown is likely to occur at the air terminal rather than at the surface of equipment within its zone of protection.

anode Within a galvanic cell, the metal that dissolves into the electrolyte to form positive charged metal ions and that provides electrons that flow through the external circuit.

appliance leakage current interrupter A circuit protection device that trips when a current mismatch between the power conductors of 30 mA or more is sensed. Functionally the same as a GFCI/RCBO but with a higher mismatch current level. This higher level is intended to protect the equipment rather than the operator, and will likely result in fewer nuisance trips.

arc fault A fault where the current flows through the air between two conductors; the arc fault may be between a conductor and ground, between a conductor of one phase and a conductor of another phase, or in-line between conductors of the same phase. Arc faults typically produce significant heat; they may be either high or low resistance. Arc faults may or may not be detected by conventional overcurrent protection.

balanced inductance The phase inductance of a set of conductors, where the current in each conductor is equal in magnitude and the sum of the currents is zero, such that the phase voltage drop between the ends of phase conductors is

Design of Shipboard Power System Grounding/Earthing, First Edition. Norbert Doerry, Mohammed M. Islam, and John Prousalidis.
© 2025 The Institute of Electrical and Electronics Engineers, Inc.
Published 2025 by John Wiley & Sons, Inc.

equal to the phase inductance times the time derivative of the phase current. The balanced inductance is a combination of the conductor self-inductance and mutual inductance with the other conductors. For most symmetrical three-phase AC cables and most DC cables, the balanced inductance is equal to the self-inductance minus the mutual inductance.

Basic Insulation Level A design voltage level for electrical equipment used to measure the ability of an insulation system to withstand surge voltages.

bearing corrosion Corrosion, in the form of pitting of the bearing surfaces, due to currents flowing through the bearing.

blocking capacitor A capacitor inserted into a circuit for the purpose of blocking DC current.

bonding The act of deliberately connecting, via a low-impedance path, two or more exposed metal parts that are not designed to carry electrical currents under normal operation. Typically, one of the metal parts is the hull of the ship. Bonding is primarily performed as a safety measure to prevent electrical shock to personnel caused by capacitively coupled voltages to the exposed metal parts, inductively coupled currents in exposed metal parts, or insulation failures.

bonding strap/bonding cable A strap or cable used to electrically connect two conductive surfaces that are not normally intended to conduct electrical power. One end of the bonding strap is typically connected either to the ship's hull or to another conductor that is electrically connected to the ship's hull; the other end is typically connected to an exposed conductive surface. Bonding straps and cables are used primarily for personnel protection by ensuring a very low voltage difference between conductors that a person can touch; they can also provide a path for common mode currents.

bus duct A set of solid conductor bars within a metal enclosure used instead of cable to distribute electrical power.

bus tie A cable, bus duct, or insulated bus pipe used to interconnect two switchboards.

cable management system Part of a shore power connection that manages the cables that connect the shore station on land to the shipboard shore power station.

cable shield A component of a multiconductor cable consisting of a conductive tape or mesh surrounding all of the insulated conductors of a cable. When installed onboard a ship, both ends of the cable shield are typically grounded to provide a path for common mode currents.

cathode Within a galvanic cell, the metal that inserts electrons into the electrolyte from the external circuit. The cathode metal is not corroded.

cathodic protection system A system that protects the ship's hull, structure, and system components in contact with the sea by altering the electrochemical environment such that all protected metals serve as cathodes. Cathodic protection is implemented either through a passive system employing sacrificial anodes, or through an active system employing impressed current cathodic protection (ICCP). Often, both methods are used at the same time; the active system provides an overall level of protection while the passive system provides additional protection in specific areas such as propellers, rudders, bilges, and sea chests.

circulating current A component of the current through a set of conductors normally having a return path through either a grounding bus or the ship's hull; circulating currents can also flow through the normal current-carrying conductors if there are multiple paths between sources and loads. The magnitude of the circulating current is equal to the common mode current.

clearance The shortest distance in air between two powered conductors, or between a powered conductor and a grounded conductor.

cold ironing When a ship is in port, the process of providing electrical power from shore (shore power) to enable shutting down shipboard generator sets. It may be referred to as Onshore Power Supply (OPS), Shore Side Electricity (SSE), or Alternative Maritime Power (AMP).

combination RCD and miniature circuit breaker A circuit protection device that combines the functionality of a Residual Current Detector and a miniature circuit breaker. Typically used in the European Union, the device trips if the current in the phase conductor and the current in the neutral conductor are mismatched by more than 10 mA.

common mode choke An inline transformer configured to provide low impedance to differential mode currents and a high impedance to common mode currents.

common mode current The instantaneous sum of the conductor currents in a set of conductors. With respect to a dq0 transformation for three-phase power, the common mode current is equal to three times the zero sequence component.

common mode grounding The act of deliberately inserting an impedance between a conductor or neutral of a power system and the hull of the ship for the purpose of providing common mode currents a path to ground. Common mode grounding typically focuses on current frequency components above three times the fundamental frequency; all frequency components, including DC, may be impacted by common mode grounding.

common mode impedance For a component with two power interfaces, each of which can have one or more conductors, the ratio of the common mode voltage between the two power interfaces to the common mode current through the two interfaces.

common mode shunt A network of circuit components configured to provide a low impedance to ground for common mode currents, a high impedance to ground for differential mode currents, and a high impedance between each conductor for differential mode currents. For DC systems, a common mode shunt is typically composed of a capacitor network. For AC systems, a common mode shunt is typically composed of a specially configured transformer in series with a capacitor network.

common mode voltage The voltage of a set of conductors with respect to a reference voltage potential (typically ground) at any instant of time, equal to the average value of the instantaneous conductor voltages with respect to the reference voltage. The neutral voltage is a common mode voltage.

conductor shield A component of a cable consisting of a conductive tape or mesh surrounding each of the insulated conductors of a cable. The purpose of the conductor shield is to control the electric field within the conductor insulation to prevent insulation breakdown. When installed onboard a ship, only one end of the conductor shield is typically grounded. The conductor-to-conductor shield capacitance contributes significantly to a power system's parasitic line-to-ground capacitance.

converter Power electronic equipment used to convert DC power or AC power of one frequency, voltage, and number of phases to DC power or AC power of potentially another frequency, voltage, or number of phases. Propulsion motor drives and variable frequency drives are examples of converters.

corrosion The deterioration of a material or its properties due to a reaction of the material within its chemical environment.

creepage The shortest distance along the surface of an insulator between two powered conductors, or between a powered conductor and a grounded conductor.

delta-delta transformer A three-phase transformer, or set of three single-phase transformers, where the three primary windings are connected in a ring (delta winding) and the three secondary windings are also connected in a ring (delta winding). The transformers are connected such that the phase sequence of the primary windings is preserved on the secondary windings.

delta-wye transformer A three-phase transformer, or set of three single-phase transformers, where the three primary windings are connected in a ring (delta winding) and the three secondary windings are connected such that one end is connected to a common neutral conductor and the other end to the power system (wye winding). The transformers are connected such that the phase sequence of the primary windings is preserved on the secondary windings; the voltages and currents on the secondary windings are shifted 30 electrical degrees from the voltages and currents on the primary windings.

dielectric constant A property of a substance (typically an insulator) that is a measure of its ability to store electrical energy in an electric field. Also known as the relative permittivity.

drain wire A conductor provided in a cable used to conduct common mode current. A drain wire is typically grounded at both ends of the cable.

down conductor A conductor in a lightning protection system used to conduct current due to a lightning strike from an air terminal to the ship's hull or ground plate.

earth In the context of shipboard power systems, the electrical potential of the seawater in which the ship is floating. It is equivalent to ground. Earth is also used to refer to the ship's metallic structure and hull in electrical contact with the seawater.

earthing The act of deliberately inserting a solid connection or an impedance between a conductor or neutral of a power system and the hull of the ship. It is equivalent to grounding.

earthing resistor Equivalent to a grounding resistor.

earthing switch A switch used to connect to or disconnect from another conductor that has a low-impedance electrical connection to earth/ground. It is equivalent to grounding switch.

electromagnetic compatibility The ability of electrical equipment and systems to function acceptably in their electromagnetic environment.

electromagnetic interference Degradation of the operation of an electronic device due to electromagnetic induction, electrostatic coupling, or electrical conduction from an external source.

emergency switchboard A switchboard used to supply power to the emergency loads. It may be powered from the normal ship service generator sets (possibly via transformers), from the emergency generators, or from energy storage systems.

EMI filter A circuit inserted in the power conductors of equipment to reduce the probability of electromagnetic interference. EMI filters often include capacitors connected between the phase conductors and ground (earth); these EMI capacitors can manifest as a low common mode impedance between the set of conductors and ground.

equipotential The condition where two conductors have equal voltages with respect to a reference voltage. Two conductors that are electrically connected with low impedance are equipotential.

equipotential bonding Equivalent to bonding (see bonding).

equipotential grounding Equivalent to bonding (see bonding).

galvanic corrosion Corrosion that occurs when dissimilar metals in electrical contact are also in contact with a fluid electrolyte.

generator set A prime mover, such as a gas turbine or diesel engine, integrated with a generator, generator exciter, speed governor, and controls. It may also include a rectifier for DC systems.

generator switchboard A switchboard directly connected to a generator set. It often includes instrumentation, fault protection relays, and synchronization equipment to enable paralleling of generators.

ground In the context of shipboard power systems, the electrical potential of the seawater the ship is floating within. It is equivalent to earth. "Ground" is also used to refer to the ship's metallic structure and hull in electrical contact with the seawater.

ground detection lamps A set of lights, one per phase, connected in wye with a normally closed switch connecting the wye point to the ship's hull. When a ground fault is not present, all the lights will be on, but dim. Upon a ground fault, the light associated with the faulted phase will either be off or very dim; the lights associated with the un-faulted phases will be bright. The test button is used to ensure none of the lights are burned out; pressing the test button should result in all the lamps being on, but dim.

ground detection system A system used to detect and alert the operator of ground faults.

ground fault An unintentional low-impedance (measured at the power system fundamental frequency for AC power systems, or DC for DC power systems) electrical connection of a power system to the ship's hull. In the presence of non-isolated power electronic converters, the low-impedance electrical connection to the ship's hull may effectively include a series voltage source.

ground fault circuit interrupter A circuit protection device for a single-phase distribution system that measures the difference in current magnitudes of the two conductors; if the difference is large enough, typically 4–6 mA, the circuit is interrupted. A "ground fault circuit interrupter" is intended to protect personnel from electrical shock.

ground fault current The current flowing through a ground fault.

ground fault detection The act of determining that a ground fault is present.

ground fault localization The process of determining which cable has a ground fault, and where along the length of the cable the ground fault is located.

ground plate An external metallic plate attached to the underwater exterior of a non-metallic hull for the purpose of providing a system ground, protective earth connection, or a ground connection for a lightning protection system.

grounding The act of deliberately inserting a solid connection or an impedance between a conductor or neutral of a power system and the hull of the ship. Equivalent to earthing.

grounding bus A conductor that provides a path for ground currents between switchboards; all grounding resistors, common mode shunts, EMI filters, and shaft brushes connect to the grounding bus. The grounding bus in turn is connected via a grounding switch to the ship's hull. The grounding switches are located in the generator switchboards. The purpose of the grounding bus is to divert a portion of the common mode current out of the ship's hull and into the grounding bus.

grounding circuit The components and interconnections used to implement resistance or solid grounding. Grounding circuit components may include, for example, grounding transformers, grounding resistors, grounding switches, and instrumentation.

grounding resistor A resistor used to provide impedance at the power system's fundamental frequency for AC systems, or DC for DC systems, between the power system and the ship's hull. In AC systems, grounding resistors are used to dampen the line-to-ground voltage transient response due to intermittent ground faults. In DC systems, the grounding resistors are used to reduce voltage stress on insulation systems when a ground fault is cleared.

grounding switch A switch used to connect or disconnect from another conductor that has a low-impedance electrical connection to earth/ground. Equivalent to earthing switch.

harmonic A frequency component with frequency equal to an integer multiple of the fundamental frequency. Normally applied to the frequency content of voltage and current waveforms.

high-resistance grounded (high-resistance grounded system) A condition where the grounding resistor has a high enough resistance to enable continued operation without tripping protective devices due to a ground fault on one phase; yet low enough to effectively dampen the line-to-ground voltage transient response due to intermittent ground faults on AC systems, or to provide sufficient protection to insulation systems on DC systems.

high voltage Voltages above 1 kV. It may also be referred to as medium voltage. In some contexts, 1 kV is also considered high voltage.

high-voltage shore connection A shore connection (shore power) system with an operating voltage greater than 1 kV.

impressed current cathodic protection An active corrosion protection system that inserts a regulated current through an anode electrode to alter the electrochemical environment such that all protected metals are cathodic. Protected metals are typically the ship's hull, structure, and system components in contact with the sea.

induced static charge Electrical charge on an aircraft parked onboard a ship or on land typically resulting from heavily charged clouds in an electrical storm creating an electrostatic field over a large area. This electrostatic field can slowly cause a charge to accumulate on an aircraft. Should the charge in the clouds dissipate, such as during a lightning strike, the electrostatic field will disappear, but the charge on the aircraft will only slowly dissipate through the tires if the aircraft does not include a protective ground. In aviation literature, a protective ground is often called a "static" ground.

insulated bus pipe An insulated solid conductive rod or solid conductive pipe used in place of cable for high-current power distribution. The conductors are normally made of copper or aluminum.

insulation monitoring device A device used in ungrounded power systems to measure the insulation resistance to ground. Insulation monitoring devices typically inject either a voltage or current signal between the power system conductors and the ground, then measuring the current or voltage impact of the injected signal.

integrated power system A power and propulsion system architecture where electric propulsion motors and ship service (hotel) loads are powered by the same set of generator sets.

inverter Power electronic equipment used to convert DC power to AC power.

isolated power conversion equipment Power conversion equipment where the input and output power interfaces are part of different separately derived systems. A ground fault on the input will not be manifest as a ground fault on the output and vice versa.

isolated system An ungrounded system.

lightning protection system A system for reducing or preventing damage from a lightning strike. Components of a lightning protection system include air terminals, down conductors, low-impedance ground system, and surge protection.

line reactor A set of series inductors installed between a power system and a load for the purpose of reducing the magnitude of differential mode current harmonics. Line reactors also reduce common mode currents, but not as effectively as a common mode choke.

load center A type of switchgear typically powered by a switchboard and used to distribute power to individual loads, circuits, distribution transformers, and power panels via circuit breakers. Load centers are usually deck mounted.

low-resistance grounded (low-resistance grounded system) A condition where a grounding resistor has a low enough resistance to provide sufficient ground fault current to reliably trip protective devices due to a ground fault on one phase.

low voltage Voltages at or below 1 kV. In come contexts, restricted to voltages below 1 kV.

low-voltage distribution system An electrical power distribution system with a nominal system voltage equal to or less than 1 kV.

low-voltage shore connection system A shore connection (shore power) system with an operating voltage less than or equal to 1 kV.

mechanical drive ship A ship with a propulsion system where the propulsors, such as propellers or waterjets, are mechanically coupled to gas turbines, diesel engines, or steam turbines.

medium voltage Voltages above 1 kV. It may also be referred to as high voltage. In some contexts, may include 1 kV.

mutual inductance The voltage difference between the ends of one conductor divided by the time derivative of the current through a different conductor, where the currents through the one conductor and all other conductors in a cable are zero. Mutual inductance is a measure of the electromagnetic coupling of two conductors.

neutral conductor A conductor that in normal operation is intended to have a voltage equal to the neutral voltage of a set of conductors. The voltage at the neutral conductor may differ somewhat from the neutral voltage during normal operation and may differ significantly during faulted conditions. The common connection point of a wye set of windings in a balanced three-phase system is a neutral conductor.

neutral earthing resistor A grounding resistor that is connected between the hull of the ship (perhaps via a grounding switch or grounding bus) and the neutral conductor of a generator or transformer. It is equivalent to a neutral grounding resistor.

neutral grounding resistor A grounding resistor that is connected between the hull of the ship (perhaps via a grounding switch or grounding bus) and the neutral conductor of a generator or transformer. It is equivalent to a neutral earthing resistor.

neutral voltage The voltage of a set of conductors with respect to a reference voltage potential (typically ground) at any instant of time equal to the average value of the instantaneous conductor voltages with respect to the reference voltage. It is equivalent to common mode voltage.

non-isolated power conversion equipment Power conversion equipment where the input and output power interfaces are part of the same separately derived system. A ground fault on the input will be manifest as a ground fault on the output and vice versa.

one-line diagram A simplified diagram of a ship's electrical system showing electrical connectivity between major electrical equipment; the multiple conductors in a cable are represented by a single line.

parasitic capacitance Unintended capacitance between the conductors of a power system and ground. Parasitic capacitance may exist in both cabling and equipment.

parasitic resistance Unintended resistance between the conductors of a power system and ground. Parasitic resistance may exist in both cabling and equipment.

partial discharge In the context of cables, an electric discharge within a void of the conductor insulation. Partial discharge occurs when the voltage on opposite sides of a void results in the electric field strength within the void exceeding the breakdown electric field strength of the air (or other gas within the void). The voltage difference results from the insulation being between the conductor at the line voltage and the conductor shield which is grounded. The voids are due to defects in the insulation and can grow due to the heat and pressure from the arcing caused by the excessive electric field strength. This breakdown is called "partial" because it occurs for only part of the path between the conductor and the shield.

passive rectifier A rectifier composed of diodes with no active regulation.

pilot contact A contact in a plug/socket pair used to signal that the connection is correctly made. The pilot contact may be designed to make connection, before, at the same time, or after the main power contact connections are made. Pilot contacts are often used as part of a safety system.

polarization curve An electrochemical characterization of a metal electrode within an electrolyte using a galvanic half-cell. The polarization curve plots voltage between the metal electrode and a hydrogen electrode against the current flowing through the metal electrode. The polarization curve is used to predict the rate of corrosion of the anode in a galvanic cell.

power panel A type of low power switchgear typically powered by a load center or another power panel and used to distribute power to individual loads, circuits, distribution transformers, and other power panels via circuit breakers. Power panels are usually bulkhead (wall) mounted. In some cases, a power panel may be powered by a switchboard. A power panel powered by another power panel may be designated a sub-power panel.

power system grounding system System composed of equipment for establishing and monitoring an impedance between the power system conductors and ground. A power system grounding system can include grounding transformers, grounding resistors, grounding switches, and instrumentation.

primary distribution system A power distribution system, either AC or DC, with a nominal system voltage above 1 kV. If a ship is provided with a primary distribution system, generator sets provide power directly to the primary distribution system via switchboards; large loads are also directly powered from the primary distribution system. Smaller loads are normally powered via transformers or power converters from the primary distribution system.

propulsion motor An electric motor connected to a propulsor such as a propeller or waterjet for the purpose of propelling the ship. A propulsion motor is usually provided power from a power electronic converter called a propulsion motor drive.

propulsion motor drive A power electronic converter configured as a variable frequency drive used to power propulsion motors. A propulsion motor drive may also provide the capability for dynamic braking by either dissipating braking power in a dynamic braking resistor or feeding electrical power back on the power distribution system.

propulsion transformer A transformer between the power system and a propulsion motor drive. The transformer may have multiple sets of secondary windings feeding separate rectifier stages for the purpose of reducing harmonic currents on the transformer primary windings. Propulsion transformers can also be used to convert power from the distribution system voltage to the voltage needed by the propulsion motor drive. Since the secondaries of the propulsion transformer and the power system are in separately derived systems, propulsion transformers also can minimize the insertion of common mode currents into the power system.

protective earth A connection between exposed conductors that are not intended to normally conduct current and ground (earth) for the purpose of preventing electrical shock to personnel. Current may flow through the protective earth due to ground faults and due to common mode currents through parasitic capacitances.

rectifier A device for converting AC power into DC power. A rectifier is typically composed of an array of semiconductor diodes, an array of silicon-controlled rectifiers, or a power electronic converter.

Residual Current Detector A circuit protection device for a single-phase distribution system that measures the difference in current magnitudes of the two conductors; if the difference is large enough, typically 10 mA, the Residual Current Detector will send a trip signal to indicate a ground fault. Normally, the trip signal is used to trip a circuit breaker to interrupt power to the faulted circuit. Residual Current Detectors are intended to protect personnel from electric shock.

resistivity A measure of the electrical resistance properties of a material in units of ohm-meters. The resistance between the ends of a component of cross-sectional area A m^2 and length L m is equal to the resistivity times $\left(\frac{L}{A}\right)$.

sacrificial anodes A component of a passive cathodic protection system; sacrificial anodes are metal bars that are anodic compared to the metals they are intended to protect. They are physically connected to the ship hull, rudder, propellers, etc. to provide cathodic protection. Most sacrificial anodes are made of alloys of zinc, magnesium, or aluminum.

low-voltage secondary distribution system A power distribution system, either AC or DC, normally serving single-phase loads with a nominal system voltage no greater than 300 V. Low-voltage secondary distribution systems typically are powered via transformers or power electronic converters from low-voltage distribution systems. Low-voltage secondary distribution systems normally provide power to only a small region of the ship.

low-voltage special circuit Circuits that normally would be powered by a low-voltage secondary distribution system, but require continued operation with a single ground fault. Low-voltage special circuits are typically used to serve applicable emergency loads, vital loads, and medical facility loads.

self-inductance The voltage difference between the ends of a conductor divided by the time derivative of the current through the conductor, where the currents through all other conductors in a cable are zero.

separately derived systems The condition where no direct electrical connection exists between two power systems; a ground fault on one system does not result in a ground fault on the other system. For example, systems only connected by a transformer with one system connected to the primary and the other system connected to the secondary are separately derived systems.

shaft brush An electrically grounded metallic brush intended to minimize the voltage between a propulsion shaft or motor shaft and ground for the purpose of preventing damage to bearings due to currents through the bearings.

shore power When a ship is in port, an electrical connection to the terrestrial power system; shore power enables shutting off shipboard generator sets to conserve fuel and reduce emissions. The use of shore power is known as cold ironing.

silicon-controlled rectifier A power electronic device that allows unidirectional current flow between its anode and cathode when triggered by a voltage pulse on its gate terminal with respect to the cathode; the device shuts off when the current through the device falls below a latching current level. Often used to regulate the output voltage of an AC to DC rectifier.

six-pulse rectifier A passive rectifier or silicon-controlled rectifier applied to a single set of three phases. The DC ripple on the voltage at the output of the rectifier has six distinct pulses for every period of the fundamental frequency.

solidly grounded A condition where a power system conductor, or a neutral conductor, is connected with a very low resistance to the hull of the ship, either directly or via a grounding bus. A ground fault results in a very high fault current which is cleared by a circuit breaker tripping.

space charge An accumulation of electric charge within cable insulation; this distribution of charge changes the electric field within the insulation to be nonuniform. The nonuniformity of the electric field in turn can lead to more space charge accumulation in addition to localized areas where the electric field is large enough to cause insulation breakdown.

split power supply A power supply consisting of two regulated power supplies providing the same magnitude voltage (equal to half the line-to-line voltage) connected in series with their common connection grounded either solidly or through a grounding resistor.

static discharge wand A grounded device used by a member of the ship's crew to discharge static electricity on a helicopter hovering over a ship; the static discharge wand brings the helicopter structure and the ship's structure to the same voltage to protect the ship's crew from electrical shock.

sub-power panel A power panel that receives its power from another power panel.

surge protection The act of reducing the probability of damage to electrical equipment from overvoltages due to transients and lightning strikes through the use of surge protective devices and equipment specification and design. Specifying an adequate Basic Insulation Level is one technique for providing surge protection within a layered approach.

surge protective device A nonlinear device that has high impedance when exposed to voltages up to its maximum continuous operating voltage; above this voltage the impedance rapidly drops such that it maintains the voltage to less than or equal to the clamping voltage.

switchboard A type of switchgear typically powered by a connected generator or another switchboard and used to distribute power to other switchboards, load centers, and large loads via circuit breakers. Switchboards are usually deck mounted. In some cases, a power panel may be powered by a switchboard.

transformer saturation A condition when the ferromagnetic core material of a transformer is operating in the saturation region; an increase in the magnetic field strength (H) does not result in an appreciable increase in the magnetic flux density (B). Transformer saturation can result from excessive voltage on the primary, low frequency operation, or DC currents. Transformer saturation can result in significant waveform distortion of the secondary voltage and overheating.

transient voltage A voltage disturbance, normally of duration on the order of milliseconds or less, but may be as long as several seconds.

triboelectric charging Buildup of static electricity that occurs when specific materials come into contact and then are separated. Examples of triboelectric charging include rubbing glass with fur or using plastic comb in hair. For aircraft, triboelectric charging can occur when aircraft fly through water droplets, dust or snow; or alternately, when the aircraft is parked onboard the ship and strong winds blow dust or snow across the aircraft. The fast rotation of helicopter blades can result in significant charge buildup. The flow of fuel through a refueling hose can also result in triboelectric charging.

twelve-pulse rectifier A passive rectifier or silicon-controlled rectifier applied to two sets of three phases; one three-phase set is 30 electrical degrees phase shifted from the other three-phase set. The DC ripple on the voltage at the output of the rectifier has 12 distinct pulses for every period of the fundamental frequency. The two sets of three phases are normally obtained from a propulsion transformer with two secondaries; one secondary is connected in wye, and the other in delta. A twelve-pulse rectifier is typically used to reduce the harmonic content of the propulsion transformer primary winding differential mode current.

ungrounded (ungrounded system) A condition where there is no intentional electrical connection between a power system conductor, or a neutral conductor, and the hull of the ship. An ungrounded power system is electrically connected to the hull of the ship via parasitic capacitances and parasitic resistances. A single line-to-ground fault on an ungrounded system results in relatively low levels of fault current.

unshielded cable Cable without conductor shields or cable shields; typically, low-voltage cables.

variable frequency drive A power electronic converter that can produce variable frequency AC power on its output, usually for the purpose of controlling the speed of an AC motor. The input power may be either AC power or DC power. A variable frequency drive is a type of variable speed drive or adjustable speed drive.

voltage spike A transient overvoltage of short duration, typically less than a millisecond.

wye-broken delta transformer A grounding transformer configuration where the primary is connected in wye with the neutral conductor solidly grounded; the secondary is connected in a ring (delta) that also includes the grounding resistor. Wye-broken delta transformers are often used in power systems with a nominal system voltage over 1 kV; a turns ratio that reduces the voltage on the secondary to a level below 1 kV (such as 450 V) enables the use of commonly available instrumentation.

wye-delta transformer A three-phase transformer, or set of three single-phase transformers, where the three primary windings are connected such that one end is connected to a common neutral conductor and the other end to the power system (wye winding) and the three secondary windings are connected in a ring (delta winding). The transformers are connected such that the phase sequence of the primary windings is preserved on the secondary windings; the voltages and currents on the secondary windings are shifted 30 electrical degrees from the voltages and currents on the primary windings. The wye-delta transformer can also be used in grounding transformer configuration where the neutral conductor of the primary is connected either directly or via a grounding resistor to the ship's hull.

wye-wye transformer A three-phase transformer, or set of three single-phase transformers, where the three primary windings are connected such that one end is connected to a common neutral conductor and the other end to the power system (wye winding) and the three secondary windings are connected such that one end is connected to a common neutral conductor and the other end to a second power system (wye winding). The transformers are connected such that the phase sequence of the primary windings is preserved on the secondary windings; the voltages and currents on the secondary windings are in phase with the voltages and currents on the primary windings.

zigzag transformer A grounding transformer configuration where each primary is connected in series with a secondary of a different phase which are then connected in wye configuration. The neutral conductor is then connected either directly or via a grounding resistor to the ship's hull. Zigzag transformers are often used in power systems with a nominal system voltage less than 1 kV because the power rating of each transformer is less than for other grounding transformer configurations.

zonal distribution A power distribution system using an architecture that is implemented by dividing the ship into multiple zones, designing low-voltage distribution system for each zone, and integrating the zones with a primary distribution system or low-voltage interzone distribution system. The connection between the primary distribution and the zone distribution system is typically either a transformer or power converter. The connection between a low-voltage interzone distribution system and the zone distribution system is typically a switchboard.

zone of protection For a lightning protection system, the volume in which equipment is likely to not be subjected to damage from a direct lightning strike.

Appendix B

Acronyms and Abbreviations

ABS	American Bureau of Shipping
AC	Alternating current
ALCI	Appliance leakage current interrupter
AMP	Alternative maritime power
AUV	Autonomous underwater vehicle
AVT	Absence of voltage tester
BIL	Basic insulation level
CCS	Combined charging system
CE	Conducted emissions
CEA	Cognizant engineering authority
CHAdeMO	CHArge de MOve
CM	Common mode
CRES	Corrosion-resistant steel
CS	Conducted susceptibility
CT	Current transformer
DC	Direct current
DM	Differential mode
EMC	Electromagnetic compatibility
EMI	Electromagnetic interference
EPR	Ethylene propylene rubber
GFCI	Ground fault circuit interrupter
HVSC	High-voltage shore connection
HRG	High-resistance ground
ICCP	Impressed current cathodic protection
IEC	International Electrotechnical Commission
IEEE	Institute of Electrical and Electronics Engineers
IMO	International Maritime Organization
IPS	Integrated power system

Design of Shipboard Power System Grounding/Earthing, First Edition. Norbert Doerry, Mohammed M. Islam, and John Prousalidis.
© 2025 The Institute of Electrical and Electronics Engineers, Inc.
Published 2025 by John Wiley & Sons, Inc.

IT	Isolation terre
LED	Light-emitting diode
LPS	Lightning protection system
LRG	Low-resistance ground
LNGC	Liquid natural gas carriers
LVAC	Low-voltage alternating current
LVSC	Low-voltage shore connection
MVAC	Medium-voltage alternating current
NFPA	National Fire Protection Association
NGR	Neutral grounding resistor
OPS	Onshore power supply
PD	Partial discharge
PE	Protective earth
RCBO	Combination RCD and miniature circuit breaker
RCD	Residual current detector
RE	Radiated emissions
RS	Radiated susceptibility
SC	Space charge
SCR	Silicon-controlled rectifier
SPD	Surge protective device
SSE	Shoreside electricity
TN-S	Terre-neutral (separated)
UAV	Unmanned aerial vehicles
VFD	Variable frequency drive
XLPE	Cross-linked polyethylene

Appendix C

Impact of Electric Current on Humans

C.1 Introduction

The main function of protective earthing is to prevent harm to personnel due to electric current. To provide context for the protective earthing system, this appendix details the impact of different types of current on the human body.

Electric shock is a living body's response to electric current flow through any part of the body. Electrocution is death caused by an electric current flowing through a body. Damage to a body can be a primary electrical injury due to the electric current (or voltage) or can be a secondary injury, such as those resulting from falls or striking immobile objects, which result from dynamic reactions of a body to the electric current.

Due to the variation in the electrical properties across a population of humans, there is also considerable variation in the levels of current necessary to produce a given physiological response; the electric current characteristics, the path of the electric current through a body, the sex of the individual, as well as environmental conditions play key roles.

The level of response is generally categorized into:

Barely perceptible: An individual is barely able to sense that an electric current is flowing in the body.

Let-go current: The maximum current an individual can grasp and "let go"; muscular control is retained.

Paralysis of respiratory muscles: The minimum current level where an individual cannot breathe.

Ventricular fibrillation threshold: The minimum current level impacting normal heart operation.

Cardiac arrest and internal organ damage: The minimum current level leading to cardiac arrest and internal organ damage.

Design of Shipboard Power System Grounding/Earthing, First Edition. Norbert Doerry, Mohammed M. Islam, and John Prousalidis.
© 2025 The Institute of Electrical and Electronics Engineers, Inc.
Published 2025 by John Wiley & Sons, Inc.

The design of protection systems usually is based on minimizing the risk to humans of body currents exceeding the let-go current threshold. Under normal operation, currents through the body should not be perceptible.

The current that flows through the body depends on the electrical resistance (impedance) of the body and the applied voltage. The resistance depends on the characteristics of the electrodes, the pressure being applied by the electrode on the skin, the skin contact area, the amount of sweat on the skin, and the location of the skin on the body. Turner (1972) indicates that variations in these dependencies may result in the body resistance ranging between 500 Ω and 100 kΩ or more. The resistance of dry and wet skin can vary by a factor of 100. Pinching an electrode, on average, results in a body resistance of 14 kΩ, while gripping an electrode, on average, results in a body resistance of 7.4 kΩ.

Fish and Geddes (2009) report that when skin is exposed to voltages above 500 V, the high resistance of the skin breaks down; the voltage is then applied directly to the internal body resistance, which is on the order of 300 Ω. The lower body resistance can result in much higher currents that can lead to significant injury or electrocution.

C.2 Low-Frequency (60 Hz) Currents

Much has been written on electric current levels necessary for different levels of response. Fish and Geddes (2009), for example, provide the following data for 60 Hz:

1 mA	Barely perceptible.
10 mA	Let-go current (sensitive female).
16 mA	Let-go current (average man).
20 mA	Paralysis of respiratory muscles.
100 mA	Ventricular fibrillation.
2 A	Cardiac arrest and internal organ damage.

Dalziel (1961) provides similar values:

0.7 mA	Barely perceptible (women).
1.1 mA	Barely perceptible (men).
6 mA	Let-go current (0.5% women).
9 mA	Let-go current (0.5% men).
10.5 mA	Let-go current (median woman).
16 mA	Let-go current (median man).

15 mA	Paralysis of respiratory muscles (women).
23 mA	Paralysis of respiratory muscles (men).
100 mA	Ventricular fibrillation – three seconds shock.
$\frac{165}{\sqrt{T}}$ mA	Ventricular fibrillation – short shocks (T in seconds).
13.6 J	High-voltage surges – energy.

Ground Fault Circuit Interrupters (GFCIs) are designed to trip with a current mismatch by more than 4–6 mA; at this level, fewer than 0.5% of women and men will lose muscular control and not be able to "let-go." Residual Current Detectors (RCDs) trip with 10 mA current mismatch; this level is less than the level that median men and women can let-go. With a trip current of 30 mA, an appliance leakage current interrupter (ALCI) will not protect personnel; an ALCI may permit currents through a body that could cause paralysis of respiratory muscles.

C.3 DC Currents

If the body current is sufficiently high, AC currents will be perceived as long as the current flows. On the other hand, until the body current is at a much higher level, DC currents will be perceived only when the current is applied and when it is removed. (Fish and Geddes 2009). In general, a higher DC current is required to achieve the same level of response compared to 60 Hz current.

Dalziel (1961) provides the following DC current thresholds for the level of responses:

3.5 mA	Barely perceptible (women).
5.2 mA	Barely perceptible (men).
41 mA	Let-go current (0.5% women).
62 mA	Let-go current (0.5% men).
51 mA	Let-go current (median woman).
76 mA	Let-go current (median man).
60 mA	Paralysis of respiratory muscles (women).
90 mA	Paralysis of respiratory muscles (men).
500 mA	Ventricular fibrillation – three seconds shock.
50 J	High-voltage surges – energy.

C.4 High-Frequency Currents

At frequencies above 100 Hz, the let-go current increases as the frequency increases. Figure C.1 depicts the let-go current for men as a function of frequency up to 10 kHz. Dalziel and Massoglia hypothesize that the let-go current for women would be roughly two-thirds that for men.

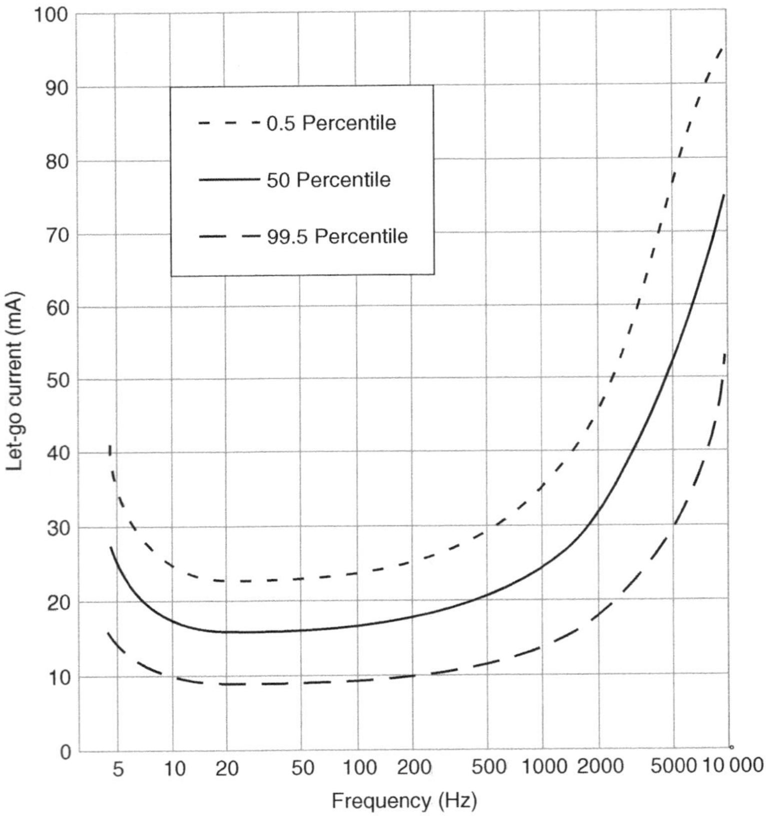

Figure C.1 Let-go current for men as a function of frequency. *Source:* Adapted from Dalziel and Massoglia (1956).

Dalziel (1961) provides the following 10 kHz current thresholds for the level of responses:

12 mA	Barely perceptible (men).
55 mA	Let-go current (0.5% men).
75 mA	Let-go current (median man).
94 mA	Paralysis of respiratory muscles (men).

While tests at 10 kHz were not performed on women, Dalziel hypothesizes that the equivalent thresholds for women would be roughly two-thirds that for men.

Dalziel (1961) also reports that between 100 and 200 kHz, the sensation changes from tingling to heat.

References

Dalziel, Charles F. and Massoglia, F.P. (1956). Let-Go Currents and Voltages. *Transactions of the American Institute of Electrical Engineers, Part II: Applications and Industry.* 75 (2): 49–56.

Dalziel, Charles F. (1961). Deleterious Effects of Electric Shock. *Proceedings of a meeting on Electrical Accidents and Related Matters sponsored by the International Labour Office, World Health Office and International Electrotechnical Commission*, Geneva, Switzerland (23–31 October 1961).

Fish, Raymond M. and Geddes, Leslie A. (2009). Conduction of Electrical Current to and Through the Human Body: A Review. *Eplasty* 9: e44.

Turner, Hebert S. (1972). Human Responses to Electricity: A Literature Review. *Prepared Under NASA Contract NSR 36-008-108*, Accession Number 73N10065, Document ID 19730001338.

Index

Design of Shipboard Power System Grounding/Earthing, First Edition. Norbert Doerry, Mohammed M. Islam, and John Prousalidis.
© 2025 The Institute of Electrical and Electronics Engineers, Inc.
Published 2025 by John Wiley & Sons, Inc.

d

e

h

i

l

Printed and bound by CPI Group (UK) Ltd, Croydon, CR0 4YY
17/09/2026
14964241-0005